Advanced Problem Solving with Maple™

A First Course

Textbooks in Mathematics
Series editors:
Al Boggess and Ken Rosen

LINEAR METHODS FOR THE LIBERAL ARTS

David Hecker and Stephen Andrilli

CRYPTOGRAPHY: THEORY AND PRACTICE, FOURTH EDITION

Douglas R. Stinson and Maura B. Paterson

DISCRETE MATHEMATICS WITH DUCKS, SECOND EDITION

Sarah-Marie Belcastro

BUSINESS PROCESS MODELING, SIMULATION AND DESIGN, THIRD EDITION

Manual Laguna and Johan Marklund

GRAPH THEORY AND ITS APPLICATIONS, THIRD EDITION

Jonathan L. Gross, Jay Yellen and Mark Anderson

A FIRST COURSE IN FUZZY LOGIC, FOURTH EDITION

Hung T. Nguyen, Carol L. Walker, and Elbert A. Walker

EXPLORING LINEAR ALGEBRA

Crista Arangala

A TRANSITION TO PROOF: AN INTRODUCTION TO ADVANCED MATHEMATICS

Neil R. Nicholson

COMPLEX VARIABLES: A PHYSICAL APPROACH WITH APPLICATIONS, SECOND EDITION

Steven G. Krantz

GAME THEORY: A MODELING APPROACH

Richard Alan Gillman and David Housman

FORMAL METHODS IN COMPUTER SCIENCE

Jiacun Wang and William Tepfenhart

ADVANCED PROBLEM SOLVING WITH MAPLE: A FIRST COURSE

William P. Fox and William C. Bauldry

Advanced Problem Solving with Maple™

A First Course

William P. Fox
William C. Bauldry

CRC Press
Taylor & Francis Group
Boca Raton London New York

CRC Press is an imprint of the
Taylor & Francis Group, an **informa** business
A CHAPMAN & HALL BOOK

Maplesoft and Maple are trademarks of Waterloo Maple Inc.

CRC Press
Taylor & Francis Group
6000 Broken Sound Parkway NW, Suite 300
Boca Raton, FL 33487-2742

First issued in paperback 2022

Version Date: 20190429

ISBN 13: 978-1-03-247554-7 (pbk)
ISBN 13: 978-1-138-60185-7 (hbk)

DOI: 10.1201/9780429469633

Publisher's Note
The publisher has gone to great lengths to ensure the quality of this reprint but points out that some imperfections in the original copies may be apparent.

Library of Congress Cataloging-in-Publication Data

Names: Fox, William P., 1949- author. | Bauldry, William C., author.
Title: Advanced problem solving with Maple : a first course / William P. Fox and William C. Bauldry.
Description: Boca Raton : Taylor & Francis, CRC Press, 2020. | Includes index.
Identifiers: LCCN 2019010023 | ISBN 9781138601857
Subjects: LCSH: Problem solving--Data processing. | Quantitative research--Data processing. | Maple (Computer file)
Classification: LCC QA401 .F69 2020 | DDC 519.0285/53--dc23
LC record available at https://lccn.loc.gov/2019010023

Visit the Taylor & Francis Web site at
http://www.taylorandfrancis.com

and the CRC Press Web site at
http://www.crcpress.com

Contents

Preface

The study of problem solving is essential for anyone who desires to use applied mathematics to solve real-world problems. We present problem-solving topics using the computer algebra system Maple^{TM} for solving mathematical equations, creating models and simulations, as well as obtaining plots that help us perform our analyses. We present cogent applications of applied mathematics, demonstrate an effective use of a computational tool to assist in doing the mathematics, provide discussions of the results obtained using Maple, and stimulate thought and analysis of additional applications. This book serves as either an introductory course or capstone course, depending on the chapters covered, to start to prepare students to apply the mathematical modeling process by formulating, building, solving, analyzing, and criticizing mathematical models. It is intended for a first course at the sophomore or junior level for applied mathematics or operations research majors that introduces these students to mathematical topics that they will revisit within their major. This text can also be used for a beginning graduate-level course or as a capstone experience for mathematics majors as well as mathematics education majors as modeling has a much bigger role in secondary education with the newest National Council of Teachers of Mathematics (NCTM) standards. We also introduce many additional mathematics topics that students may study more in depth later in their majors.

Although calculus (either engineering or business) is the prerequisite material, many sections and chapters, especially in Volume II, require multivariable calculus. In addition, the use of linear algebra is required in some chapters. For students without the necessary background, these chapters can be omitted for a specific course. We realize that there are more chapters in this text than could ever be covered within one semester. The increased number of topics and chapters provides flexibility for designing a course appropriate to the background of your students.

Goals and Orientation

This course bridges the study of mathematics topics and the applications of mathematics to various fields of mathematics, science, and engineering. This text affords the student an early opportunity to see how assumptions drive the models as well as an opportunity to put the mathematical modeling process together. The student investigates real-world problems from a variety of disciplines such as mathematics, operations research, engineering, computer science, business, management, biology, physics, and chemistry. This book provides introductory material to the entire modeling process. Students will find themselves applying the process and enhancing their problem-solving capabilities, becoming competent, confident problem solvers for the 21st century. Students are introduced to the following facets of problem solving:

- CREATIVE PROBLEM SOLVING. Students learn the problem-solving process by identifying the problem to be solved, making assumptions and collecting data, proposing a model (or building a model), testing their assumptions, refining the model as necessary, fitting the model to the data if appropriate, and analyzing the mathematical structure of the model to appraise the sensitivity of the results when the assumptions are not strictly met.

- PROBLEM ANALYSIS. Given a model, students will learn to work backward to uncover the assumptions, access how well those assumptions fit the scenario, and estimate the sensitivity of the results when the assumptions are not strictly met.

- PROBLEM RESEARCH. The students investigate a specific area to gain understanding of behavior and to learn how to apply or extend what has been already created or developed to new scenarios.

Course Content

We introduce problem solving early. Volume I is a typical applied mathematics or introduction to operations research topics of ODES, mathematical programming, data fitting with regression, probabilistic problem solving and simulation. Volume II contains discrete dynamical system, both constrained and unconstrained optimization, linear systems, advanced regression, game theory, and multi-attribute decision making.

Organization of Text

Volume I covers introductory topics. Chapter 1 of Volume I is repeated in Volume II, introduces Maple and its basic command structure as well as introduces the problem-solving process. Because the book uses Maple as the tool in mathematical modeling, the chapter provides the foundation or cornerstone of using technology in the modeling process. In Volume I, Chapter 2 introduces ordinary differential equations, and Chapter 3 covers systems of ordinary differential equations. Chapter 4 covers linear, integer, and mixed integer programming as well as the Simplex method. Chapter 5 covers model fitting concentrating on regression methods to fit data. Chapter 6 covers statistical and probabilistic problem solving; Chapter 7 extends these ideas into Monte Carlo simulations. Scenarios are developed within the scope of the problem-solving process. Student thought and exploration is required.

Volume II covers more advanced topics. Chapter 1, the introduction to problem solving and to Maple, is repeated. Chapter 2 covers discrete dynamical systems, and is complementary to Chapters 2 and 3 in Volume I. Chapters 3 and 4 cover both unconstrained and constrained optimization in single variable and multi-variable topics. Chapter 5 deals with solving problems from engineering, economics, and chemistry with linear systems. Chapter 6 continues with regression but moves into more advanced topics such as non-linear regression, logistic regression, and Poisson regression. Chapter 7 covers game theory and relies heavily on linear and non-linear programming methods. Chapter 8 completes Volume II discussing multi-attribute or multi-criteria decision making with methods such as AHP, TOPSIS, and SAW.

Student Projects

The backbone of this course is the student project. In each project, students apply the mathematical modeling process and the mathematical tools they have learned. Each chapter and many sections have sets of student projects. We have seen significant student growth in project submissions from their first project to the final submission. Student projects take time to apply the modeling process, so we typically do not assign more than one project to a student. Most of these projects are designed to be group projects with two or three students working together, although they can be done as individual projects with strong students.

Technology

Technology is fundamental to serious mathematical modeling. The computer algebra system Maple was chosen as our technology for this text, but any technology could be used — from graphing and symbolic calculators through spreadsheets.

Emphasis on Numerical Approximations

Numerical solutions techniques are used in the dynamical systems, explicative modeling with some numerical analysis approaches, and in optimization search procedures. These are the methods most easily employed in iterative and recursive formulas. Early on, the student is exposed to numerical techniques. These numerical procedures are algorithmic and iterative.

Focus on Algorithms

All algorithms are provided with step-by-step formats to aid students in learning to do mathematical modeling with these methods and Maple. Examples follow the summary to illustrate an algorithm's use and application. Throughout, we emphasize the process and interpretation, not the rote use of formula.

Problem Solving and Applications

Each chapter includes examples of models, real-world applications, and problems and projects. Problems are modeled, formulated, and solved within the scenario of the application. These models and applications play an important role in student growth in working in today's complex world.

Exercises

Many exercises are provided at the end of each section and chapter so the student can practice the solution techniques and work with the mathematical concepts discussed. Review problems are given at the end of each chapter, some of which combine elements from several chapter sections. Projects are also provided at the end of each chapter to enhance student understanding of the concepts and their application to real-world problems.

Computer Usage

Maple is the computer algebra system used for this text. Tutorial labs are widely available for learning Maple's syntax and command structure. Our emphasis is on providing the student with the ability to use Maple to assist with mathematical modeling. We illustrate graphing in both two and three dimensions. We provide Maple packages *PSM* and *PSMv2* containing programs, functions, and data to use with each volume.

Acknowledgments

We need to begin by thanking Frank R. Giordano for being a mentor and choosing to involve each of us in mathematical modeling.

William Fox. I am indebted to my colleagues who taught with me over the years and who were involved with problem and project development: Jack Pollin, David Cameron, Rickey Kolb, Steve Maddox, Dan Hogan, Paul Grimm, Rich West, Chris Fowler, Mike Jaye, Mike Huber, Jack Piccuito, Jeff Appleget, Steve Horton, and Gary Krahn. A special thanks to William Hank Richardson, from Francis Marion University, who assisted in developing and refining many of our Maple programs used within this text.

William Bauldry. I am also indebted to my colleagues who have shared their expertise and wisdom over the years and shaped my approach to modeling, to teaching, and to technology: Wade Ellis, Joe Fiedler, Rich West, Jeff Hirst, and most recently Mike Bossé. The folks at Maplesoft have been incredibly supportive and helpful from the beginnings when I used Maple running as tool in the *Macintosh Programmer's Workshop* to today's Maple 2019.

William P. Fox
William C. Bauldry

1

Introduction to Problem Solving and Maple

Objectives

(1) **Understand the nature of problem solving.**

(2) **Understand the use of Maple commands.**

(3) **Understand the Maple Applications Center and its uses.**

1.1 Problem Solving

What do we mean by *problem solving*? We interpret this as having a real problem whose understanding and solution requires quantitative analysis and one or more solution techniques using mathematics. To put into context, we say we need a well-defined problem. After we have a well-defined problem, we must brainstorm variables and assumptions that might impact the problem. We build or select a known model and choose a solution technique or combinations of techniques to obtain answer. We solve and perform sensitivity analysis. We interpret all results, implement, and if necessary, refine the entire process.

In many ways, this process is very similar to the mathematical modeling processes described in other texts: Albright [A2011], Giordano *et al* [GFH2013], and Meerscheart [M2007] to name a few. Readers may want to examine these texts for a more detailed approach. As a co-author of the Giordano text, my approach is most similar to the approach we describe in that text.

There are four- and five-step processes for problem solving. We present and describe a simple five-step method.

Step 1. Define and understand the problem.

Step 2. Develop strategies to solve the problem. This includes a problem-solving formulation including a methodology to obtain a solution. If data is available, examine the data, plot it, and look for patterns.

Step 3. Solve the problem formulated in Step 2.

Step 4. Perform a self-reflection of your process. You want to make sure the solution answers the problem from Step 1. You also want to ensure the results pass the "commonsense" test. If not go back to Step 2 and reformulate the strategy.

Step 5. If necessary, extend the problem.

We will not concentrate on the modeling portion but on the selection of the model and the solution technique processes including the use of technology in the solution process.

One key point is that our results must pass the "commonsense" test. For example, we were conducting spring-mass experiments in a classroom on the 3rd floor of our mathematics and science building. The simple purpose was to investigate Hooke's Law. The springs were small and the weights varied from a fraction to about 50 grams. After the experiments, we asked the students to calculate the stretch of their spring if it were attached to a seat, and they sat on the seat. Every student found an answer but none said the spring would most likely break long before it stretched that far.

Let's preview a problem we will see in Volume I and II. We have data for time (t) and an index (y) from $[0, 100]$. Our plot shows a negative linear trend. We compute the correlation which is -0.94, and is interpreted as a strong negative linear relationship. We use linear regression to build a regression equation which has some very good diagnostics, but one questionable diagnostic from the residual plot. The main goal is predicting the future, which is why the problem is being solved in the first place. The answer for y comes out negative which is not a possible answer for y. So we continue our problem solving and correct the residual plot issue by adding a quadratic term to the regression equation. Again, our diagnostics are all excellent this time. We attempt to use the model to predict but our answer does not pass the commonsense test as it is too large. A simple plot shows that for the time value in the future we are on the increasing past of the quadratic polynomial. If we cannot use our regression equation then our work is useless. Now, we continue on the nonlinear regression and use an exponential function to fit our data. Finally, not only are all the diagnostics excellent, but our use of the new regression equation passes the commonsense test.

We also believe that in the 21st-century, technology is a key element in all problem solving. Technology does not tell you what to do, but its use provides insights and the ability to check out possibilities. In this book our technology of choice is the computer algebra system, Maple^{TM}.[1]

1.2 Introduction to Maple

Maple is a symbolic computation system or computer algebra system (CAS) that manipulates information in a symbolic or algebraic manner. You can use these symbolic capabilities to obtain exact, analytical solutions to many mathematical problems. Maple also provides numerical estimates to whatever precision is desired when exact solutions do not exist.

Maple 2019 and higher is different from previous versions of Maple. With Maple 2019 you can create profession quality documents, presentations, and custom computational tools. You can access the power of the Maple computational engine through a variety of interfaces: standard worksheet, classical worksheet, command line version, graphing calculator (Windows only), or Maple applications. Although you type in the commands in a very similar manner as in previous versions of Maple, the statement appears in a "pretty print" format on the screen; that is, the statement appears more like typeset mathematics.

Standard Worksheet

This is a full-featured graphical user interface offering features that help to create documents that show all assumptions, the calculations, and any margin of error in your results. You can even hide the computations to focus on problem setup and final results.

Classic Worksheet

The basic worksheet environment works best for older computers with limited memory.

Command-Line Version

Command-line interface, without graphical user interface features, is used for solving very large, complex problems or batch processing.

[1]Maple is a trademark of Waterloo Maple, Inc.

Maplesoft™ Graphing Calculator

The graphical interface to the Maple computational engine allows you to perform simple computations and create customizable, zoomable graphs. The Graphing Calculator is only available in the Windows version.

Maple Applications

The graphical user interface containing windows, textbox regions, and other visual features, gives you point-and-click access to the power of Maple. It allows you to perform calculations and plot functions without using the worksheet or command-line interfaces.

Maple's extensive mathematical functionality is most easily accessed through all these interfaces. Previous older versions relied on its advanced worksheet-based graphical interface. A worksheet is a flexible document for exploring mathematical ideas or mathematical alternatives and even creating technical reports.

Experimental mathematical modeling, a natural stepping stone to statistical analysis, has an obvious coupling with computers, which can quickly solve equations, and plot and display data to assist in model test and evaluation. The software computer algebra system, Maple, is a powerful tool to assist in this process. When dealing with real-world problems, data requirements can be immense. When evaluating immigration trends, for example, and the political, social, and economic effects of these trends, thousands of data points are used; in some cases, millions of data points. Such problems cannot be analyzed by hand, effectively or efficiently. The manipulation required to plot, curve fit, and statistically analyze with goodness of fit techniques, cannot feasibly be done without the assistance of a computer software system.

The Maple system is easy to learn and can be applied in many mathematical applications. While these demonstrate its versatility, Maple is also an extremely powerful software package. Maple provides over 5000 built-in definitions and mathematical functions, covering every mathematical interest: calculus, differential equations, linear algebra, statistics, and group theory to mention only a few. The statistical package reduces many standard time-consuming statistical questions into one-step solutions, including mean, median, percentile, kurtosis, moments, variance, standard deviation, and so forth. There are many references for Maple, and a short list would include:

- Maple Quick Reference
- Maple Flight Manual
- Maple Language Reference Manual
- Maple Library Reference Manual
- Maple "What's New" Release Notes

- Maple "Getting Started" tutorials and worksheets

This chapter presents a quick review of some basic Maple commands. It is not intended to be a self-contained tutorial for Maple, but will however, provide a quick review of the basics prior to more sophisticated commands in the modeling chapters. There are many good references for those new to Maple. Additionally, there is a collection of self-contained tutorials accessed by clicking the "Getting Started" button on Maple's opening screen. See Figure 1.1.

1.3 The Structure of Maple

Maple is an example of a Computer Algebra System (CAS). It is composed of thousands of commands, to execute operations in algebra, calculus, differential equations, discrete mathematics, geometry, linear algebra, numerical analysis, linear programming, statistics, and graphing. It has been logically designed to minimize storage allocation, while remaining user friendly. Maple allows a user to solve and evaluate complicated equations and calculations, analytically or numerically, such as optimization problems, least square solutions to equations, and solving equations that involve special functions.

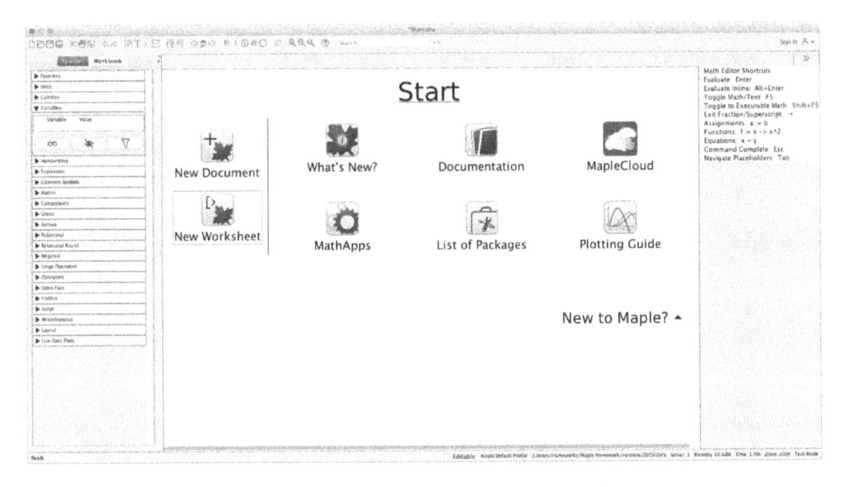

FIGURE 1.1: Maple 2019 Opening Screen

Clicking on the icon (small triangle) to the right of "New to Maple?" in the opening screen will expand the section showing videos, tutorials, and links to more information as seen in Figure 1.2.

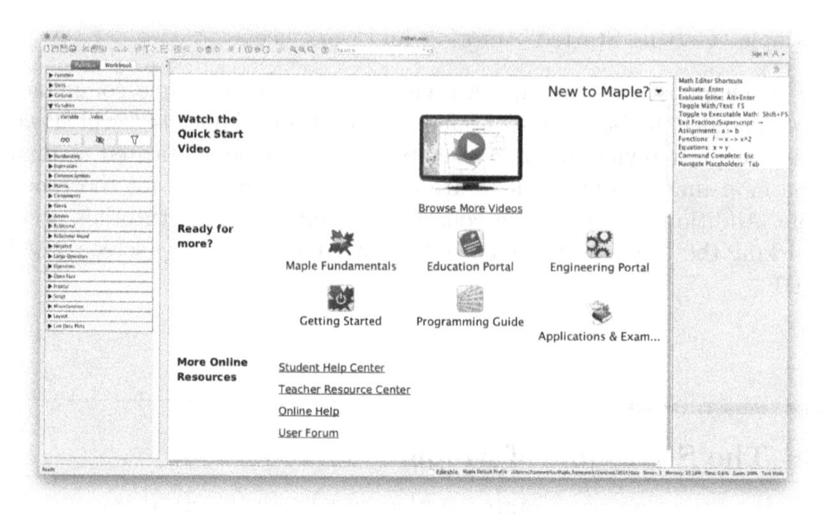

FIGURE 1.2: "New to Maple" Section Expanded

Commands

To begin using Maple from the opening screen, we suggest clicking on the "New Worksheet" button shown in Figure 1.1. Enter commands at Maple's *command prompt*: >. See Figure 1.3. A *Worksheet* shows command prompts, while a *Document* does not.

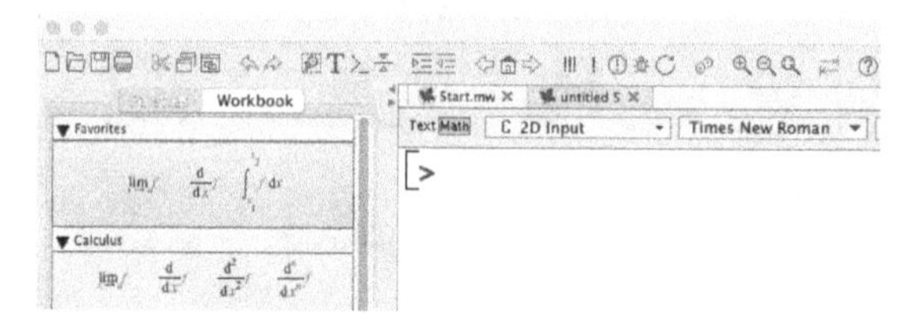

FIGURE 1.3: Maple's Command Prompt

Notation and Conventions

Throughout this book, different types of fonts and styles are used to distinguish between Maple commands, Maple output, and other information. Maple commands are copied directly from Maple 2019 and the output will immediately follow the Maple commands. In the first example below, the variable a has been assigned an expression as its value. Notice that the symbol := is used to indicate this assignment. In the second line of the example, the value of a, the expression, is differentiated with respect to x.

Type the command in either the Worksheet or the Document mode as

`a := 2*x^3` [→] `- 5*x/6.`

(The [→] indicates pressing the "right arrow" cursor key to exit superscript mode.) You will type the statement this way, but the screen version appears as follows:

$> a := 2 \cdot x^3 - \frac{5 \cdot x}{6}$

$$a := 2x^3 - \frac{5}{6}x$$

$> diff(a, x)$

$$6x^2 - \frac{5}{6}$$

The screen version shows Maple's mathematical interpretation of what you have typed.

1.4 General Introduction to Maple

Maple has many types of windows: the worksheet or document window, help window, 2-D plot window, 3-D plot window, and the animation window. Many "Maple Assistants" use a "maple" window. In this book, the worksheet and help windows are used most often. We provide a brief explanation of each.

The worksheet is where all interaction between Maple and the user occurs. Within a worksheet, commands (input) and text (remarks for clarification) are entered by the user, and results, numerical or symbolic (output and graphics) are produced by Maple. The user may manipulate these interactions to create a flowing document that can be saved by clicking File, then clicking *SaveAs* followed by naming the document. The name of the document can be any word or group of letters and/or numbers, which has a length of less than nine characters. Once a document has initially been saved, it can be retrieved by clicking File, clicking Open, and entering the document name. Then it can be re-saved after modification by clicking File, followed by clicking Save. We will use Worksheet mode rather than Document mode since Worksheets show a command prompt, but Documents do not.

The input and text regions of the worksheet can be modified to change a document, but the graphic and output regions cannot be modified once Maple inserts them into a document. A command must be edited and re-executed to alter its associated output. The input region is identified by the > prompt which precedes all command entries into Maple. (Note: Maple only recognizes the Maple generated > prompt. If the symbol > itself is typed by the user, Maple does not respond to it as an input prompt, but as the "greater than" relation.) The input commands, output characters and text regions are all of different font size and color to assist the user in distinguishing between them. Text regions assist in documentation and explanation of the input/output regions of the document and they may be placed anywhere in a document. Graphic regions, once they are generated by input commands, can be copied and pasted into a worksheet or into another document. Once the graphic is pasted into the worksheet, it can no longer be edited or manipulated. The output regions are generated by the user's input commands and cannot be manipulated once they appear in a document, although the user is allowed to delete these results.

The Maple menu bar is located at the top of the screen on a Macintosh, and immediately below the Maple title in Windows. The menu bar provides easy access and *collocation* (collocation is defined as a sequence of words or terms which co-occur more often than would be expected by chance) of many commonly used options. The menu bar includes File, Edit, View, Insert, Format, Plot, Tools, Window, and Help, much like any software. Enter *"? worksheet,reference,standard Menubar"* for detailed information on all menus. Immediately below the Maple window's title bar (Macintosh) or menu bar (Windows) is the Maple tool bar. The tool bar provides accelerated access to the most commonly used options; see Figure1.4. Enter *"? WorksheetToolbar"* for detailed information.

FIGURE 1.4: Maple's Tool Bar

Maple syntax uses either a semicolon or a colon to end a statement. A single command ends with a semicolon implicitly. Multiple statements may be on the same line, but each must have its own colon or semicolon. The colon suppresses output of the command, while the semicolon signals that the results are to be printed to the screen immediately after the enter key is pressed. There is a large set of commands either readily available in Maple memory or stored separately in Maple packages, which assist in more efficient memory storage. Standard commands such as addition and multiplication are *built-in*, not contained in packages. Enter *"? inifcns"* to see the complete list of functions always available.

The Calling Sequence

A "package" is a collection of related definitions and functions that can be brought into a Maple session using the *with* command. The syntax for *with* is

[> *with(<package_name>);*

When using commands stored in packages, such as graphing commands in the *plots* package, the command "*with(plots)*:" is issued prior to using any of the commands in that package. The *with* command is required only once. Several of the Maple packages will be used in this text. Specifically, *plots* will be used extensively for creating graphs, *LinearAlgebra* for linear algebra operations, and *Statistics*, for statistical and linear regression commands. Other important with packages include *DETools* for differential equations and *MultivariableCalculus*, *Optimization*, and *simplex* for optimization.

The Help Command

The Maple help database can provide all the information found in the Maple Library Reference Manual. See Figure 1.5. However, help can be obtained immediately to assist the user in solving problems without leaving the document. Help can be acquired by a number of methods: click on Help in the menu bar, type "help" at the > prompt, or type a "?" at the > prompt. By using the ? or the *help* at the > prompt, the user must type the keyword for the help search. If the specific syntax of a command is in question, for example, the syntax for differentiation, type *? differentiate* at the > prompt. This procedure is perhaps the most convenient one for help on syntax. "Quick Help," the last item on the tool bar, gives quick feedback.

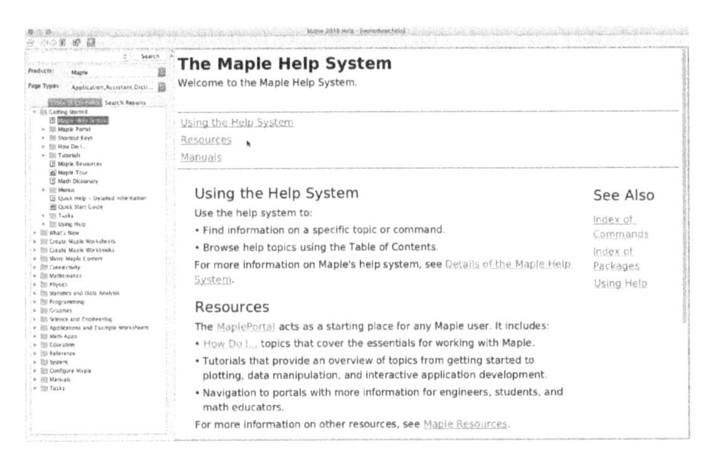

FIGURE 1.5: The Maple Help System

Begin typing a command, say *dif*, then press esc, the escape key. A pop-up menu appears with a list of commands beginning with or related to *dif* appears. Use the mouse to select the desired choice. This technique also works with variables or names that have been defined. Enter and execute *myVariable* := 10. Now type *myVar*, pause, and press esc. Maple fills in the rest of the name. If more than one possibility exists, a pop-up menu will appear with the choices.

Maple's help database also includes a mathematical encyclopedia. Try entering

```
> ? Definition, integral
```

Data Entry

Commonly, a set of data will describe a process. Entering the data is the first step to analyzing the data. Maple provides a convenient method for manually entering data via a list. A list is a group of data to which Maple's many operations are applied. Suppose a group of five college students' weights are known to be 185, 202, 225, 195, and 145. We demonstrate the command required to enter the data into Maple.

```
> weights := [185, 202, 225, 195, 145];
                    weights := [185, 202, 225, 195, 145]
```

The use of brackets in the command indicates a list. The brackets will maintain the initial ordering of the data and allows for duplicate values, and may be manipulated by the methods described later in this book. Commas are required between the list elements. Any alphanumeric element may be included in a list; integers, rational numbers, decimals, strings, variable names, and expressions are all allowed. If any data should be missing, a place holder can be entered in its place, such as the letter x. If there was a sixth student in the group with an unknown weight, a symbol could be used to represent the sixth student's weight.

```
> weights := [185, 202, 225, 195, 145, x];
                    weights := [185, 202, 225, 195, 145, x]
```

Data Entry and Verification

Figure 1.6 illustrates entering, verifying, and naming data pertaining to the length and weight of bass caught during a fishing derby. In subsequent chapters, several models for predicting the weight of a snook fish as a function of length of the fish is suggested. In Figure 1.6 the data is entered in rows and the data is printed to verify correct entry.

```
> length_inches := [12, 14, 12.5, 16, 21.5];
                    length_inches := [12, 14, 12.5, 16, 21.5]
> weight_oz := [15, 21, 10, 33, 41];
                    weight_oz := [15, 21, 10, 33, 41]
```

Correcting Erroneous Entries

In the above illustration, Maple displayed the snook fish data immediately after its input. This is done to help the user verify that all elements have been entered correctly. Reentry of each command is typically the best method for correcting an erroneous entry with many errors.

Transformation and Functions

As described, the symbol := is used to assign the value on the right-hand side of the statement to the name on the left-hand side of the statement. An example, *x:=3*; assigns the value 3 to x, to unassign x, use the command x='x'. Functions can be defined using the mapping arrow symbol $\rightarrow$. The function can be evaluated numerically or symbolically, we provide a short example.

> $f := x \rightarrow x^2 - 3 \cdot x + 4.5;$

$$f := x \mapsto x^2 - 3x + 4.5$$

> $f(5);$

$$14.5$$

> $f(x-2);$

$$(x-2)^2 - 3x + 10.5$$

A problem solution may suggests transforming a variable. Perhaps we want to transform both x and y by taking natural logarithms (ln) of each. In the two-dimensional case, the model suggests a functional relationship between a dependent and an independent variable. Given values of an independent variable, a function can transform the given data to yield predicted values for the dependent variable. Transforming data requires the understanding of algebraic operations and functions that are used in Maple. Table 1.1 presents the regular arithmetic operations that Maple recognizes.

TABLE 1.1: Maple Arithmetic Functions.

Symbol	Operation
+	Addition
−	Subtraction
*	Multiplication
/	Division
^	Exponentiation

Many other functions are also recognized by Maple. Table 1.2 has an abbreviated listing.

We illustrate by taking the natural logarithm of our length and weight

TABLE 1.2: Maple Algebraic Functions

Command	Operation
abs	absolute value of real or complex argument
arg	argument of a complex number
ceil	least integer $\geq x$
conjugate	conjugate of a complex number
exp	the exponential function: $e^x = \exp(x) = \sum_{k=0}^{\infty} x^k/k!$
factorial, !	the factorial function, factorial$(n) = n!$
floor	greatest integer $\leq x$
ln	natural logarithm (with base $e = 2.718...$)
log[b]	logarithm to arbitrary base b
log 10	log to the base 10
max, min	maximum/minimum of a list of real numbers
RootOf	function for expressing roots of algebraic equations

data.

```
> ln_length := map(x -> evalf(ln(x)), length_inches);
    ln_length := [2.484906650, 2.639057330, 2.525728644, 2.772588722,
                  3.068052935]
> ln_weight := map(x -> evalf(ln(x)), weight_oz);
    ln_weight := [2.708050201, 3.044522438, 2.302585093, 3.496507561,
                  3.713572067]
```

Some other functions that are available will be discussed in a later chapter when considering statistical operations that may be performed on columns of data: sums, mean, standard deviation, and so forth. Table 1.3 presents a few examples of data transformations, with the Maple commands.

TABLE 1.3: Examples of Maple Commands in the Worksheet

Expression	Typed Command
x^2	x^2
$2x^2 + 2.5 + 9$	2 * x^2 + 2.5 * x + 9
$(x^2 + 2)^{0.5}$	(x^2 + 2)^0.5

Column Operations

In this section, the operations that can be performed directly on worksheet columns are present. The operations require the *LinearAlgebra* package to be loaded prior to use. The following demonstrates these commands with six examples using two columns of data; $c1 := [1, 2, 3, 4]$ and $c2 := [5, 6, 7, 8]$.

$> with(LinearAlgebra):$

$> c1 := \langle 1|2|3|4 \rangle;$

$$c1 := [1, 2, 3, 4]$$

$> c2 := \langle 5|6|7|8 \rangle;$

$$c2 := [5, 6, 7, 8]$$

Summing the vectors $c1$ and $c2$

$> VectorAdd(c1, c2);$

$$[6, 8, 10, 12]$$

$> c1 + c2;$

$$[6, 8, 10, 12]$$

To add a constant to the entries of $c1$, use '+~'

$> 10 +\sim c1;$

$$[11, 12, 13, 14]$$

To multiply $c2$ by constant, use either

$> VectorScalarMultiply(c2, 0.5);$

$$[2.50000000000000000 \quad 3.0 \quad 3.50000000000000000 \quad 4.0]$$

$> 0.5 \cdot c2;$

$$[2.5 \quad 3.0 \quad 3.5 \quad 4.0]$$

To apply a function to the elements of a vector use *map* or '~'.

$> map(\ln, c1);$
$evalf(\%);$

$$[0 \quad \ln(2) \quad \ln(3) \quad 2\ln(2)]$$

$$[0.0 \quad 0.6931471806 \quad 1.098612289 \quad 1.386294361]$$

$> \ln\sim(c2);$

$$[\ln(5) \quad \ln(6) \quad \ln(7) \quad 3\ln(2)]$$

$> map(x \to x^2, c2);$

$$[25 \quad 36 \quad 49 \quad 64]$$

Arrays and Matrices

Arrays and matrices are structured devices used to store and manipulate data. An array is a specialization of a table; a matrix is a two-dimensional array. Both *array* and *matrix* are part of the linear algebra package, and require the *with(LinearAlgebra)*: command prior to use. Table 1.4 presents a few examples of the use of the *array* and *matrix* commands.

TABLE 1.4: Array and Matrix Commands

Command	Output
$with(LinearAlgebra)$:	(load the *LinearAlgebra* package)
$a := array([1,2,2,3,4]);$	$a := [\ 1\ 2\ 2\ 3\ 4\ ]$
$b := array([[1,2],[3,5]]);$	$b := \begin{bmatrix} 1,2 \\ 3,5 \end{bmatrix}$
$c := Vector[row]([9,3,1,8,3]);$	$c := [\ 9\ 3\ 1\ 8\ 3\ ]$
$d := Matrix([[7,3],[2,3]]);$	$d := \begin{bmatrix} 7,3 \\ 2,3 \end{bmatrix}$

Saving and Printing a Worksheet

To start Maple from Windows or MacOS, launch Maple by double-clicking on the Maple icon. Once Maple has been started, it will automatically open an empty worksheet, with a flashing cursor to the right of a character prompt >. To save the worksheet, click on File and then click on *SaveAs* and then specify a name for the worksheet. Once the worksheet has been named and saved, click on File and then click on Save to re-save the worksheet. To open a previously saved worksheet, click on File and then click on Open, then specify the name of the worksheet.

After re-opening a document, the document will contain the commands and display the results, but not have any values defined. After a command is executed, the result is stored in memory. If the document is closed and then reopened, Maple recovers the commands, but does not recall the results.

After the completion of a document, involving commands and results, the document can be printed by choosing File ▶ Print (which calls up the standard printing dialog).

To quit Maple, choose File ▶ Exit (Windows) or Maple 2019 ▶ Quit (Macintosh); that is, choose Exit from the File menu (Windows) or Quit from the Maple 2019 menu (Macintosh). Saving your work is prompted when quitting

Maple. (In the Command-line version type "quit," "done," or "stop" at the Maple prompt. CAUTION: these are the only commands that do not require a trailing semicolon in the Command-line version; there is no opportunity to save your work when using these commands.)

Procedures

The **proc** command is very useful. The following comes directly from the Maple Help page in Maple 2019 and explains and provides an example for a procedure.

Procedures

Calling Sequence	Evaluation Rules
Parameters	Notes
Description	Examples
Implicit Local Variables	Details
The Operands of a Procedure	

Calling Sequence

proc (parameterSequence) :: returnType; **local** localSequence; **global** globalSequence; **option** optionSequence; **description** descriptionSequence; **uses** usesSequence; statementSequence **end proc;**

Parameters

parameterSequence	- formal parameter declarations
returnType	- (optional) assertion on the type of the returned value
localSequence	- (optional) names of local variables
globalSequence	- (optional) names of global variables used in the procedure
optionSequence	- (optional) names of procedure options
descriptionSequence	- (optional) sequence of strings describing the procedure
usesSequence	- (optional) names of modules or packages the procedure uses
statementSequence	- statements comprising the body of the procedure

Description

- A procedure definition is a valid expression that can be assigned to a name.

That name may then be used to refer to the procedure in order to invoke it in a function call.

- The parenthesized parameterSequence, which may be empty, specifies the names and optionally the types and/or default values of the procedure's parameters. In its simplest form, the parameterSequence is just a comma-separated list of symbols by which arguments may be referred to within the procedure.

- More complex parameter declarations are possible in the parameterSequence, including the ability to declare the type that each argument must have, default values for each parameter, evaluation rules for arguments, dependencies between parameters, and a limit on the number of arguments that may be passed. See Procedure Parameters for more details on these capabilities.

- The closing parenthesis of the parameterSequence may optionally be followed by ::, a returnType, and a ;. This is *not* a type declaration, but rather an assertion . If kernelopts(assertlevel) is set to 2, the type of the returned value is checked as the procedure returns. If the type violates the assertion, then an exception is raised.

- Each of the clauses local localSequence;, global globalSequence;, option optionSequence;, description descriptionSequence;, and uses usesSequence; is optional. If present, they specify respectively, the local variables reserved for use by the procedure, the global variables used or modified by the procedure, any procedure options, a description of the procedure, and any modules or packages used by the procedure. These clauses may appear in any order.

- Local variables that appear in the local localSequence; clause may optionally be followed by :: and a type. As in the case of the optional returnType, this is not a type declaration, but rather an assertion. If kernelopts(assertlevel) is set to 2, any assignment to a variable with a type assertion is checked before the assignment is carried out. If the assignment violates the assertion, then an exception is raised.

- A global variable declaration in the global globalSequence clause cannot have a type specification.

- Several options that affect a procedure's behavior can be specified in the option optionSequence; clause. These are described in detail on their own page.

- The description descriptionSequence; clause specifies one or more lines of description about the procedure. When the procedure is printed, this description information is also printed. Even library procedures, whose body is generally elided when printing, have their description (if any) printed. The descriptionSequence is also used when information about the procedure is printed by the Describe command.

- The optional uses usesSequence; clause is equivalent to wrapping the statementSequence with a use statement. In other words,
  proc ... uses LinearAlgebra; ... end proc
  is equivalent to:
  proc ... use LinearAlgebra in ... end use; end proc
- The statementSequence consists of one or more Maple language statements, separated by semicolons (;), implementing the algorithm of the procedure.
- A procedure assigned to a name, f, is invoked by using f(argumentSeq). See Argument Processing for an explanation of argument passing.
- The value of a procedure invocation is the value of the *last* statement executed, or the value specified in a return statement.
- In both 1-D and 2-D math notation, statements entered between proc and end proc must be terminated with a colon (:) or semicolon (;).

Implicit Local Variables

- For any variable used within a procedure without being explicitly mentioned in a local localSequence; or global globalSequence; the following rules are used to determine whether it is local or global:
  The variable is searched for amongst the locals and globals (explicit or implicit) in surrounding procedures, starting with the innermost. If the name is encountered as a parameter, local variable, or global variable of such a surrounding procedure, that is what it refers to.
  Otherwise, any variable to which an assignment is made, or which appears as the controlling variable in a 'for' loop, is automatically made local.
  Any remaining variables are considered to be global.
- **Note:** Any name beginning with _Env is considered to be an environment variable, and is not subject to the rules above.

The Operands of a Procedure

- A Maple procedure is a valid expression like any other (e.g., integers, sums, inequalities, lists, etc.). As such, it has sub-parts that can be extracted using the op function. A procedure has eight such operands:

  op 1 is the parameterSequence,

  op 2 **is the** localSequence,

  op 3 **is the** optionSequence,

  op 4 **is the** remember table,

op 5 **is the** descriptionSequence**,**

op 6 **is the** globalSequence**,**

op 7 **is the lexical table (see note below), and**

op 8 **is the returnType** (if present).

- Any of these operands will be NULL if the corresponding sub-part of the procedure is not present.
- **Note:** The lexical table is an internal structure used to record the correspondence between undeclared variables and locals, globals, or parameters of surrounding procedures. It does not correspond to any part of the procedure as written.

Evaluation Rules

- Procedures have special evaluation rules (like tables) so that if the name f has been assigned a procedure then:
  f evaluates to just the name f,
  eval(f) yields the actual procedure, and
  op(eval(f)) yields the sequence of eight operands mentioned above (any or all of which may be NULL).
- Within a procedure, during the execution of its statementSequence, local variables have *single level evaluation*. This means that using a variable in an expression will yield the current value of that variable, rather than first evaluating that value. This is in contrast to how variables are evaluated outside of a procedure, but is similar to how variables work in other programming languages.

Notes

- Remember tables (option remember) should not be used for procedures that are intended to accept mutable objects (e.g., rtables or tables) as input, because Maple does not detect that such an object has changed when retrieving values from remember tables.

Examples

```
> lc := proc( s, u, t, v )
    description "form a linear combination of the arguments";
    s · u + t · v;
    end proc;
    lc := proc(s, u, t, v)
          description "form a linear combination of the arguments";
          s * u + t * v
     end proc
```

```
> lc(π, x − I, y)
```

$$\pi x - \mathrm{I}y$$

```
> Describe(lc)
# form a linear combination of the arguments
lc( s, u, t, v )
> lc
                                        lc
> eval(lc)
  proc (s, u, t, v)
       description "form a linear combination of the arguments";
       s * u + t * v
   end proc
> op(1, eval(lc))
                                      s, u, tv
> addList := proc(a::list, b::integer)::integer;
    local x,i,s;
    description "add a list of numbers and multiply by a constant";
    x:=b;
    s:=0;
    for i in a do
      s:=s+a[i];
    end do;
    s:=s*x;
  end proc;
  addList := proc(a :: list, b :: integer) :: integer;
  localx, i, s;
  description)"add a list of numbers and multiply by a constant";
  x := b; s := 0; for iin ado s := s + a[i]end do; s := s · x;
  end proc
> sumList := addList([1, 2, 3, 4, 5], 2)
                                  sumList := 30
```

Details

For details on defining, modifying, and handling parameters, see Procedure Parameters.

See Also

_nresults, assertions, Function Calls, kernelopts, Last-Name Evaluation, Procedure Options, ProcessOptions, procname, Reading and Saving Procedures, remember, return, separator, Special Evaluation Rules, use

---

We add an example procedure for using Newton's Method to find the roots of a differentiable function. First, recall the iterating formula for Newton's root-finding algorithm:

$$x_{new} = x_{old} - \frac{f(x_{old})}{f'(x_{old})}$$

We iterate the formula, finding new values of x_{new}, until the absolute difference $|x_{new} - x_{old}| < tolerance$ or $|f(x_{new})| < tolerance$.

The Maple procedure: Newton's Method for Finding Roots of a polynomial.
Assumptions:

1. The function must be differentiable.
2. You have a fair guess at a starting point x_0 for the method. We suggest plotting the function and estimating a value near the desired root.

Algorithm:

1. Pick a tolerance, *tol* (*tol* is small), and a maximum number of iterations, $MaxN$, allowed.
2. Pick an initial value $xold$.
3. Iterate $xnew := xold - f(xold)/f'(xold)$.
4. Stop when either $|xnew - xold| < t$ or $|f(x(new))|$ approximately equals 0.

Maple Procedure:

```
> Newton := proc(f, x0, tol, MaxN)
    local df, xold, x_new, i;
    df := D(f);
    xold := x0;
    print(xold);
    for i from 1 to MaxN do
      xnew := xold - f(xold)/f'(xold);
      if |f(xnew)| < tol or |xnew - xold| < tol then
        return(xnew);
        end if;
      xold := xnew;
      print(xold);
      end do;
    return(cat(MaxN, " iterate "), xold);
    end proc :
```

Let's consider an example.

$> f := x \to (x-5)^2 - 3;$

$$f := x \mapsto (x-5)^2 - 3;$$

We plot the function to be able to estimate the roots for the procedure.

$> plot(f(x), x = -1..10, thickness = 3, color = black);$

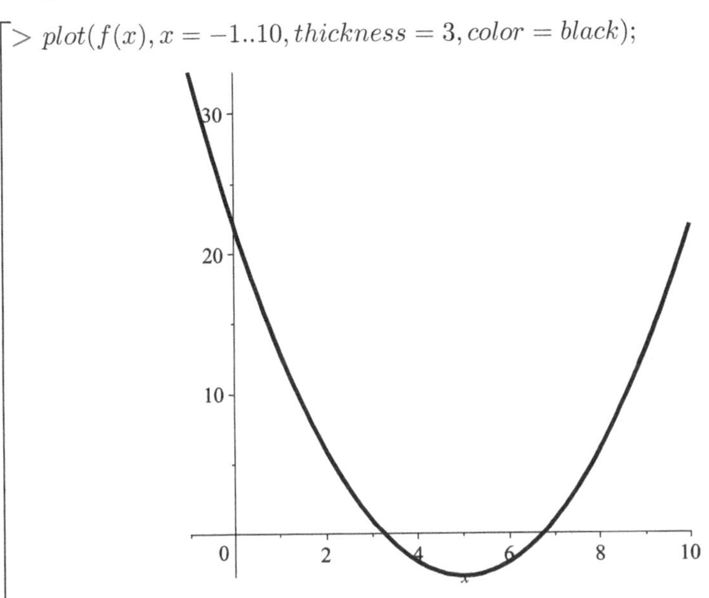

$> Newton(f, 2.0, 10^{-9}, 30);$

$$2.0$$
$$3.000000000$$
$$3.250000000$$
$$3.267857143$$
$$3.267949190$$
$$3.267949192$$
$$3.267949192$$

$> f(3.267949192);$

$$1.\ 10^{-9}$$

$> Newton(f, 7.0, 10^{-9}, 30);$

$$7.0$$
$$6.750000000$$
$$6.732142857$$
$$6.732050810$$
$$6.732050808$$
$$6.732050808$$

$> f(6.732050808);$

$$1.\ 10^{-9}$$

We found approximations to the two roots as $x = 3.2679449192$ and $x = 6.732050808$.

We will often create **proc** s in both volumes of this text to solve problems using specialized techniques.

A Quick Review of Key Commands

Assignments and Basic Mathematics

For example, let's compute $(1.2^{4.1} + 4.3(9.8))/34$.

> $(1.2^{4.1} + 4.3 \cdot (9.8))/34;$

$$1.301522146$$

All Maple statements are entered after the > prompt. Again note that Maple commands end with a semicolon; in the document interface, semicolons are optional. Maple output, printed in blue, is centered in the page.

To assign a label or name to a number or an expression, we use :=. For example, let's use two Maple statements to assign a the value 11.5 and b the value 9.

> $a := 11.5;$

$$a := 11.5$$

> $b := 9;$

$$b := 9$$

We can enter multiple commands on the same line or in the same input cell. If we separated the two commands with a colon instead of a semicolon, only the second command would be displayed in the output even though both commands were executed. A colon suppresses output.

> $a := 11.5 : b := 9;$

$$b := 9$$

Once the assignments have been made we can perform arithmetic operations. For example, let's compute $a^2 + b^3$, $a^2 \cdot b^3$, and $\sqrt{a^2 + b^3}$.

> $a^2 + b^3;$
> $a^2 \cdot b^3;$
> $sqrt(a^2 + b^3);$

$$861.25$$
$$96410.25$$
$$29.34706118$$

A very useful command is *evalf*. This command produces the decimal equivalent of a given expression. It's mnemonic is *eval*uate as *f*loating point (decimal).

> $c := evalf(a^2/b^3);$

$$0.1814128944$$

For expressions, we assign y using the same assignment operator :=.

> $y := x^2 + 21.6 \cdot x - 1$

$$y := x^2 + 21.6x - 1$$

To evaluate this type of expression for specific values of x, we use the *subs* command (substitution) or the *eval* command (evaluate). For example, we want to substitute 3 for x.

> $subs(x = 3, y);$

$$72.8$$

> $eval(y, x = 3);$

$$72.8$$

To use functional notation, such as $f(3)$, we must start with a different form of assignment. To create a function assignment f we use the *arrow operator* as follows. (Type '–', then '>' for the arrow operator; the image changes to '→'.)

> $f := x \rightarrow x^2 + 21.6 \cdot x - 1;$

$$f := x \mapsto x^2 + 21.6x - 1$$

> $f(3);$

$$72.8$$

We can even substitute variables such as $(x + h)$ for x;

> $f(x + h);$

$$(x + h)^2 + 21.6x + 21.6h - 1$$

The two forms, expressions (objects) and functions (operations), are very important in both programming and plotting as we shall see later. Creating a function from an expression is easy using the *unapply* command.

> $y := x^3 + 3 \cdot \cos(x) - 4;$

$$x^3 + 3\cos(x) - 4$$

> $f := unapply(y, x);$

$$f := x \mapsto x^3 + 3\cos(x) - 4$$

> $f(x);$

$$x^3 + 3\cos(x) - 4$$

> $f(2);$

$$4 + 3\cos(2)$$

Maple can easily handle functions of more than one variable. For example, consider a surface area defined by $\pi(x^2y^3 + 3)$. Suppose we want to evaluate the surface area at the point $(2, 5)$.

$> s := (x, y) \rightarrow \text{Pi} \cdot (x^2 \cdot y^3 + 3)'$

$$s := (x, y) \mapsto \pi(x^2 \cdot y^3 + 3)'$$

$> s(2, 5);$

$$503\,\pi$$

$> evalf(\%)$

$$1580.221105$$

The expression *evalf*(%) contains the symbol '%' that means insert the result of the last expression evaluated. *(This may not be the result just above in your Maple document if you've re-executed another statement.)*

Algebra and Calculus

Let's return to our expression $y = x^2 + 21.6x - 1$.

$> y = x^2 + 21.6x - 1;$

$$y = x^2 + 21.6x - 1;$$

Let's factor y. We can use the *factor* command or the *solve* command. The *solve* command has more utility.

$> factor(y);$

$$(x + 21.64619749)(x - 0.04619749036)$$

$> solve(y = 0);$

$$0.04619749036, -21.64619749$$

Let's consider the function $q(x) = ax^2 + bx + c$. We use the *solve* command and obtain the result:

$> restart;$

$> q := x \rightarrow ax^2 + bx + c$

$$q := x \mapsto ax^2 + bx + c$$

$> solve(q(x) = 0, x);$

$$\frac{-b + \sqrt{-4ac + b^2}}{2a}, \frac{-b + \sqrt{-4ac + b^2}}{2a}$$

Note the results are the quadratic formula. We also used the command *restart*. A *restart* forgets all previous assignments to f, a, b, c, and anything else we defined.

In calculus, we can differentiate and integrate in one and many variables. The commands for differentiation and integration are:

diff & *Diff*: differentiation or partial differentiation and 'indefinite' differentiation

int & *Int*: definite and indefinite integration

For example, let's differentiate an expression $y = 2x^2 + 24.1x - 1$, and then find the area under the curve in general and from $x = 1$ to $x = 4$.

> $y := 2x^2 + 24.1 \cdot x - 1$

$$y := 2x^2 + 24.1x - 1$$

> $Diff(y, x)$

$$\frac{d}{dx}\left(2x^2 + 24.1x - 1\right)$$

> $diff(y, x)$

$$4x + 24.1$$

> $Int(y, x)$

$$\int \left(2x^2 + 24.1x - 1\right) dx$$

> $int(y, x)$

$$0.6666666667x^3 + 12.05000000x^2 - x$$

> $Int(y, x = 1..4)$

$$\int_1^4 \left(2x^2 + 24.1x - 1\right) dx$$

> $int(y, x = 1..4)$

$$219.7500000$$

Note the difference between *diff* and *Diff*. The capital letters indicate the "inert form" of the command.

Often, we want to find critical points of the first derivative (where $y' = 0$). We can use the *solve* command as follows,

> $solve(diff(y, x) = 0, x);$

$$-6.025000000$$

Plotting and Graphs

Maple has an extremely detailed and developed *plot* command, which provides graphs in both two and three dimensions. For modeling purposes, 2-D plots will typically be used. We suggest loading the *plots* package, via *with(plots)*, prior to plotting. The syntax for *plot* is *plot(y, hr, vr, options)*, where y is the expression to be plotted, *hr* is the horizontal range and *vr* is the vertical range. Additionally, many other options can be added after the vertical range to control a variety of items. Table 1.5 presents a short list of options (defaults values are in the third column). See the Maple Help topic "plot/options" for the full list.

TABLE 1.5: Plot Command Options

Option	Description	Default
scaling =	*constrained* or *unconstrained*	*unconstrained*
style =	*point, line, patch* or *patchnogrid*	*line*
title =	"*a title*"	no title
thickness =	0, 1, 2, or 3	1
axes =	*framed, boxed, normal* or *none*	*normal*
view =	[*xmin..xmax, ymin..ymax*]	entire graph

Continuous Plots

The command, *plot(y):* will generate a 2-D plot of y with a default horizontal range of $-10..10$ and a vertical range that shows the curve. No other command information is required; however, the axes ranges and options may be specified to generate a specific plot. The default range can be specified as a finite range or an infinite range. By constraining the scale, equal units occur in both the x and y directions. However, a plot is generally easier to see when the scale is unconstrained, although it would be distorted (e.g., a circle would appear as an ellipse). Maple automatically scales the axes to spread the data over as large a space as possible, but this procedure does not imply that the area of interest will be plotted most effectively. As a result, the *view* option must be employed to ensure that the correct portion of the plot is best displayed. To demonstrate the plot command with a variety of options, a few examples are provided of $\sin(x)$ below, the first with the default ranges, the second with an infinite range.

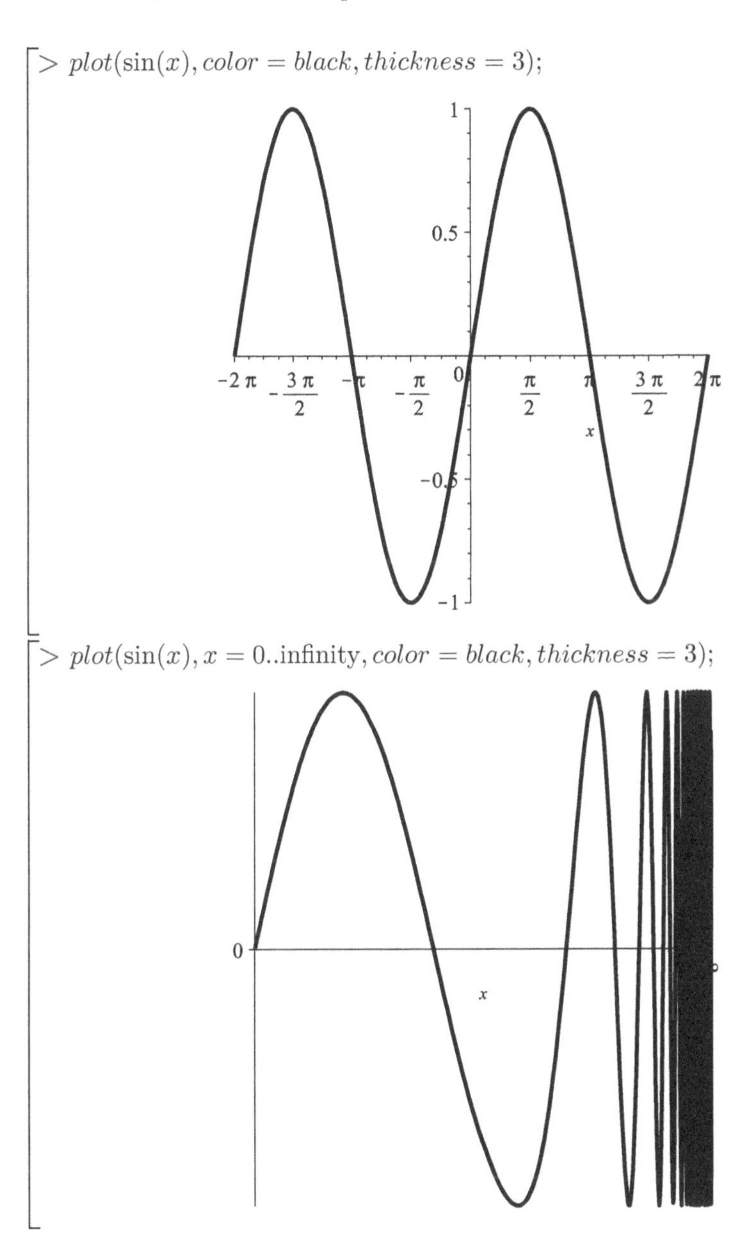

Scatterplots

The previous plots demonstrate functions with continuous x-values; either defaulted to $-10..10$ or selected by the user. However, Maple can also plot discrete sets of data. Using the data provided in Table 1.6, we present an example of plotting the ages of five people versus their respective weights.

TABLE 1.6: Age-Weight Data

Age (years)	1	5	13	17	24
Weight (lbs)	15	40	90	160	180

```
> age := [1, 5, 13, 17, 24] :
  wt := [15, 40, 90, 160, 180] :
> agewt := {seq([age[k], wt[k]], k = 1..5)}
          agewt := {[1, 15], [5, 40], [13, 90], [17, 160], [24, 180]}
> plot(agewt, style = point, symbol = diamond, symbolsize = 14)
```

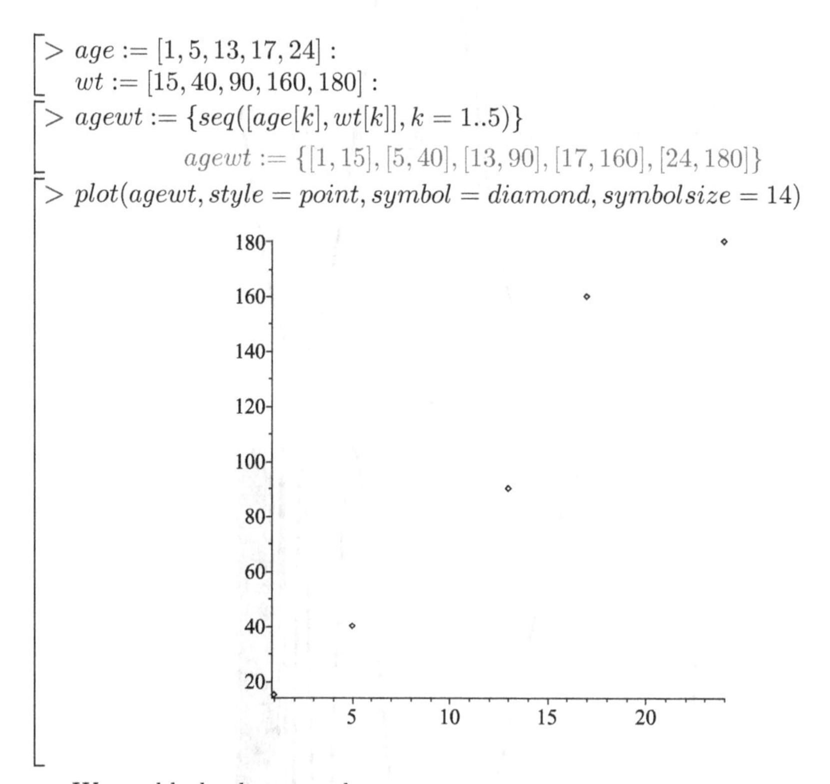

We could also have used

```
> with(plots) :
  pointplot(agewt, symbol = diamond, symbolsize = 14);
```

Multiple Plots

It is also an option to plot multiple functions on one set of axes. One method requires that both functions have the same domain, presented in Figure 1.11. The second method uses the *display* command which requires the *plots* package be loaded via *with(plots):*. This method does not restrict the domain, as presented below.

> $myOptions := (scaling = constrained, axes = framed, thickness = 2):$
> $plot([\sin(x), x^{0.5}], x = 0..5, myOptions);$

> $with(plots):$

> $curve := plot(30 \cdot \sin(x), x = 0..30):$

> $points := pointplot(agewt, symbol = diamond, symbolsize = 14):$

> $display(curve, points);$

An intriguing method of adding curves to an existing graph is to select the output (blue font) expression desired, then "drag and drop" that selection onto an existing plot image. Try it!

1.5 Maple Training

A Maple training video of the basic features is provided at the website:
https://www.maplesoft.com/support/training/quickstart.aspx.
The website has a self-contained video tutorial "Maple Quick Start." Maple's basic features covered include:

- Numeric Computations
- Symbolic Computations
- Programming Basic Maple Procedures
- Visualization
- Calculus
- Linear Algebra
- Student Packages
- Other Maple Packages

This training serves as a basic refresher of Maple commands. The website also has a "QuickStart" PDF reference.

1.6 Maple Applications Center

The *Maple Applications Center* found at
https://www.maplesoft.com/applications/
contains a large collection of worksheets available to students and practitioners alike. The authors have several worksheets available on the site.

Exercises

Perform the following operations in Maple:

1. $\sqrt{2.3^5(4.5)}$
2. $11.3^3 + 5.1^2$
3. $\sqrt[3]{21.6}$
4. Let $a = 8$, $b = 7$, then compute $\left(a^2 - b^2\right)$.
5. $2(11.5) + 6.2^2(0.7)$

Enter the following functions. Obtain a graph. Find roots and solve for the intercepts.

6. $f(x) = -x^2 + 3x + 3$
7. $f(x) = x^2 - 3x - 1$
8. $f(x) = -0.1213x^4 + 3.462x^3 - 29.22x^2 + 64.68x + 97.69$
9. $g(x) = x^3 - 2x^2 - 5x + 6$
10. $s(x) = 2x^3 - 3x^2 - 11x + 7$

Enter the following date sets into Maple and obtain a scatter plot for each:

11.

x	1	3	8	10
y	0.7	5	15.2	36

12.

t	7	14	21	28	35	42
P	8	41	133	250	280	297

13.

x	29	48	72.7	92	118	140	165	199
y	0.49	0.82	1.23	1.54	1.97	2.34	2.74	3.30

Perform the required function in Maple.

14. $\dfrac{d}{dx}\left(1.104x - 0.542x^2\right)$
15. $\displaystyle\int 1.104x - 0.542x^2\,dx$
16. $\displaystyle\int_1^5 1.104x - 0.542x^2\,dx$

References

Modeling References

[A2011] Brian Albright, *Mathematical Modeling with Excel*, Jones & Bartlett Publishers, 2011.

[GFH2013] Frank Giordano, William P. Fox, and Steven Horton, *A First Course in Mathematical Modeling*, Nelson Education, 2013.

[M2007] Mark Meerscheart, *Mathematical Modeling*, Third Edition. 2007.

Maple References and Help

[T2017] Ian Thompson, *Understanding Maple*, 2017.

[MUM2018] Maplesoft, *Maple User Manual*, 2018. https://www.maplesoft.com/support/help/category.aspx?CID=2317

[MPG2018] Maplesoft, *Maple Programming Guide*, 2018. https://www.maplesoft.com/support/help/category.aspx?CID=2318

[MPG2012] L. Bernardin, P. Chin, P. DeMarco, K. O. Geddes, D. E. G. Hare, K. M. Heal, G. Labahn, J. P. May, J. McCarron, M. B. Monagan, D. Ohashi, and S. M. Vorkoetter, *Maple Programming Guide*, 2012.

[EM2007] Richard H. Enns and George McGuire, *Computer Algebra Recipes: An Advanced Guide to Scientific Modeling*, 2007.

2

Introduction, Basic Concepts, and Techniques in Problem Solving with First-Order, Ordinary Differential Equations

Objectives

(1) Define and build a differential equation for a real problem.

(2) Find a closed-form solution and a numerical solution.

(3) Graph a solution.

(4) Understand equilibrium values and stability.

(5) Solve both linear and nonlinear systems of differential equations.

(6) Solve systems of differential equations.

(7) Use numerical approaches to obtain good approximate solutions to differential equations and systems of differential equations.

2.1 Introduction

Consider a sports parachutist jumping from a plane. We want to build a model on how far the parachutist free falls before they open their parachute. We might initially assume the person is falling from rest and choose to neglect all resistive forces. We might choose to neglect a variable because we want to investigate the problem without the influence of that variable. For instance, we might want to study the falling body by neglecting buoyancy but considering the effects of drag within the earth's atmosphere. We can simplify the model for the resistive force by making reasonable assumptions that the drag force equals some constant times the speed of the falling body or we might decide to try a constant times the speed squared of the falling body. Next, we want to construct a model for propulsion, drag, and buoyancy forces. The propulsion force acting on the falling body from rest is due to gravity. This gravitational attraction in turn depends on the mass of the falling body and its distance above the surface of the earth. Thus

$$\begin{aligned} \textit{Propulsion force} &= F_p \\ &= \textit{gravitational attraction} \\ &= f(\textit{mass, distance}) \end{aligned}$$

Next, consider the resistive force, which is the sum of the drag and buoyancy forces:

$$\begin{aligned} \textit{Drag} &= F_d \\ &= f(\textit{speed, air density, body's cross-sectional area, body's shape}) \end{aligned}$$

and

$$\begin{aligned} \textit{Buoyancy} &= F_b \\ &= f(\textit{air density, body's density}) \end{aligned}$$

A free body diagram displays all the forces acting on the body with a coordinate system indicating directions. From the free body diagram in Figure 2.1,

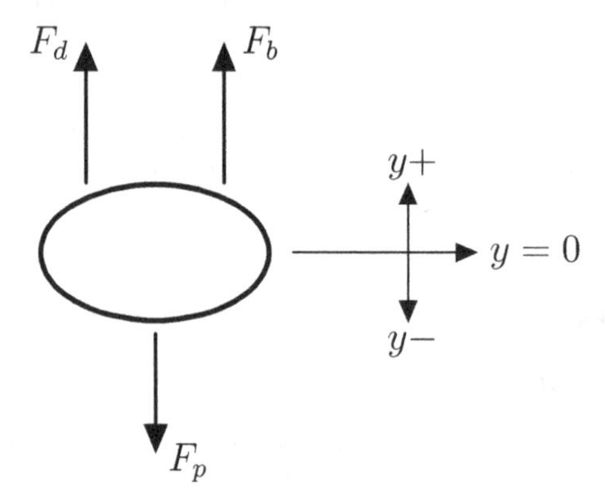

FIGURE 2.1: A Free Body Diagram

we see that the total force F acting on the body is given by

$$F = F_p + F_d + F_b$$

We assume that Newton's second law applies, $F = ma$ (*itself a mathematical model*). If we assume we can neglect all resisting forces we find

$$F = ma = -mg + 0 + 0$$

or

$$my'' = -mg,$$

where y'' is a second derivative and subject to initial conditions $y(0) = h, y'(0) = 0$.

Here m is the mass of the body, g is the acceleration due to gravity, and y is the distance of the body measured from the surface of the earth. The acceleration a is the second derivative of y with respect to time. It is this idea of change that led to a differential equation. Since the propulsion force F_p is acting down, it is negative.

If we were to include the force due to air resistance as a constant times speed then $F_r = kv$. We then would have the model

$$\begin{aligned} F &= F_p + F_d + F_b \\ F &= ma = my'' \\ &= -mg + kv \end{aligned}$$

with the same initial conditions $y(0) = h, y'(0) = 0$.

In this chapter, we will consider a variety of models, concepts, and techniques that give the reader some of the basic tools needed in solving and analyzing first-order differential equations. Since differential equations are used often in solving problems related to continuous change, we devote a more thorough approach in this chapter.

Section 2.2 provides some models that we will solve and analyze. These models come from a variety of disciplines in science and engineering, including chemistry, physics, biology, fluids mechanics, Newtonian Mechanics, environmental engineering, and financial mathematics. Modeling techniques are discussed that help determine the necessary coefficients in the models.

Section 2.3 introduces slope fields. Slope fields provided a qualitative view of solutions of first-order differential equations. Recall that we said this book discusses the modeling of change. Slope fields provide information about rates of change since they are derived from information given by the derivative.

Section 2.4 covers the analytical method of solution with Maple. Existence and uniqueness are addressed from conceptual and theoretical bases. The focus is on the initial-value problem and the question as to whether or not the initial value problem has a unique solution, and whether new functions can be defined by the initial-value problem.

Section 2.4 investigates the special class of ODEs called *linear differential equations.*

Section 2.5 is devoted to various numerical techniques. Euler, Improved Euler, and Runge-Kutta techniques are introduced as methods to approximate solutions. Since problems may arise when using numerical methods, some errors are discussed, as well as methods to compare long-term behavior of the solutions to determine the qualitative accuracy of generated solutions.

2.2 Applied First-Order Differential Equations and Solution Methods

In this section, we introduce many methods to solve differential equations across a variety of disciplines. Our emphasis in this section is building and solving the mathematical model and its associated differential equation that will be solved later in the chapter. Recall we have previously discussed the problem-solving process. In this section, we will confine ourselves to the first three steps: (1) identifying the problem, (2) assumptions and variables, and (3) setting up the differential equation problem.

Example 2.1. Newton's Law of Cooling.
You are working as an assistant for the Crime Scene Investigators at a potential homicide. A body was found in a trash dumpster and you were called to the site immediately. The ambient temperature is approximately 68°F. The body temperature is 77 degrees. After one hour, the body temperature is 74°F.

Problem.
Build a differential equation relating a dead body's cooling temperature to time in order to find the time of death.

Assumptions:
A simplified list could include normal body temperature, average seasonal temperature outdoors, constant indoor temperature, and no other outside sources of heat or cooling.

Mathematical Problem:
Newton's Law of Cooling states that the surface temperature of an object changes at a rate proportional to its relative temperature. That is, the difference between its temperature and the surrounding environment temperature A. Let $T(t) =$ the temperature of the object at time t. According to Newton's Law,

$$T(t) = k \cdot (A - T(t))$$

where k is the proportionality constant.

Example 2.2. Compound Mixtures.
Consider an initial amount of a substance, say barley measured in pounds, is dissolved in a large vat in order to make beer. The solution is pumped in at one rate, well-mixed, and pumped out at the same rate. We want to know the concentration of the substance in the vat after time.

Problem:
Relate the concentration of barley in the vat to time t.

Assumptions:
We know the size of the vat, 300 gallons. The initial amount dissolved is 50 pounds. The pumping rate is constant both into and out of the vat. No other substances are interacting with the vat and the pumps.

Model Selection:

The rate of change of concentration is measured as the difference between the rate pumped in and the mixture being pumped out of the vat. Let $C(t)$ be the concentration or amount of barley in the vat after time t, then

$$C'(t) = r_{in} - r_{out}$$

Example 2.3. Population Growth or Decay.
Consider an experiment measuring the growth of a yeast culture. An analysis yields the growth rate as approximately 0.6. We might try as our ODE $Y'(t) = Y_0 \cdot e^{0.6t}$.

Note that this model will predict a population that increases forever. We might think that is not possible.

Model Refinement: Modeling Births, Deaths, and Resources

Both births and deaths during a period are proportional to the population, then the change in population itself should be proportional to the population, as was illustrated above. However, certain resources (food, for instance) can support only a maximum population level rather than one that increases indefinitely. As these maximum levels are approached, growth should slowly move towards the carrying capacity.

Suppose we estimate the carrying capacity to be 665. Nevertheless, as p_n approaches 665, the change does slow considerably. Since $665 - -p_n$ does get smaller as p_n approaches 665, consider the following analysis:

If we accept the proportionality argument, we can estimate the slope of the line approximating the data to be about $k \sim 0.00082$, which gives the model:

$$p_{n+1} - p_n = 0.00082\, p_n \cdot (665 - p_n).$$

Converting to an ODE yields

$$p'(t) = 0.00082\, p(t) \cdot (665 - p(t)).$$

Example 2.4. The Spread of a Contagious Disease.
Suppose there are 400 students in a college dormitory and that one or more of the students has a severe case of the flu. Let i_n represent the number of infected students after n time periods. Assume some interaction between those infected and those not infected is required to pass the disease. If all are susceptible to the disease, $(400 - i_n)$ represents those susceptible but not yet infected. If those infected remain contagious, we may model the change in infected as proportional to the product of those infected by those susceptible but not yet infected, or

$$i_{n+1} - i_n = k \cdot i_n \cdot (400 - i_n).$$

In this model the product $i_n \cdot (400 - i_n)$ represents possible interactions between those infected and those not infected. A fraction k of those interactions would now become infected. There are many refinements to the above model. For example, we can consider that a segment of the population is not susceptible to the disease, that the infection period is limited, that infected students are removed from the dorm to prevent interaction with uninfected students, and so forth.

TABLE 2.1: Some Classical First-Order Equations in Differential Equations

Classical Model	Differential Equation
Decay	$\frac{dQ}{dt} = -kQ, \quad k > 0$
Growth	$\frac{dQ}{dt} = -kQ, \quad k < 0$
Newton's Law of Cooling	$\frac{dT}{dt} = k(T - T_0), \quad k > 0$
Chemical Mixtures, Drug Dissemination, Dialysis	$\frac{dA}{dt} = R_{in} - R_{out}$
L-R Series Circuit	$L\frac{dI}{dt} + RI = E(t)$
Falling Bodies, Newton's Law $\sum$ Forces $= 0$, $(k > 0)$	$m\frac{dV}{dt} = F_{weight} + F_{resistance} = mg - kv$
Chemical Reactions	$\frac{dX}{dt} = k(\alpha - X) \cdot (\beta - X) \cdot \ldots$
Spread of a Disease or Rumor	$\frac{dN}{dt} = kN(L - N), \quad k > 0$ and $L > 0$
Tank with cross-sectional area A and height h, filling from the top at rate K, while draining from the bottom through an orifice of area α.	$\frac{dh}{dt} = \frac{k - \alpha a\sqrt{2gh}}{A}$

2.3 Slope Fields and Qualitative Assessments of Autonomous First-Order ODEs

We have used the modeling process to create a number of models that are important to differential equations. However, sometimes we can predict outcomes of models graphically without actually solving them. This is true when we are looking for behavior rather than specific values. For example, will a specific population survive is a question that can be answered graphically.

In calculus, we used derivative information to analyze a function or just to gain more information. Knowing where a function is increasing or decreasing, locations of extrema, concavity, and so forth can be important information.

Interpreting the derivative as the slope of the line tangent to the graph of the function is useful in gaining information about the solution to the differential equation. From this information, it is possible to sketch qualitatively the solution curves to the equations from any real initial condition. These graphs provide considerable information regarding the solution and these graphs provide practical benefits as well. Since most real differential equations cannot be solved analytically, other techniques, such as graphical analysis are needed to analyze the behavior of solutions.

Creating a Slope Field

Suppose the first-order differential equation $dy/dt = f(t, y)$ is given. Each time an initial value $y(t_0) = y_0$ is specified, the solution curve is required to pass through the coordinate point (t_0, y_0) and we also know the value of slope at that point, $m = \frac{dy}{dt} = f(t_0, y_0)$. Therefore, it is possible to draw a short line segment with the correct slope through each point (t, y) in the t-y plane. The resulting image is known as the *slope field of the first-order differential equation.* We can say that the slope field of a differential equation is the set of all line segments with the correct slope or an array of mini-tangent lines.

Let's sketch the possible solution curves of the differential equation, without solving the DE. Consider the information in Table 2.2 below that provides specific points in the plane with the slope at each point.

TABLE 2.2: Points with Their Slopes

Point	Slope	Point	Slope
$(0, 1)$	$-1/2$	$(-2, 0)$	-1
$(0, 2)$	-1	$(1, 1)$	0
$(1, 0)$	$1/2$	$(-1, 1)$	-1
$(2, 0)$	1	$(2, 2)$	0
$(-1, 0)$	$-1/2$	$(1, 2)$	$-1/2$

It would be extremely tedious — and error prone — to calculate and graph sufficiently many slopes to have a useful image. So, we turn to technology.

In Maple, first load the DEtools package via entering *with*(*DEtools*). Then we enter the differential equation, and the commands to get a specific slope field plot. The command sequence to produce a slope field for $dy/dt = (t-y)/2$ over the region $[-2,2] \times [0,2]$ is:

$[> with(DEtools):$

$[> deq := diff(y(t),t) = (t - y(t))/2;$

$[> DEplot(deq, y(t), t = -2..2, y = 0..2);$

A complete slope field over the region $[-2,2] \times [0,2]$ is seen in Figure 2.2.

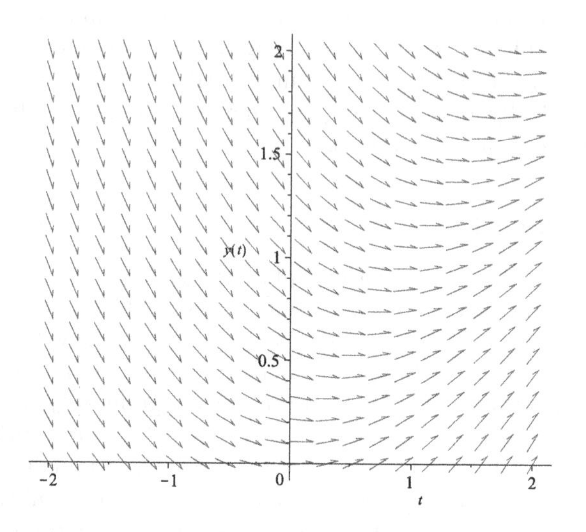

FIGURE 2.2: Slope Field for $dy/dt = (t - y)/2$

We use Maple to produce the slope field plot. Although knowledge of how to graph slope fields is important, you will not want to graph complete slope fields without technology. A solution curve is then tangent to the slope line at each point through which the curve passes. Thus, the slope field gives a visual representation of a family of possible solution curves of the differential equation. Figure 2.3 shows two solution curves, with starting points $(0, 2)$ and $(0, 900)$, added the slope field of a logistic differential equation.

Examining the slope field, we notice that the solution curves do not cross tangent line segments — the segments are tangent to the solution.

The Maple commands to obtain slope fields and possible solution curves for several first-order differential equations appear below. Remember to first enter *with*(*DEtools*).

$[> SlopeField1 := diff(y(t),t) = y(t) - t^2;$

$$SlopeField1 := \frac{d}{dt}\, y(t) = y(t) - t^2;$$

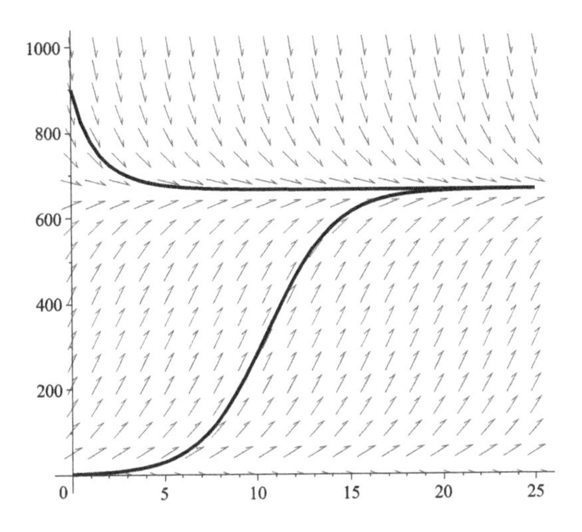

FIGURE 2.3: Slope Field for $dP/dt = 0.0008271 \cdot P \cdot (665 - P)$ with Starting Points $(0, 2)$ and $(0, 900)$

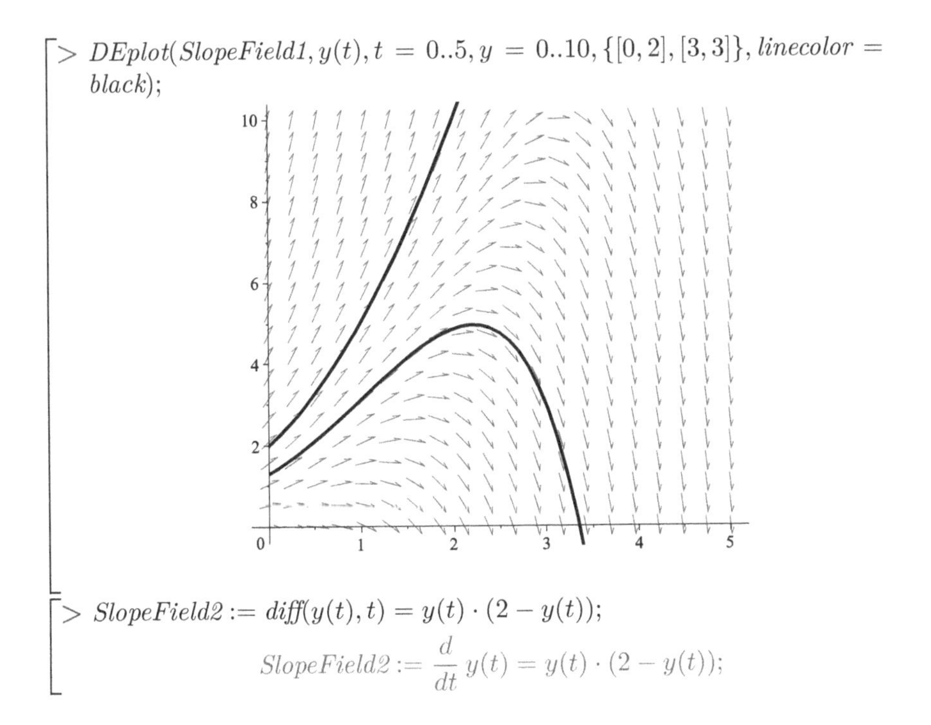

> $DEplot(SlopeField1, y(t), t = 0..5, y = 0..10, \{[0, 2], [3, 3]\}, linecolor = black);$

> $SlopeField2 := diff(y(t), t) = y(t) \cdot (2 - y(t));$

$$SlopeField2 := \frac{d}{dt}\, y(t) = y(t) \cdot (2 - y(t));$$

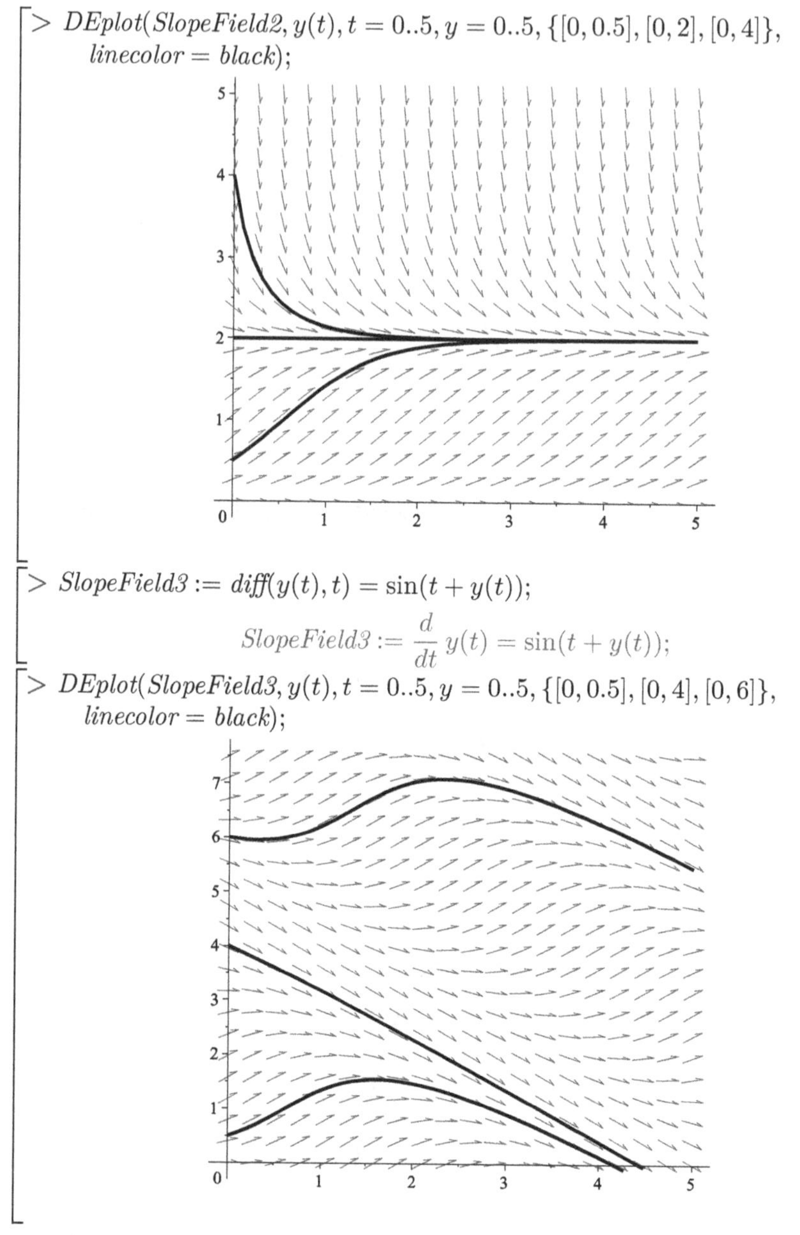

> $DEplot(SlopeField2, y(t), t = 0..5, y = 0..5, \{[0, 0.5], [0, 2], [0, 4]\},$
> $linecolor = black);$

> $SlopeField3 := diff(y(t), t) = \sin(t + y(t));$

$$SlopeField3 := \frac{d}{dt}\, y(t) = \sin(t + y(t));$$

> $DEplot(SlopeField3, y(t), t = 0..5, y = 0..5, \{[0, 0.5], [0, 4], [0, 6]\},$
> $linecolor = black);$

Slope fields provide us with an informative display of possible solution curves and their qualitative behavior for a differential equation.

Exercises

1. Construct a slope field for the following differential equations and sketch in a solution curve.

   a. $\dfrac{dy}{dx} = y$

   b. $\dfrac{dy}{dx} = x$

   c. $\dfrac{dy}{dx} = x + y$

   d. $\dfrac{dy}{dx} = x - y$

   e. $\dfrac{dy}{dx} = xy$

   f. $\dfrac{dy}{dx} = y^{-1}$

2. Sketch a number of solutions to the following equations showing the correct slope, concavity, and any points of inflection.

   a. $\dfrac{dy}{dx} = (y+2)(y-3)$

   b. $\dfrac{dy}{dx} = y^2 - 4$

   c. $\dfrac{dy}{dx} = y^3 - 4$

   d. $\dfrac{dy}{dx} = x - 2y$

3. Analyze graphically the equation $\dfrac{dy}{dt} = ry$, when $r < 0$ What happens to any solution curve as t becomes large?

4. Develop graphically the following models. First graph $\dfrac{dP}{dt}$ versus P, then obtain various graphs of P versus t by selecting different initial values $P(0)$ (as in our population example in the text). Identify and discuss the nature of the equilibrium points in each model.

   a. $\dfrac{dP}{dt} = a - bP$ with $a, b > 0$

   b. $\dfrac{dP}{dt} = P \cdot (a - bP)$ with $a, b > 0$

   c. $\dfrac{dP}{dt} = k(M - P)(P - m)$ with $k, M, m > 0$

d. $\frac{dP}{dt} = k(m-P)(P-m)$ with $k, m > 0$

5. The Department of Fish and Game in a certain state is planning to issue deer hunting permits. It is known that if the deer population falls below a certain level, m, then the deer will become extinct. It is also known that if the deer population goes above the maximum carrying capacity of the environment, M, the population will decrease to M. M and m are the *carrying capacity* and *threshold*, respectively.

   (a) Discuss the reasonableness of the model for the growth rate of the deer population as a function of time: $\frac{dP}{dt} = kP \cdot (M-P)(P-m)$ where P is the population of the deer and k is a constant of proportionality. Include a graph of $\frac{dP}{dt}$ versus P as part of your discussion.

   (b) Explain how this growth rate model differs from the logistic model $\frac{dP}{dt} = kP \cdot (M-P)$. Is it better or worse than the logistic model?

   (c) Show that if $P > M$ for all t, then the limit of $P(t)$ as $t \to \infty$ is M.

   (d) Discuss what happens if $P < m$ for all t.

   (e) Assuming that $m < P < M$ for all t, explain briefly the steps you would use to solve the differential equation. Do not attempt to solve the differential equation.

   (f) Graphically discuss the solutions to the differential equation. What are the equilibrium points of the model? Explain the dependence of the equilibrium level of P on the initial conditions. How many deer hunting permits should be issued?

2.4 Analytical Solution of First-Order Ordinary Differential Equations

2.4.1 Separable Ordinary Differential Equations

A differential equation of the form $\frac{dy}{dt} = f(t, y)$ is called *separable*, if $f(t, y) = h(t) \cdot g(y)$. That is, $\frac{dy}{dt} = h(t) \cdot g(y)$. You need to able to write the function $f(t, y)$ as the product of two functions $h(t)$ of t only and $g(y)$ of y only.

Example 2.5. A Simple Separable ODE.

$$\frac{dy}{dt} = t \cdot y$$

This ODE is separable because the function $f(t, y) = t \cdot y$ can be written as the product of $h(t) = t$ and $g(y) = y$.

Example 2.6. A Simple Nonseparable ODE.

$$\frac{dy}{dt} = t + 9y$$

This ODE is not separable since $f(t, y) = t + 9y$ cannot be written as the product of two functions $h(t)$ of t only and $g(y)$ of y only.

Sometimes it's is more complicated to see whether an ODE is separable or not as in the following example.

Example 2.7. A More Complex Separable ODE.

$$\frac{dy}{dt} = 8ty - 12y - 2t + 3$$

This ODE can be written as $\frac{dy}{dt} = (2t-3)\cdot(4y-1)$ and is seen to be separable when in this form.

In order to solve separable ODEs, we perform the following steps:

1. Rewrite the differential equation into a clearly separable form of

$$\frac{dy}{dt} = h(t) \cdot g(y)$$

2. Separate the differential equation as

$$\frac{dy}{g(y)} = h(t)\, dt$$

3. Integrate both sides of the expression, and include a constant of integration C,

$$\int \frac{1}{g(y)}\, dy = \int h(t)\, dt$$

to obtain the general solution $y(t)$ (usually in implicit form)

$$G(y) = H(t) + C$$

4. If you are given an initial value problem (IVP), use the initial conditions to find its particular solution; i.e., to determine C.

Example 2.8. Solving a Simple Separable ODE.
Solve $\frac{dy}{dt} = t \cdot y(t)$.

1. This ODE is separable with $h(t) = t$ and $g(y) = y$.

2. Separate:

$$\frac{dy}{y} = t\,dt$$

3. Integrate:

$$\begin{aligned}
\int \frac{1}{y}\,dy &= \int t\,dt \\
\ln(y) + C &= \frac{t^2}{2} + C \\
e^{\ln(y)} &= e^{t^2/2+C} \\
y &= c\,e^{t^2/2}
\end{aligned}$$

If the desired particular solution of the ODE passes through the point $(1,1)$, we use $(t,y) = (1,1)$ as the initial condition. Then $y = e^{(t^2-1)/2} \approx 0.60653e^{t^2/2}$.

Figure 2.4 shows a graph of the solution passing through the initial value.

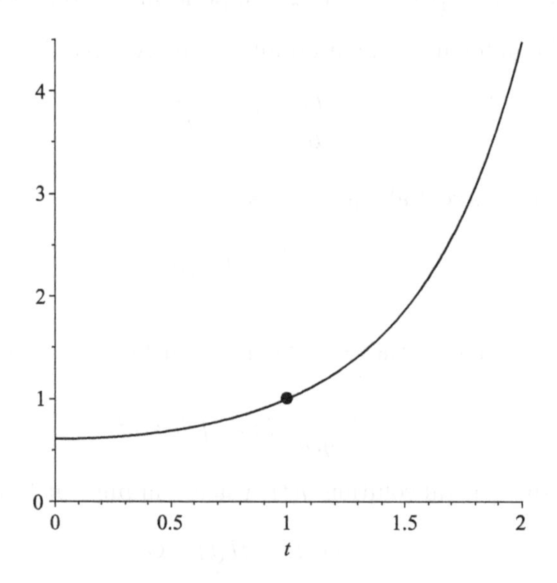

FIGURE 2.4: The Particular Solution to $\dfrac{dy}{dt} = t \cdot y(t)$ Passing through $(1,1)$

Example 2.9. Radioactive Decay.
Consider the radioactive decay model, $dQ/dt = -kQ$ where $k > 0$, with a radioactive isotope that has a half-life of 16 days. You wish to have 30 grams at the end of 25 days.

First, solve the separable ODE.

$$\frac{dQ}{dt} = -kQ$$

separates and integrates to become

$$\begin{aligned} \ln(Q) &= -kt + C \\ Q &= c\, e^{-kt} = Q_0 e^{-kt} \end{aligned}$$

Since the half-life is 16 days, then

$$\frac{1}{2} Q_0 = Q_0 e^{-16k}$$

so that

$$k = \frac{\ln(2)}{16} \approx 0.043322$$

Therefore

$$Q = Q_0 e^{-0.043322\, t}$$

We want 30 grams after 25 days, so

$$\begin{aligned} 30 &= Q_0 e^{-0.043322(25)} = Q_0 e^{-1.083042} \\ Q_0 &= 88.61 \end{aligned}$$

Initially, we should start with 88.61 grams of the radioactive substance. This problem is common in oncology treatments.

The desired model is $Q = 88.61\, e^{-0.043322\, t}$.

Example 2.10. Malthusian Population Model.
Model the population of the US provided in Table 2.3. (Data source: U.S. Census Bureau.)

TABLE 2.3: U.S. Population 1990 2000

Year	1990	1991	1992	1993	1994	1995
Pop (millions)	249.6	253.0	256.5	259.9	263.1	266.3

Year	1996	1997	1998	1999	2000
Pop (millions)	269.4	272.6	275.9	279.0	282.2

The model that we will initially use is our simple growth model $dP/dt = kP$, $k > 0$, where t is the year past 1990. Solve this separable ODE with our standard procedure:

1. Separate:
$$\frac{dP}{P} = k\,dt$$

2. Integrate:
$$\int \frac{1}{P}\,dP = \int k\,dt$$
$$\ln(P) = kt + C$$
$$P = c\,e^{kt} = P_0\,e^{kt}$$

The initial population in 1990 was 249.6 (million), so $P(0) = P_0 = 249.6$. Therefore
$$P = 249.6\,e^{kt}$$

Now, we need to find a good estimate for the growth parameter k. The estimate we'll use comes from substituting the values for the middle year, 1995, in our model, then solving for k. (Later, we'll investigate better methods for approximating k.) We find the estimate is $k \approx 0.010535$. Then
$$P = 249.6\,e^{0.010535t}$$

Predicting the population in 2020 yields $P(20) = 249.6\,e^{0.010535\cdot 20} = 308.14$ million people. The U.S. Census Bureau's projection for 2020 is 334.5 million people.

There are two problems with our model. First, it is not realistic to assume that populations tend to infinity as an exponential model must. Second, the predictions of this model always underestimate the real populations. We could use a refinement to this model which will be suggested later that does not yield the slow exponential growth toward infinity.

Example 2.11. Newton's Law of Cooling.
Suppose a cold can of soda is taken from a refrigerator and placed in a warm classroom. The temperature is measured periodically. The temperature of the soda is initially 40° Fahrenheit; room temperature is a cozy 72°F. Model the temperature of the soda.

Using Newton's Law of Cooling gives
$$\frac{dT}{dt} = -k(T_a - T), \quad T(0) = 40.$$

Assume that we have measured $k = 0.05$. Substituting our values for k and room temperature, T_a, gives
$$\frac{dT}{dt} = -0.05(72 - T), \quad T(0) = 40.$$

Solve this separable IVP with our standard method.

1. Separate:

$$\frac{dT}{72-T} = -0.05\,dt$$

2. Integrate:

$$\begin{aligned}\int \frac{dT}{72-T} &= \int -0.05\,dt\\ \ln(72-T) &= -0.05\,t + C\\ 72 - T &= c\,e^{-0.05t}\\ T &= 72 - c\,e^{-0.05t}\end{aligned}$$

3. Determine c by substituting the initial condition $T(0) = 40°\text{F}$.

$$\begin{aligned}40 &= 72 - c\,e^{-0.05\cdot 0}\\ c &= 32\end{aligned}$$

Our model is $T = 72 - 32e^{-0.05t}$.

After 10 minutes the soda's temperature is $T(10) = 72 - 32e^{-0.05\cdot 10} = 52.591$. After 30 minutes the soda's temperature is $T(30) = 72 - 32\,e^{-0.05\cdot 30} = 64.860$.

Note the temperature of the soda warms and eventually will reach the temperature of the room, a cozy 72° Fahrenheit (but not cozy for drinking a soda).

Example 2.12. Chemical Mixtures.
Initially 50 pounds of salt is dissolved in a tank containing 300 gallons of water. A brine solution is pumped into the tank at a rate of three gallons per minute. A well-stirred mixed solution is then pumped out at a rate of three gallons per minute. Determine the amount of salt present at any time t if the concentration of the entering solution is two pounds per gallon.

Write a model for the problem as

$$\frac{dA}{dt} = R_{in} - R_{out}$$

where A represents the amount of salt (pounds) in the tank at time t. Now, determine the Rs, the rate salt enters/leaves (lb/min), respectively, from

$$\begin{aligned}R_{in} &= \frac{3\text{ gallons}}{1\text{ minute}} \cdot \frac{2\text{ pounds}}{1\text{ gallon}}\\ &= 6\,\frac{\text{pounds}}{\text{minute}}\end{aligned}$$

and

$$R_{out} = \frac{3 \text{ gallons}}{1 \text{ minute}} \cdot \frac{A \text{ pounds}}{300 \text{ gallon}}$$

We have the IVP

$$\frac{dA}{dt} = 6 - \frac{A}{100}, \quad A(0) = 50$$

Rewriting as

$$\frac{dA}{dt} = 0.06(600 - A), \quad A(0) = 50$$

makes it easy to see the ODE is separable.

Once again, apply our standard technique to solve the IVP.

1. Separate:

$$\frac{dA}{600 - A} = 0.06\, dt$$

2. Integrate:

$$\begin{aligned} \int \frac{dA}{600 - A} &= 0.06\, dt \\ -\ln(600 - A) &= 0.01t + C \\ A &= 600 + c\, e^{-0.01t} \end{aligned}$$

3. Use the initial value to find that $c = -550$.

Our model for the amount of salt in the tank at time t is $A(t) = 600 - 550\, e^{-0.01t}$.

Example 2.13. A Refined Population Model.
A *logistic growth model* defined for blue crabs in Venezuela is

$$\frac{dB}{dt} = 0.25B(10 - B), \quad B(0) = 4$$

where B is in millions of bushels. Determine the maximum sustainable amount (millions of bushels).

The logistic DE is separable, so we can apply our standard technique.

1. Separate:

$$\frac{dB}{B(10 - B)} = 0.25\, dt$$

2. Integrate:

$$\int \frac{dB}{B(10 - B)} = \int 0.25\, dt = 0.25t + C$$

The integral on the left will require *partial fractions* to evaluate.

$$\frac{1}{B(10-B)} = \frac{a_1}{B} + \frac{a_2}{10-B}$$

Use your favorite technique to find $a_1 = 1/10$ and $a_2 = 1/10$, so that

$$\begin{aligned}\int \frac{dB}{B(10-B)} &= \int \frac{1}{10B} + \frac{1}{10(10-B)}\, dB \\ &= \frac{1}{10}\big[\ln(B) - \ln(10-B)\big]\end{aligned}$$

Putting the results together gives

$$\frac{1}{10}\big[\ln(B) - \ln(10-B)\big] = 0.25t + C$$

3. Determine that $C = -0.0405465$ by using $B(0) = 4$.

Using the laws of logarithms and exponents, we obtain

$$\frac{B}{10-B} = 0.666667 \cdot e^{2.5t}$$

We simplify to

$$B = \frac{6.666667 \cdot e^{2.5t}}{1 + 0.666667 \cdot e^{2.5t}} \quad \text{or} \quad B = \frac{6.666667}{0.666667 + e^{2.5t}}$$

The result is known as a *logistic curve.* It approaches the maximum sustainable amount of 6.666 million bushels as t goes to infinity.

2.4.2 Linear Differential Equations

Let's consider a class of differential equations of the form

$$a_1(t)\frac{dy}{dt} + a_0(t)y = g(t)$$

which can be simplified by dividing every term by $a_1(t)$ to obtain

$$\frac{dy}{dt} + P(t)y = f(t).$$

We seek a solution to this differential equation. Notice that the left-hand side $(dy/dt) + P(t)y$ is similar to the implicit derivative of a product

$$\big[g(t) \cdot y(t)\big]' = g(t)\,\frac{dy}{dt} + g'(t)\,y.$$

An ODE of this form where the left-hand side is such a product is called *exact.* The natural question is, "Can we find a function to multiply our DE by to turn it into one that is exact?" We look for a function $u(t)$ such that

$$\big[u(t)\cdot y\big]' = u(t)\cdot y' + u'(t)\cdot y = u(t)\cdot \frac{dy}{dt} + [u(t)P(t)]\cdot y$$

Examining the equation above leads us to solving for u in the simple separable DE

$$\frac{du}{dt} = uP(t).$$

Solving gives us the desired multiplier $u(t)$, called an *integrating factor*,

$$u(t) = e^{\int P(t)\,dt}.$$

We now have a method for solving linear DEs.

Step 1. Put the linear differential equation in the form $\dfrac{dy}{dt} + P(t)y = f(t)$.

Step 2. Find the integrating factor $u(t) = e^{\int P(t)\,dt}$.

Step 3. Multiply the differential equation by the integrating factor u

$$e^{\int P(t)\,dt}\cdot\frac{dy}{dt} + e^{\int P(t)\,dt}\cdot P(t)\,y = e^{\int P(t)\,dt}\cdot f(t)$$

which, by the choice of u, must simplify to

$$\Big[e^{\int P(t)\,dt}\cdot y(t)\Big]' = e^{\int P(t)\,dt}\cdot f(t).$$

Step 4. Integrate both sides, and include a constant of integration,

$$\int \Big[e^{\int P(t)\,dt}\cdot y(t)\Big]'\,dt = \int e^{\int P(t)\,dt}\cdot f(t).$$

$$e^{\int P(t)\,dt}\cdot y(t) = \int e^{\int P(t)\,dt}\cdot f(t)\,dt + C.$$

Step 5. Solve for y.

Example 2.14. Solving a Linear Differential Equation.
Solve the linear DE from the slope field shown in Figure 2.2,

$$\frac{dy}{dt} = \frac{t-y}{2}.$$

Step 1. We put the linear differential equation in the proper form.

$$\frac{dy}{dt} + \frac{1}{2}y = \frac{1}{2}t$$

identifying $P(t) = 1/2$ and $f(t) = t/2$.

Step 2. Calculate the integrating factor u.

$$u(t) = e^{\int 1/2\,dt} = e^{t/2}$$

Step 3. Multiply by the integrating factor and simplify to obtain

$$\left[e^{t/2}y\right]' = e^{t/2} \cdot \frac{t}{2}.$$

Step 4. Integrate (using integration by parts on the right).

$$\begin{aligned} e^{t/2}y &= (t-2)e^{t/2} + C \\ y &= t - 2 + C\,e^{t/2} \end{aligned}$$

Example 2.15. Newton's Law of Cooling.
We know that Newton's Law of Cooling gives the temperature T of a mass being cooled by the ODE

$$\frac{dT}{dt} = k(T_a - T), \quad k > 0$$

where T_a is the *ambient temperature* or the constant temperature of the surroundings, and $k > 0$ is called the *heat transfer coefficient.*

Consider a mass cooling from an initial temperature of $T(0) = 300°$ F in a room with ambient temperature $70°$ F and the heat transfer coefficient $k = 0.190$. Then

$$\frac{dT}{dt} = 0.190(70 - T), \quad T(0) = 300.$$

Observe that the ODE from Newton's Law of Cooling is both linear and separable. We'll find T using the integrating factor technique on this linear ODE.

Step 1. We put the linear differential equation in the proper form

$$\frac{dT}{dt} + 0.19T = 13.3$$

identifying $P(t) = 0.19$ and $f(t) = 13.3$.

Step 2. Calculate the integrating factor u.

$$u(t) = e^{\int 0.190\, dt} = e^{0.19t}$$

Step 3. Multiply by the integrating factor and simplify to obtain

$$\left[e^{0.19t} y\right]' = e^{0.19t} \cdot \frac{t}{2}.$$

Step 4. Integrate.

$$e^{0.19t} T = 70.0\, e^{0.19t} + C$$
$$T = 70.0 + Ce^{-0.19t}.$$

Use the initial condition $T(0) = 300$ to find that $C = 230$, so

$$T = 70.0 + 230e^{-0.19t}$$

We observe that the $\lim_{t\to\infty}\left(70.0 + 230e^{-0.19t}\right) = 70$ (the ambient room temperature) as expected.

2.5 First-Order Ordinary Differential Equations and Maple

We illustrate the commands to both solve an ODE and to plot an ODE using Maple. We'll use Maple's differential equation toolkit, the *DEtools* package, which contains *DEplot* and many other tools. We also introduce the *dsolve* command, which can solve many first-order ordinary differential equations both with and without initial conditions. The basic form of *dsolve* requires two parameters: (1) *ODE*, which gives the ordinary differential equation in functional form with its variables, and (2) *vars*, a variable or set of variables using functional notation. If there are initial conditions *IC*, Maple requires the set $\{ODE, IC\}$. For an ODE describing y in terms of x, i.e., $y(x)$, the basic forms are

$$dsolve(ODE, y(x), options)$$
$$dsolve(\{ODE, IC\}, y(x), options)$$

Enter "\$ *dsolve*" for more information. We'll use Maple to re-solve many of our previous examples.

Get ready by entering

```
> restart; #to forget any previous definitions and start fresh
```

First, we will solve an ODE using Maple without initial conditions, then we'll redo the problem with initial conditions.

Example 2.16. Basic *dsolve*.

Solve the ODE $\dfrac{dy}{dt} = y \cdot t$.

```
> deqn := diff(y(t), t) = y(t) · t;
```

$$deqn := \frac{d}{dt}\, y(t) = y(t) \cdot t$$

```
> dsolve(deqn, y(t));
```

$$y(t) = \_C1\, e^{\frac{t^2}{2}}$$

Note that Maple introduced $\_C1$ for the constant of integration.

We solve again, this time adding the initial condition $y(1) = 1$.

```
> inits := y(1) = 1;
```

$$y(1) := 1$$

```
> dsolve({deqn, inits}y(t));
```

$$y(t) = \frac{e^{t^2/2}}{e^{1/2}}$$

Next, we plot the solution over $[-1, 1]$.

> $plot(\exp(t^2/2)/\exp(1/2), t = -1..1, 0..1);$

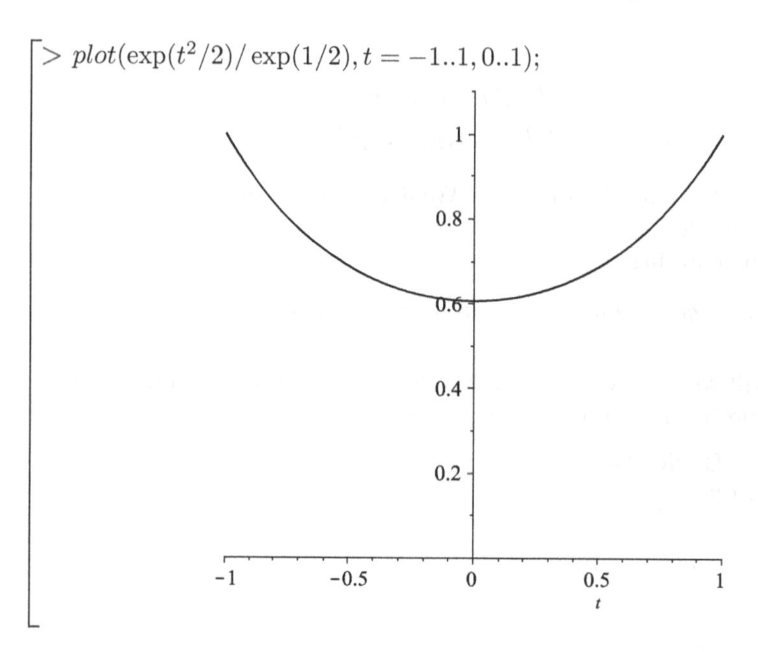

Example 2.17. Solving the Refined Population Model with Maple. Use Maple to solve the logistic growth model defined for blue crabs in Venezuela

$$\frac{dB}{dt} = 0.25B(10 - B), \quad B(0) = 4$$

where B is in millions of bushels.

> $deqn := diff(B(t), t) = 0.25 \cdot B(t) \cdot (10 - B(t));$

$$deqn := diff(B(t), t) = 0.25 \cdot B(t) \cdot (10 - B(t))$$

> $inits := B(0) = 4;$

$$B(0) := 4$$

> $dsolve(\{deqn, inits\}B(t));$

$$B(t) = \frac{20}{2 + 3\,e^{\left(-\frac{5t}{2}\right)}}$$

Now a plot of $B(t)$.

> $plot(20/(2+3\cdot\exp(-5\cdot t/2)), t=-1..5);$

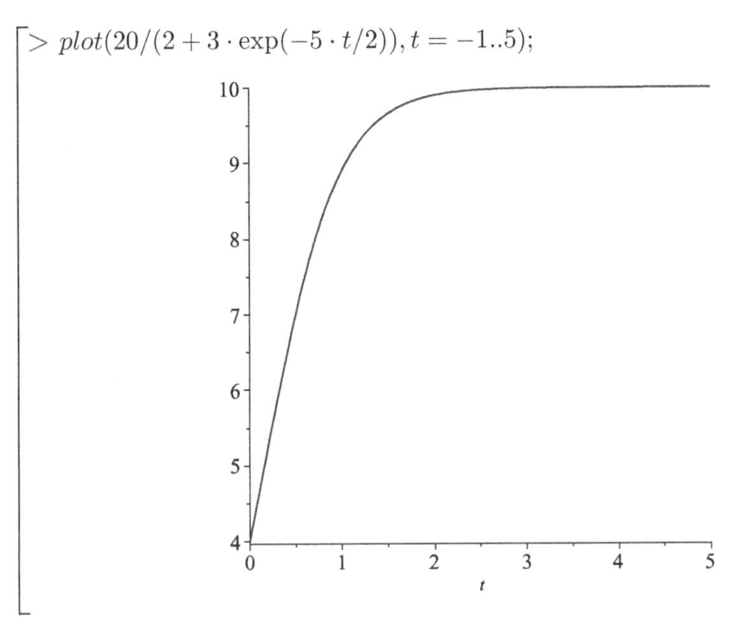

Not all ODEs are in fact solvable in closed-form. If Maple does not return a solution, there might not be a closed-form analytical solution for the ODE.

> $deqn := diff(c(t),t) = (c(t)+2*t)/(t^2+c(t)^3);$

$$deqn := \frac{d}{dt}\,c(t) = \frac{c(t)+2t}{t^2+c(t)^3}$$

> $dsolve(deqn, c(t));$

Maple didn't find a closed form solution. That doesn't indicate that a solution doesn't exist, just that a closed form *may* not be able to be found. We might try a numerical method as we will discuss in the next section.

2.6 Numerical Methods for Solutions to First-Order Ordinary Differential Equations

As mentioned above, not all first-order differential equations are solvable in closed form. In this section, we briefly introduce three numerical techniques for computing discrete values of a solution: Euler's Method (1768), Improved Euler's Method (Heun 1886–1900)[1], and Runge-Kutta (1895). We will discuss

[1] Like many named items in math, e.g., Newton's method or Taylor series, Heun didn't invent what we call Heun's method. It was actually developed by Runge.

each technique and its algorithm, and we will illustrate the Maple commands with our examples.

Euler's Method (1768)

From the point of view of a mathematician, the *ideal* form of the solution to an initial value problem would be a **formula** for the solution function. After all, if this formula is known, it is usually relatively easy to produce any other form of the solution you may desire, such as a **graphical solution**, or a **numerical solution** in the form of a table of values. You might say that a formulaic solution contains the recipes for these other types of solution within it. Unfortunately, as we have seen in our studies already, obtaining a formulaic solution is not always easy, and in many cases is absolutely impossible.

So we often have to "make do" with a numerical solution, i.e., a table of values approximating points which lie along the solution's curve. This table of values can be a perfectly usable form of the answer in many applied problems, but before we go too much further, let's make sure that we are aware of the shortcomings of this form of solution.

By its very nature, a numerical solution to an initial value problem consists of a **table of values** which is **finite** in length. On the other hand, the true solution of the initial value problem is most likely a whole continuum of values, i.e., it consists of an *infinite* number of points. Obviously, the numerical solution is actually leaving out an infinite number of points. The question might arise, "With this many holes in it, is the solution good for anything at all?" To make this comparison a little clearer, let's look at a very simple specific example starting with Euler's Method.

Euler's method, named after Leonhard Euler, is a numerical procedure for solving ODEs with a given initial value. It is the most basic of the explicit methods for numerical integration for ordinary differential equations, and is equivalent to a left-endpoint Riemann sum for approximating an integral. Euler developed the method in 1768 to prove that certain initial value problems had solutions.

The Algorithm. Given the differential equation

$$\frac{dy}{dt} = g(t, y) \text{ and } y(t_0) = y_0 \text{ for } t_0 \leq t \leq b,$$

find $y(b)$.

Euler's Method Algorithm

Step 1. Pick the *step size*, h so that the interval, $(b - t_0)/h = n$ is divided evenly, (or choose the number of steps n and let $h = (b - t_0)/n$).

Step 2. Start at $y(t_0) = y_0$ and let $k = 1$.

Step 3. Compute $t_k = t_{k-1} + h$.

Step 4. Compute $y_k = y_{k-1} + h \cdot g(t_{k-1}, y_{k-1})$.

Step 5. If $k = n$, then STOP. Otherwise, set $k = k + 1$ and return to Step 3.

Since $b = t_0 + n \cdot h$ and $y_k \approx y(t_k)$, we have calculated the sequence of points (t_0, y_0), (t_1, y_1), $\ldots$, $(t_n, y_n) = (b, y_n)$ giving a numerical solution to the ODE.

Example 2.18. Euler's Method Applied to a Simple IVP.
Estimate the solution from $y(0)$ to $y(3)$ to the IVP

$$\frac{dy}{dt} = 0.25 \cdot y \cdot t, \quad y(0) = 2$$

using Euler's Method.

The interval for the problem is $[0, 3]$ and $g(t, y) = 0.25 \cdot y \cdot t$. Let's calculate a few steps by hand.

Step 1. Take a step size of $h = 1$. Then $(b - t_0)/h = (3 - 0)/1 = 3$ steps. Let $k = 1$.

Step 2. The initial condition $y(0) = 2$ gives the point $(t_0, y_0) = (0, 2)$.

Step 3. Calculate $t_1 = t_0 + h = 0 + 1 = 1$.

Step 4. Calculate $y_1 = y_0 + h \cdot g(t_0, y_0) = 2 + 1 \cdot g(0, 2) = 2.0$. Our new point is $(1, 2.0)$.

Step 5. Now $k = 1 < 3$, so set $k = 1 + 1 = 2$, and return to Step 3 with $k = 3$.

Step 3. Calculate $t_2 = t_1 + h = 1 + 1 = 2$.

Step 4. Calculate $y_2 = y_1 + h \cdot g(t_1, y_1) = 2 + 1 \cdot g(1, 2) = 2.0$.
Our new point is $(2, 2.5)$.

Step 5. Now $k = 2 < 3$, so set $k = 2 + 1 = 3$, and return to Step 3 with $k = 3$.

Step 4. Calculate $t_3 = t_2 + h = 2 + 1 = 3$.

Step 5. Calculate $y_3 = y_2 + h \cdot g(t_2, y_2) = 2.5 + 1 \cdot g(2, 2.5) = 3.75$.
Our new point is $(3, 3.75)$.

Step 6. Now $k = 3 = n$, so STOP.

The sequence we have calculated is $(0, 2)$, $(1, 2)$, $(2, 2.5)$, $(3, 3.75)$.
The Euler's method solution is shown geometrically in Figure 2.5.

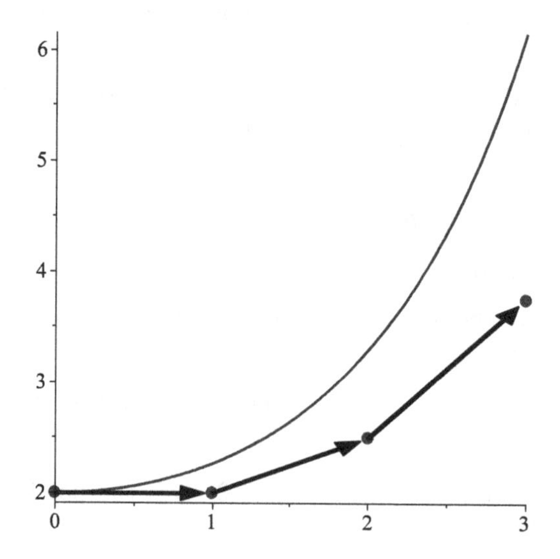

FIGURE 2.5: Euler's Method Solution ($h = 1$) and Analytic Solution

Note that we have underestimated every value between $t = 0$ and $t = 3$.
The exact solution is $y(t) = 2\,e^{t^2/8}$.
Let's make a table and see how we did.

TABLE 2.4: Euler's Method v. Exact Values for $h = 1$

t	Approx. y_a	Exact y_e	$Error = \|y_e - y_a\|$	$\%\ Error = 100 \cdot \left\| \frac{y_e - y_a}{y_e} \right\|$
0	2	2	0	0%
1	2.00	2.266	0.266	11.73%
2	2.50	3.297	0.797	24.17%
3	3.75	6.160	2.410	39.12%

Notice that as we move farther away from the initial condition $y(0) = 2$, the worse our estimate becomes.

Can we do better?

Maybe; we might improve by reducing the step size. Let's make the step size 0.5 instead of 1.

We will use Maple to obtain our numerical output since we can give it any step size and quickly generate the estimates. Here we use the new step size of $h = 0.5$.

```
> IVP := {diff(y(t), t) = 0.25 · t · y(t), y(0) = 2};
```

$$IVP := \left\{ \frac{d}{dt} y(t) = 0.25 \cdot t \cdot y(t), y(0) = 2 \right\}$$

```
> ans2 := dsolve(IVP, numeric, method = classical[foreuler],
    output = Array([0.0, 0.5, 1.0, 1.5, 2.0, 2.5, 3.0]),
    stepsize = 0.5
```

$$\left[\begin{array}{c} \left[t \quad y(t) \right] \\ \left[\begin{array}{cc} 0. & 2. \\ 0.5 & 2. \\ 1. & 2.12500000000000 \\ 1.5 & 2.39062500000000 \\ 2. & 2.83886718750000 \\ 2.5 & 3.54858398437500 \\ 3. & 4.65751647949219 \end{array} \right] \end{array} \right]$$

Let's make a table and see how we did:

TABLE 2.5: Euler's Method v. Exact Values for $h = 0.5$

t	Approx. y_a	Exact y_e	*Error* $= \lvert y_e - y_a \rvert$	% *Error*
0.	2.	2.	0.	0.%
0.5	2.	2.063487	0.063487	3.077%
1.	2.125	2.266297	0.141297	6.235%
1.5	2.390625	2.649570	0.258945	9.773%
2.	2.838867	3.297443	0.458576	13.907%
2.5	3.548584	4.368402	0.819818	18.767%
3.	4.657516	6.160434	1.502918	24.396%

The new Euler's method solution with $h = 0.5$ is shown geometrically in Figure 2.6.

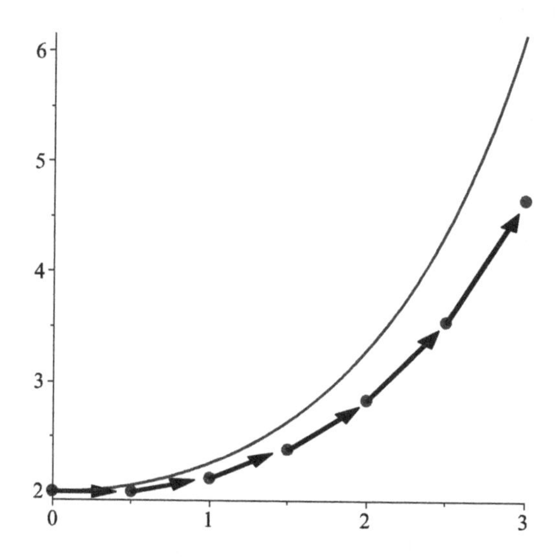

FIGURE 2.6: Euler's Method Solution ($h = 0.5$) and Analytic Solution

We are still underestimating at all points, but the values are closer to the exact solution. We can do better by making this step size even smaller. (Note: it is not always true that reducing the step size improves the estimates.)

We will make a template for using Maple to investigate an initial value problem.

Maple Initial Value Problem Template

1. Set up Maple.

   > *restart*; # Forget all old definitions and start fresh

   > *with(plots)* : # Load *display* and *pointplot*

   (Everything on a line after a hashtag '#' is a comment and is ignored by Maple.)

2. Enter your IVP.

   > $IVP := \{diff(y(t), t) = 0.25 \cdot t \cdot y(t), y(0) = 2\};$

   $$IVP := \left\{\frac{d}{dt}\, y(t) = 0.25t\, y(t), y(0) = 2\right\}$$

3. Solve the IVP analytically, if possible, and plot the solution

   > $Soln := dsolve(IVP, y(t));$

   $$y(t) = 2\,\mathrm{e}^{\left(\frac{t^2}{8}\right)}$$

   > $plot(rhs(Soln), t = 0..3, thickness = 3);$

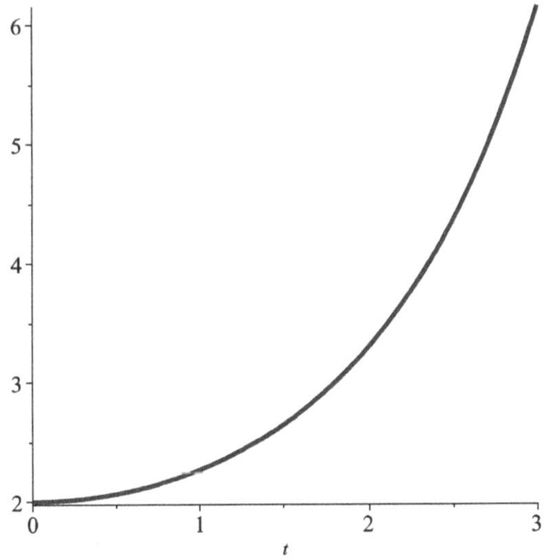

4. Use Maple's Euler's method implementation.
   We need to specify both the desired *stepsize* and the t-values we want to have listed in the output array; other values may be calculated, they're just not listed in the output array.

$> NumericSoln := dsolve(IVP, numeric, stepsize = 0.5,$
$method = classical[foreuler],$
$output = Array([0., 0.5, 1.0, 1.5, 2.0, 2.5, 3.0]);$

$$NumericSoln := \begin{bmatrix} [t \quad y(t)] \\ \begin{bmatrix} 0. & 2. \\ 0.5 & 2. \\ 1.0 & 2.12500000000000 \\ 1.5 & 2.39062500000000 \\ 2.0 & 2.83886718750000 \\ 2.5 & 3.54858398437500 \\ 3.0 & 4.65751647949219 \end{bmatrix} \end{bmatrix}$$

The option *method=classical[foreuler]* indicates that Euler's method is one of the "classical" techniques. The *foreuler* version of Euler's method that Maple used is to "look forward" along the tangent line to the next point.

5. Numerical Output for Analysis: Graphical and Percent Error.

   (a) Have *dsolve* return the numerical output as a procedure:

   $> FunctionSoln := dsolve(IVP, numeric, stepsize = 1,$
   $method = classical[foreuler], output = listprocedure);$
   $FunctionSoln := [t = \mathbf{proc}(t)...\mathbf{end\ proc}, y(t) =$
   $\mathbf{proc}(t)...\mathbf{end\ proc}]$

   (b) Prepare for function evaluation by using the $y(t)$ procedure shown in *FunctionSoln* above.

   $> fy2 := eval(y(t), FunctionSoln)$

   $$fy2 := \mathbf{proc}(t)...\mathbf{end\ proc}$$

   (c) Compare the function's output to the output array above. (Remember: the step size matters!)

   $> seq(fy2(i), i = 0..3);$

   $$2., 2., 2.50000000000000, 3.75000000000000$$

   Note the extended number of digits from *dsolve*'s procedure.

   (d) Define the exact solution function.

   $> actual\_y2 := t \to evalf\left(2 \cdot \exp\left(\frac{t^2}{8}\right)\right)$

   $$actual\_y2 := t \mapsto evalf\left(2 \cdot \exp\left(\frac{t^2}{8}\right)\right)$$

   (e) Compare the actual solution's output to the output above.

   $> seq(actual\_y2(i), i = 0..3);$

   $$2., 2.266296906, 3.297442542, 6.160433698$$

(f) Build an array that lists [t, *Numerical* $y(t)$, *Actual* $y(t)$, *Percent Error*].

```
> Array([[t, Numerical_y(t), Actual_y(t), Percent_Error],
  seq([i, fy2(i), evalf(actual_y2(i)),
  evalf(100 · abs(fy2(i) − actual_y2(i))/actual_y2(i))], i = 0..3)]);
```

t	Numerical_y (t)	Actual_y (t)	Percent_Error
0	2.0	2.0	0.0
1	2.0	2.266296906	11.7503097363360212
2	2.50000000000000000	3.297442542	24.1836675497104139
3	3.75000000000000000	6.160433698	39.1276623719293255

6. Graphical Comparisons.
First, we plot the actual solution (if we have one) then the numerical solution. Second, we will overlay the images on one graph. Otherwise, compare the numerical solution to either qualitative plots or slope field plots.

```
> curve := plot(actual_y2(t), t = 0..3, thickness = 2);
```

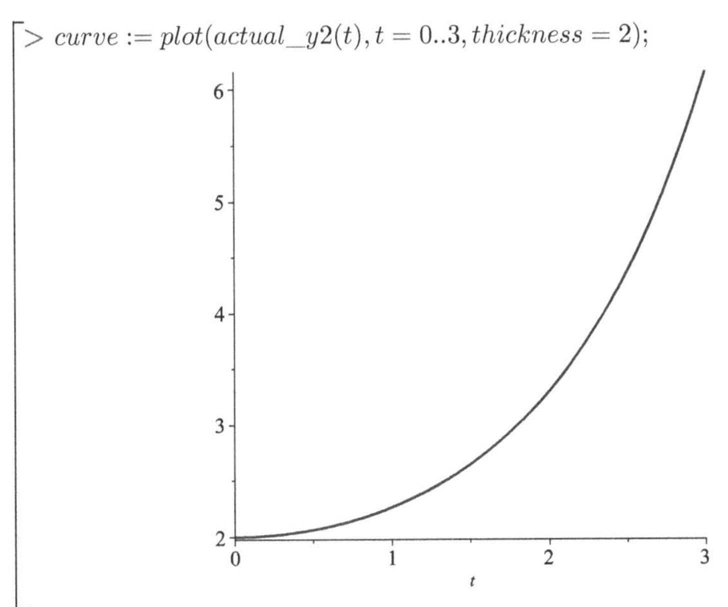

Generate a plot of the numerical solution.

```
> data := seq([i, fy2(i)], i = 0..3) :
```

> $pointplot(data, symbol = soliddiamond, symbolsize = 14, color = black);$

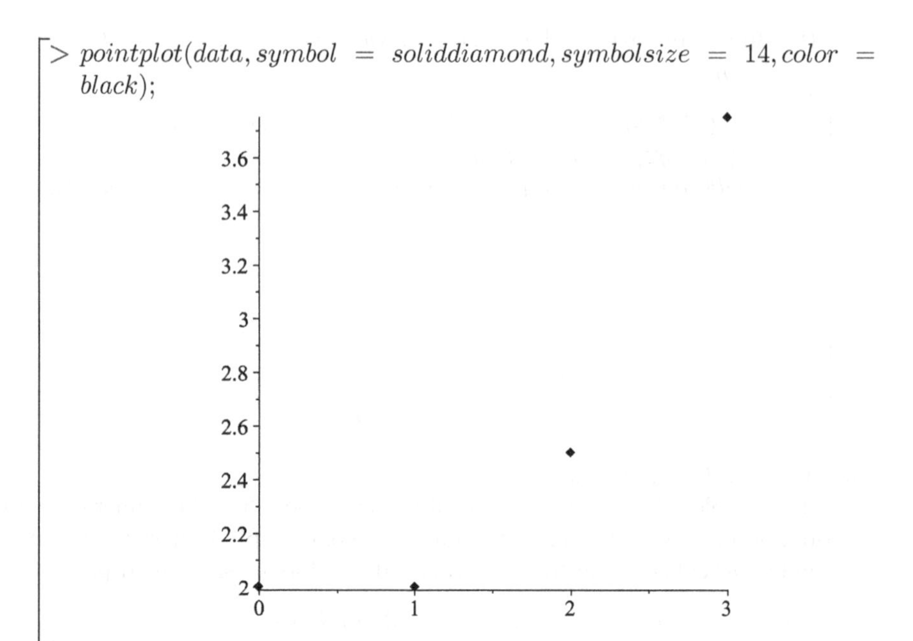

Combining the plots into a single graph provides easy-to-see comparisons.

> $display(curve, points);$

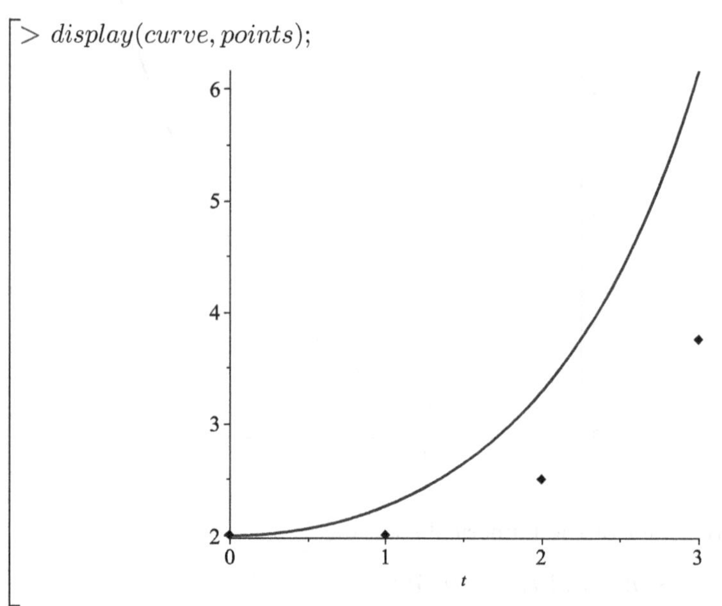

The second numerical technique to discuss is the Improved Euler's method, also known as Heun's method.

Improved Euler's Method or Heun's Method (1886 – 1900)

Euler's method takes a step along the tangent line to predict a new point. To improve the prediction, we go to the new point, and look back along its tangent line to see where we would have come from the new point's perspective. Averaging the slopes of these two tangent lines and taking a step along this new line gives us a better approximation to the true solution. This averaging technique forms the Improved Euler's method. In contrast to Euler's method being equivalent to a right endpoint Riemann sum, Improved Euler's method is equivalent to a Trapezoid Rule sum.

Let's apply Improved Euler's method to the IVP we've been working with:

$$\left\{\frac{dy}{dt} = 0.25 \cdot y(t) \cdot t, \quad y(0) = 2\right\}.$$

> *IVP*;

$$\left\{\frac{d}{dt}\, y(t) = 0.25\, y(t)\, t, y(0) = 2\right\}$$

We already have the analytic solution $y(t) = 2\, e^{t^2/8}$.

As before, we need to specify the numerical method, this time *method=classical[heunform]* for the Improved Euler's method, the output array with the desired domain points, and the step size.

> $IEMSoln := dsolve(IVP, numeric, method = classical[heunform],$
> $output = Array([0., 0.5, 1.0, 1.5, 2.0, 2.5, 3.0]),$
> $stepsize = 0.5);$

IEMSoln :=

t	$y(t)$
0.	2.
0.5	2.06250000000000
1.0	2.26391601562500
1.5	2.64418315887451
2.0	3.28457126766443
2.5	4.33666050183820
3.0	6.08148875062466

We've already graphed the actual solution (again, we do this whenever we have one) above in Item 6. Graphical Comparisons. Now, we'll plot the numeric solution. Then, we will overlay the plots on one graph. If no analytic solution were available, we would compare a numerical solution with either qualitative plots or slope field plots.

Pull the predicted values from *IEMSoln*. The [2, 1] element of *IEMSoln* is the solution array. Converting the array to a list of points makes it easy to plot.

```
> IEMdata := convert(IEMSoln[2, 1], listlist);
```

IEMdata := [[0., 2.], [0.5, 2.06250000000000], [1.0, 2.26391601562500],
[1.5, 2.64418315887451], [2.0, 3.28457126766443],
[2.5, 4.33666050183820], [3.0, 6.08148875062466]]

```
> IEMptplot := pointplot(IEMdata, symbol = soliddiamond,
    symbolsize = 14, color = black);
```

And combine the graphs.

```
> display(curve, IEMptplot);
```

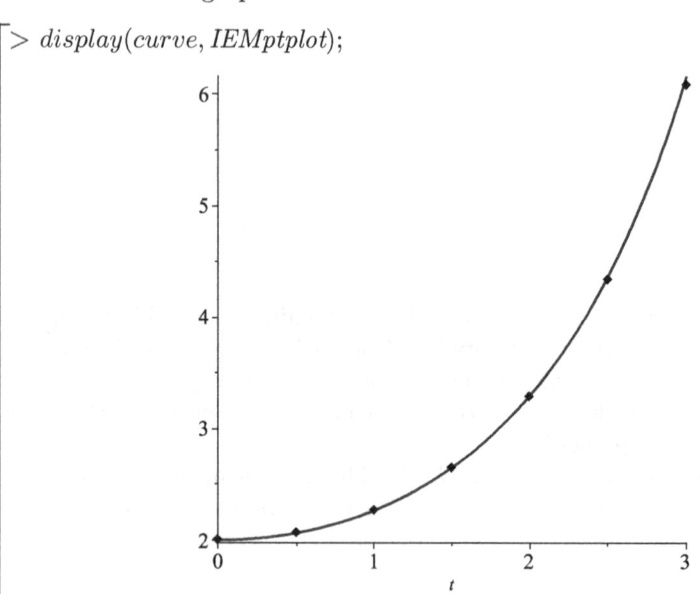

Improved Euler's method really does improve Euler's method!

As we did with Euler's method, we'll use *dsolve* to produce a function for calculating numerical values, and then analyze the results with a table. Note the change in stepsize to $h = 1$.

(a) Use *dsolve* to generate the function.

$$> IEMFcnSoln := dsolve(IVP, numeric, stepsize = 1, method = classical[heunform], output = listprocedure);$$

$$IEMFcnSoln := [t = \textbf{proc}(t)...\textbf{end proc}, y(t) = \textbf{proc}(t)...\textbf{end proc}]$$

(b) Prepare for function evaluation by using the $y(t)$ procedure shown in *IEMFcnSoln* above.

$$> IEMy2 := eval(y(t), IEMFcnSoln);$$

$$IEMy2 := \textbf{proc}(t)...\textbf{end proc}$$

(c) The exact solution function has already been defined as $actual\_y2(t)$.

(d) Compare the actual solution's output to the output above.

$$> seq(actual\_y2(i), i = 0..3);$$

$$2., 2.266296906, 3.297442542, 6.160433698$$

(e) Build an array that lists $[t, Numerical\ y(t), Actual\ y(t), Percent\ Error]$.

$$> Array([[t, Numerical\_y(t), Actual\_y(t), Percent\_Error], seq([i, IEMy2(i), evalf(actual\_y2(i)), evalf(100 \cdot abs(IEMy2(i) - actual\_y2(i))/actual_y2(i))], i = 0..3)]);$$

t	Numerical_y (t)	Actual_y (t)	Percent_Error
0	2.0	2.0	0.0
1	2.25000000000000000	2.266296906	0.719098453378023583
2	3.23437500000000000	3.297442542	1.91261989243784658
3	5.86230468750000000	6.160433698	4.83941594236763883

(f) For our graphical comparison, we'll use *DEplot* from the *DEtools* package specifying Improved Euler's numerical solution method to see the Improved Euler method solution on the DE's slopefield.

$$> with(DEtools): \quad \#\ \text{to load } DEplots$$

$$> deqn := IVP[1]:$$

> $DEplot(deqn, [y], t = 0..3, y = 0..6, method = classical[heunform],$
> $[[y(0) = 2]], linecolor = black);$

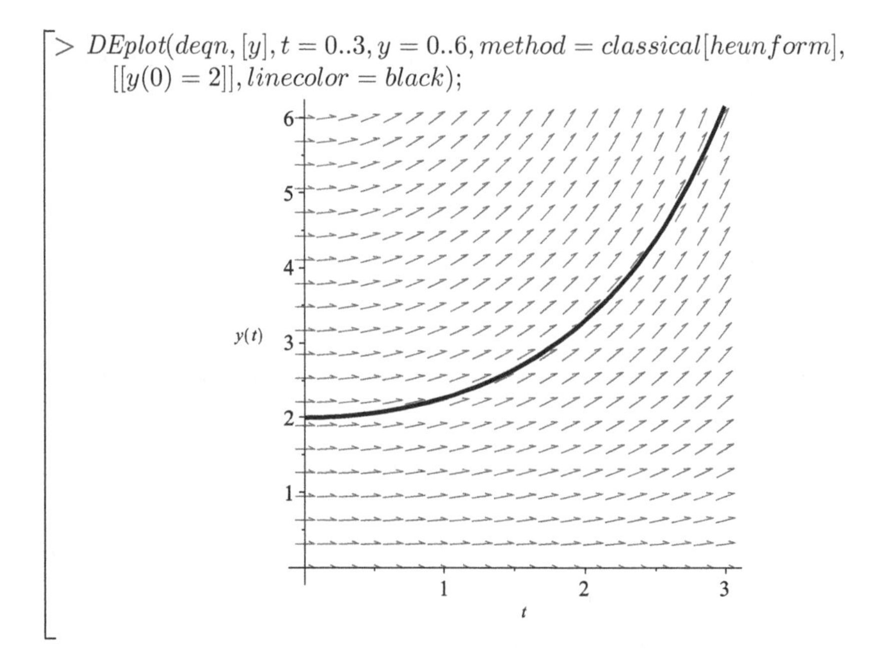

Runge-Kutta Fourth-Order Method (RK4) (1895)

Following the natural progression of numerical integration techniques from the left endpoint Riemann sum to Trapezoid Rule to Simpson's Rule leads us to our next method, a Runge-Kutta method. Runge began studying extending Euler's work in 1895. Kutta devised what we now call RK4 in 1901. We will use Maple's RK4 implementation through specifying the option *method=classical[rk4]* in *dsolve* and *DEplot.*

Let's apply the RK4 method to the IVP we've been working with:

$$\left\{\frac{dy}{dt} = 0.25 \cdot y(t) \cdot t, \quad y(0) = 2\right\}.$$

> $IVP;$

$$\left\{\frac{d}{dt}\, y(t) = 0.25\, y(t)\, t, y(0) = 2\right\}$$

We already have the analytic solution $y(t) = 2\, e^{t^2/8}$. Its graph appears in Item 6. Graphical Comparisons.

As before, we need to specify the numerical method, this time *method=classical[rk4]* for the RK4 method, the output array with the desired domain points, and the step size.

$> RK4Soln := dsolve(IVP, numeric, method = classical[rk4],$
$output = Array([0., 0.5, 1.0, 1.5, 2.0, 2.5, 3.0]), stepsize = 0.5);$

$$RK4Soln := \begin{bmatrix} [t \quad y(t)] \\ \begin{bmatrix} 0. & 2. \\ 0.5 & 2.06348673502604 \\ 1.0 & 2.26629595383808 \\ 1.5 & 2.64956434702467 \\ 2.0 & 3.29741947215258 \\ 2.5 & 4.36830972892537 \\ 3.0 & 6.16009786020672 \end{bmatrix} \end{bmatrix}$$

Now, we plot the numeric solution. Then, we combine the plots on one graph. As usual, when no analytic solution is available, we would compare a numerical solution with either qualitative plots or slope field plots.

Recover the predicted values from *RK4Soln*. The [2, 1] element of *RK4Soln* is the solution array. Converting the array to a list of points makes it easy to graph.

$> RK4data := convert(RK4Soln[2, 1], listlist);$

$> RK4ptplot := pointplot(RK4data, symbol = soliddiamond,$
$symbolsize = 14, color = black);$

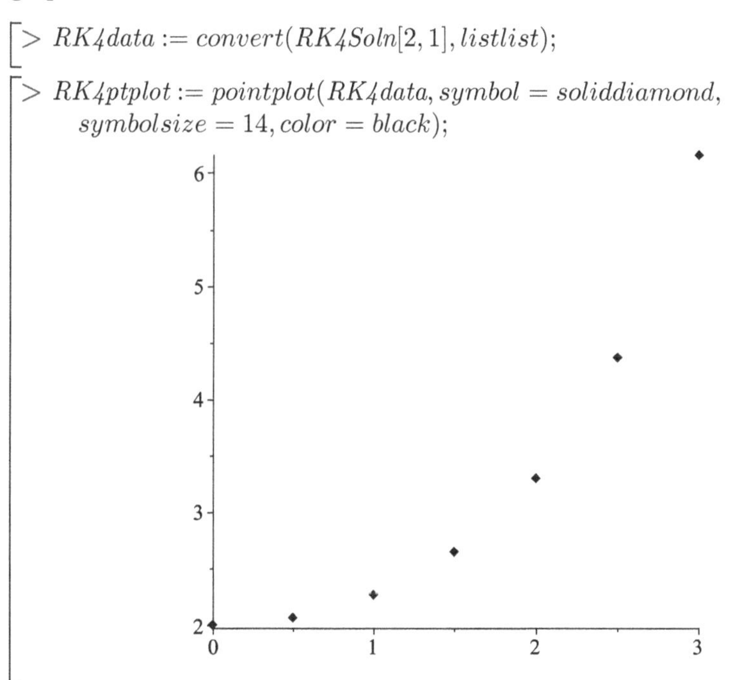

And combine the graphs.

> $display(curve, RK4ptplot);$

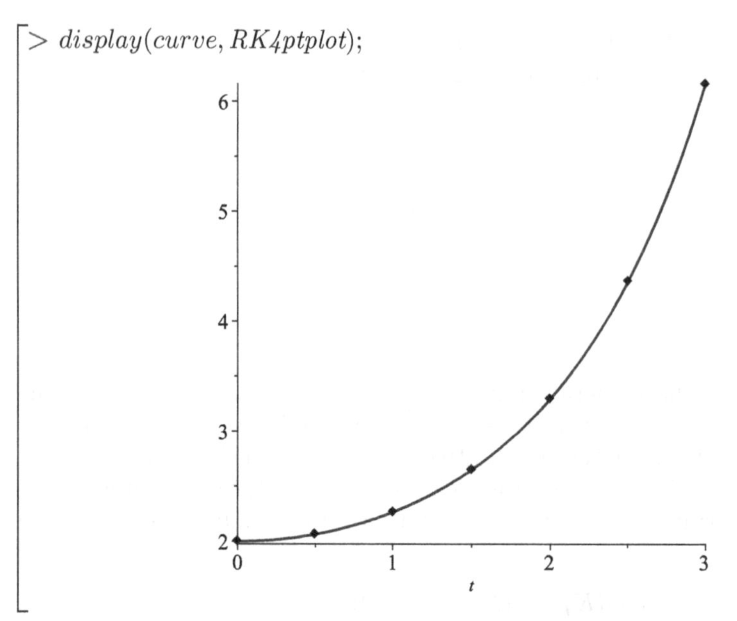

The RK4 method improves Improved Euler's method even more!

As we did with both Euler's and Improved Euler's method, we'll use *dsolve* to produce a function for calculating numerical values, and then analyze the results with a table. Note as before, we change step size to $h = 1$.

(a) Use *dsolve* to generate the function.

> $RK4FcnSoln := dsolve(IVP, numeric, stepsize = 1,$
> $method = classical[heunform],$
> $output = listprocedure);$

$$RK4FcnSoln := [t = \mathbf{proc}(t)...\mathbf{end\ proc}, y(t) = \mathbf{proc}(t)...\mathbf{end\ proc}]$$

(b) Prepare for function evaluation by using the $y(t)$ procedure shown in *RK4FcnSoln* above.

> $RK4y2 := eval(y(t), RK4FcnSoln);$

$$RK4y2 := \mathbf{proc}(t)...\mathbf{end\ proc}$$

(c) The exact solution function has already been defined as $actual\_y2(t)$.

(d) Compare the actual solution's output to the output listed above.

> $seq(actual\_y2(i), i = 0..3);$

$$2., 2.266296906, 3.297442542, 6.160433698$$

(e) Build an array that lists $[t, Numerical\ y(t), Actual\ y(t), Percent\ Error]$.

```
> Array([[t, Numerical_y(t), Actual_y(t), Percent_Error],
    seq([i, RK4y2(i), evalf(actual_y2(i)),
    evalf(100 · abs(RK4y2(i) − actual_y2(i))/actual_y2(i))], i = 0..3)]);
```

t	Numerical_y (t)	Actual_y (t)	Percent_Error
0	2.0	2.0	0.0
1	2.2500000000000000	2.266296906	0.719098453378023583
2	3.2343750000000000	3.297442542	1.91261989243784658
3	5.8623046875000000	6.160433698	4.83941594236763883

We leave plotting the results to the reader.

Comparing the Methods

To compare Euler's method, Improved Euler's method, and RK4, create a table showing the values each produced.

```
> Array([[t, "Euler's method", "Improved Euler's", "RK4", "Actual y"],
    seq([i, fy2(i), IEMy2(i), RK4y2(i), evalf(actual_y2(i))], i = 0..3)]) :
    evalf(%, 8);
```

t	"Euler's method"	"Improved Euler's"	"RK4"	"Actual y"
0.0	2.0	2.0	2.0	2.0
1.0	2.1250000	2.2639160	2.2662960	2.2662970
2.0	2.8388672	3.2845713	3.2974195	3.2974426
3.0	4.6575165	6.0814888	6.1600979	6.1604336

Comparing the values shows the better approximations with each improvement to Euler's method. In real applications with large problems, the speed and simplicity of calculation must be balanced with the accuracy needed. More accurate techniques are slower and require more computations. Each technique has its advantages and drawbacks that must be carefully weighed in the real world.

Exercises

1. Calculate Euler estimates for $y(3)$ using the DE $y' = y + t$ for step sizes of $h = 0.1$, 0.05, and 0.01. Compute the percent error for each step size.

2. Use Euler's method, computing by hand, for the IVP $\{y' = y+1, y(0) = 1\}$

with step size of $h = 0.25$ to estimate $y(0.5)$. Then repeat Euler's Method using Maple. Compare your results to make sure you have the correct solution. Compute the percent error since you can find the exact solution to this IVP. Use Maple to estimate $y(0.5)$ with the step sizes $h = 0.1$ and 0.05.

3. Consider $\{v' = 32 - -1.6v, v(0) = 0\}$. Using Euler's method in Maple and a step size of $h = 0.05$ estimate $v(2)$. What is the terminal velocity? Keep stepping out until you approximate the terminal velocity to 3 decimal places of accuracy.

4. Try **by hand** $\{y' = y + 1, y(0) = 1\}$ with a step size of $h = 0.25$ to estimate $y(.5)$. Then use Maple with the Improved Euler's Method and compare your results to make sure you have the correct solution. Compute the percent error since you can find this exact solution.

5. Consider $\{v' = 32 - -1.6v, v(0) = 0\}$. Using *heunform* in Maple and a step size of $h = 0.05$, estimate $v(2)$. What is the terminal velocity? Keep stepping out until you approximate the terminal velocity to 3 decimal places of accuracy.

6. Consider $\{P' = P \cdot (15 - 3 \cdot P), P(0) = 2\}$.

   (a) Use *heunform* in Maple with step sizes of $h = 1$, 0.5, and 0.1.

   (b) Discuss the results as compared to the qualitative solution and a slope field.

7. Try **by hand** $\{y' = y + 1, y(0) = 1\}$ with step size of $h = 0.25$ to estimate $y(0.5)$. Then do the RK4 method using Maple; compare your results to make sure you have the correct solution. Compute the percent error since you can find this exact solution.

8. Consider $\{v' = 32 - -1.6v, v(0) = 0\}$. Using RK4 in Maple and a step size of $h = 0.05$, estimate $v(2)$. What is the terminal velocity? Keep stepping out until you approximate the terminal velocity to 3 decimal places of accuracy.

9. Consider $\{P' = P \cdot (15 - 3 \cdot P), P(0) = 2\}$.

   (a) Use *rk4* in Maple with step sizes of $h = 1$, 0.5, and 0.1

   (b) Discuss the results as compared to qualitative solution and a slope field.

Chapter 2 Projects

Project 2.1. The Spread of a Contagious Disease

The logistic ODE model for the spread of a communicable disease is

$$\frac{dN}{dt} = 0.25N \cdot (10 - N), \quad N(0) = 2$$

where N is in 100's.

Let's completely analyze the behavior of this ODE.

(a) Since this is an autonomous ODE, perform a complete graphical analysis:

 i. Plot dN/dt versus N. Find and label all *rest points* (*equilibrium points*).

 ii. Find the value where the rate of change of the disease is the fastest. Why did you provide this value?

 iii. Plot N versus t using the set of initial conditions $N(0) = 2$, $N(0) = 7$, and $N(0) = 14$.

 iv. Describe the *stability* of each rest point (equilibrium point).

(b) Obtain a slope field plot of this ODE from Maple. Briefly discuss how you would analyze the slope field plot. Compare it to the qualitative plot in part (a).

(c) Solve this ODE using the technique for separable equations. (Hint: You will need to use a partial fraction decomposition.) Insure you find the value of the arbitrary constant C using the initial condition $N(0) = 2$.

(d) Simplify your solution in (a) and plot N versus t.

(e) Compare your actual plot with the graphical solution in (a) and the slope field plot in (b).

(f) From your graph of the analytical solution in (d), estimate the solution values of $N(0.5)$ and $N(5)$.

(g) Find the actual values of the two solutions $N(0.5)$ & $N(5)$ using your analytical solution from part (c).

(h) Compute the time t when N is changing the fastest for the solution using the initial condition $N(0) = 2$. Recall, you have found the value of N already; compare your results.

(i) Use Euler's Method with step sizes of $h = 0.1$ and then $h = 1.0$ to approximate the solutions to the ODE for $N(0.5)$ & $N(5)$. Find the absolute differences and relative error.

(j) i. Plot Euler's approximations for $h = 0.1$ and $h = 1.0$ and compare these graphs to your graphical analysis in part (a) and your other plots. Briefly discuss these plots; i.e., compare and contrast these plots.

ii. What happened with Euler's Method? (Any opinions?)

Project 2.2. Chemical Reactions

Purpose: To model the changing amounts of salt dissolved in brine as an initial value problem, and to use the computer to obtain graphical and numerical solutions.

Background: A brine solution is a solution of salt in water. If the brine is in a tank equipped with fill and drain pipes, then the total amount of dissolved salt in the tank varies as the concentration in the inflow stream changes and the inflow and outflow rates are adjusted. The amount of salt in the tank can be modeled by appeal to the

Balance law : *New rate of change = rate in − rate out*

The term "rate in" refers to the rate at which salt is added to the brine by means of the inflow stream and the term "rate out" is the rate at which salt leaves the tank through the outflow pipe. We make the simplifying assumption throughout that the inflow stream is instantaneously mixed with the brine in the tank so that at any time the concentration of salt in the tank is uniform. Each of these terms ('rate in' and 'rate out') is a product of the appropriate brine flow rate with corresponding salt concentrations.

Consider a tank with a capacity of 4000 liters holding 2000 liters of brine that contains 50 kg of dissolved salt. Brine with a salt concentration of 0.2 kg/liter is piped into the tank at a rate of 40 liters/min. Well-mixed brine is drawn off at a constant rate of a liters/min. The Balance Law then gives

$$\frac{dx}{dt}\frac{\text{kg}}{\text{min}} = 0.2\,\frac{\text{kg}}{\text{liter}}\cdot 40\,\frac{\text{liter}}{\text{min}} + \frac{x}{2000+(40-a)\cdot t}\,\frac{\text{kg}}{\text{liter}}\cdot a\,\frac{\text{liter}}{\text{min}}$$

or

$$\frac{dx}{dt} = 8 + \frac{a\,x}{2000+(40-a)\cdot t} \qquad \text{(Brine Eqn)}$$

Requirement 1. Show that (Brine Eqn) above is the general ODE. Find the appropriate initial condition for this Brine problem.

Requirement 2. Solve the ODE leaving a as a parameter.

Requirement 3. Solve the ODE and obtain plots when the rate out is $a = 30$, 40, then 50.

Requirement 4. (Bulk Salt) Suppose that 200 kg of bulk salt is placed in a tanking holding 2000 liters of brine already containing x_0 kg of dissolved salt. Suppose that the saturation level of brine is 300 kg of salt. The model representing this situation is

$$\frac{dx}{dt} = 0.01(200-x)(300-x), \quad x(0) = x_0$$

Solve this ODE, and obtain a plot for various choices of $x_0 = 25$, 50, 100, and 200.

Requirement 5. Bimolecular Chemical Reactions.
Background. In a bimolecular chemical reaction, two species interact and create one or more products. Consider the reaction,

$$A + B \to C + D$$

In this case, the Law of Conservation is

$$\frac{dC}{dt} = \frac{dD}{dt} = -\frac{dA}{dt} = -\frac{dB}{dt}.$$

The Law of Mass Action is: The rate of an elementary reaction is proportional to the product of the concentrations of the reactants.

The associated ODE is

$$\frac{dx}{dt} = k_1(a + c - x)(b + c - x), \quad x(0) = c$$

Consider the following reversible chemical reaction

$$\frac{dx}{dt} = (0.7 - x)(0.4 - x) - 0.1x^2, \quad x(0) = c$$

Solve qualitatively (autonomous). Solve this ODE (at least numerically). Show the curve and discuss the equilibrium values. Let c = 0.2. This reaction is only valid for $0 \leq c \leq 0.4$.

Project 2.3. Harvesting a Species

Consider harvesting the blue crab in South Carolina. There have been many reports about the declining populations of blue crabs and the difficulty in harvesting these crabs. Let's model the situation and analysis using some "what ifs."

The basic Balance law for Harvesting is:

$$\frac{dP}{dt} = r\,P(t) \cdot \left(1 - \frac{P(t)}{k}\right) - H(t)$$

where r is the intrinsic rate coefficient (growth rate if greater than 0) and k represents the carrying capacity (or saturation level).

Requirement 1. In the absence of harvesting ($H(t) = 0$), the ODE is *autonomous.* Let $r = 0.3$, $k = 12$, and $P(0) = 5$.

(a) Perform a qualitative assessment of this situation since it is autonomous. List and classify each equilibrium value.

(b) Solve this ODE.

(c) Plot the solution to this ODE and compare to your qualitative solution. Briefly discuss similarities and/or differences.

(d) Solve numerically using Euler's Method and obtain a plot of the numerical solution. Compare to the analytical solution.

Requirement 2. Consider light harvesting where $H(t) = 0.4$ (Assume r and k are the same values as above.)

(a) Obtain a slope field plot and discuss any equilibrium values observed.

(b) Find all equilibrium values of this ODE.

(c) Solve this ODE. Assume initial conditions of $P(0) = 19$, $P(0) = 8$, and $P(0) = 0.8$.

(d) Solve numerically using Euler's Method with $P(0) = 5$.

Requirement 3. Now consider a heavier harvesting where $H(t) = 1.5$.

(a) Obtain a slope field plot and discuss any equilibrium values observed.

(b) Find all equilibrium values of this ODE.

(c) Solve this ODE. Assume initial conditions of $P(0) = 19$, $P(0) = 8$, and $P(0) = 0.8$.

(d) Solve numerically using Euler's Method with $P(0) = 5$.

Requirement 4. Determine which of the models is more likely to represent the current situation for blue crabs in the Chesapeake Bay.

Project 2.4. Rural Water Supplies

Consider a cylindrical water tank discharging into a pipe located in the middle of its base. The potential energy given by the pressure of the column of liquid is converted into kinetic energy of the stream of water, as it exists in the tank. If the level of the water in the tank is h ft, then it follows that the exit velocity in ft/sec is $v = \sqrt{2gh}$ where $g = 32.2$ ft/sec^2 is the acceleration due to gravity. Since a convergent flow is set up from all sides to the orifice, inertia from the particles in the jet forces them to overshoot the edge and to converge to a smaller cross section termed the *vena contracta.* The ratio of the diameter of the vena contracta to the diameter of the orifice is called the contraction coefficient α which ranges from 0.5 to 1.0 depending on the smoothness of the edge of the orifice. The diameter of the vena cross-section and exit velocity of the stream determines the volume rate of the discharge of the stream. Thus, if the tank has cross section A and is being filled from the top at a rate K, while draining for the bottom through an orifice of area a, the height in the tank is governed by the differential equation

$$\frac{dh}{dt} = \frac{K - a\alpha\sqrt{2gh}}{A}.$$

Now, consider we are in the mountains outside of Stowe, VT. The water for our cabin is supplied from an aquifer some 125 feet below the surface at a maximum rate of 7 gallons per minute. Since the flow rate is insufficient to meet daily needs, you must design a more complicated system. In your design, this water is pumped to a cylindrical tank with diameter 10 feet and height 35 feet which is located on a nearby rise about 75 feet from the cabin, with a base 25 ft above the level of the cabin floor. The water is stored in this gravity tank until needed (water is turned on by opening the spigot).

Assume that the coefficient of contraction at the exit is $\alpha = 0.63$ and that the water main consists of 6-inch-diameter PVC pipe, so that the friction losses are less than 2 psi.

Required:

(a) Determine the minimum flow rate in the system and the minimum water pressure during the first two hours.

(b) The local fire code demands that water supplies maintain a pressure of at least 20 psi at a flow rate of 30 gallons per minute for a minimum of two hours. Will your design pass inspection?

(c) At continuous maximum flow, how long would it take for the level of the top of the water in the tank to fall to 10 ft above the base?

Project 2.5. Falling Bodies

In bridge jumping, a participant attaches one end of a bungee cord to himself or herself, attaches the other end to a bridge railing, and then drops off the bridge. In this project, the jumper will be dropping off of the Royal Gorge Bridge, a suspension bridge that is 1053 feet above the floor of the Royal Gorge in Colorado. The jumper will use a 200-foot-long bungee cord. It would be nice if the jumper has a safe jump, meaning that the jumper does not crash into the floor of the gorge or run into the bridge on the rebound. In this project, you will analyze the fall.

Assume the jumper weighs 160 lb. The jumper will free fall until the bungee cord begins to exert a force that acts to restore the cord to its natural (equilibrium) position. In order to determine the spring constant of the bungee cord, you found that a mass weighing 4 lb. stretches the cord 8 feet. Hopefully, this spring force will help slow down the descent sufficiently so that the jumper does not hit the bottom of the gorge.

Throughout this project we will assume that *down* is the *Positive* direction.

Requirement 1.

Before the bungee cord begins to retard the fall of the jumper, the only forces that act on the jumper are his weight and the force due to wind (air) resistance.

(a) If the force due to the wind resistance is 0.9 times the velocity of the jumper, then use Newton's Second Law ($\sum F = m\,a$) to write a differential equation that models the fall of the jumper. Be sure to include the initial conditions for the jumper for your differential equation. (Hint: This problem can be formulated as a second-order DE in position, or as a first-order DE in velocity — use the first-order DE model in this project.)

(b) Solve this differential equation and find a function that describes the jumper's velocity (as a function of time). Knowing integral calculus, find a function that describes the jumper's position (as a function of time).

(c) What is the velocity of the jumper after the jumper has fallen 200 feet?

(d) What is the terminal velocity of the jumper, if any?

Requirement 2.

After a bit more research, you have found that the force due to wind resistance is not linear, as assumed above. Apparently, the force due to wind resistance is more closely modeled by $0.9v + 0.0009v^2$.

(a) Write a new differential equation governing the velocity of the jumper (prior to the bungee cord coming into effect).

(b) You should notice that this DE is no longer easy to solve. Nonetheless, you are determined to find the velocity of the jumper after 4 seconds by using a numerical technique. Use Euler's method with a step size of $h = 0.5$, and estimate the velocity of the jumper after 4 seconds.

(c) Find the terminal velocity, if any exists.

(d) How do your results compare with those found in requirement one under the linear assumption?

(e) What is the velocity of the jumper after 200 feet? (Hint: you can find this from your numerical table)

Further Reading

[AB2014] Martha L. Abell and James P. Braselton, *Differential Equations with Maple V®*, Academic Press, 2014.

[Barrow1997] David Barrow, Art Belmonte, Jack Bryant, Kirby Smith, Mike Stecher, Al Boggess, Jeff Morgan, Maury Rahe, and Tom Kiffe, *Solving Differential Equations with Maple V Release 4*, Brooks/Cole Publishing Co., 1997.

[GW1991] Frank R. Giordano and Maurice D. Weir, *Differential Equations: A Modeling Approach*, Addison-Wesley, 1991.

[GFH2013] Frank Giordano, William P. Fox, and Steven Horton, *A First Course in Mathematical Modeling*, Nelson Education, 2013.

[Zill2017] Dennis G. Zill, *A First Course in Differential Equations with Modeling Applications*, 11 ed., Cengage Learning, 2017.

3

Introduction, Basic Concepts, and Techniques in Problem Solving with Systems of Ordinary Differential Equations

Objectives

(1) Define and build a system of differential equations for a real problem.

(2) Find a solution.

(3) Graph a solution.

(4) Understand equilibrium values and stability.

(5) Solve both linear and nonlinear systems of differential equations.

(6) Solve systems of differential equations.

(7) Use numerical approaches to obtain good approximate solutions to differential equations and systems of differential equations.

3.1 Systems of Differential Equations

Interactive situations occur in the study of economics, ecology, electrical engineering, mechanical systems, control systems, systems engineering, and so forth. For example, the dynamics of population growth of various species is an important ecological application of applied mathematics.

Your uncle recently retired and bought farm land in South Carolina. His desire is to have a fishing pond; his favorite fish to catch are rainbow trout and brown trout. He finds he has a fair size fresh water pond on his land, but it contains no fish. He takes a water sample to the local fish and game authority. They analyze his water and conclude that the pond can sustain a fish population. He visits the local fish hatchers where they provide him the growth rates of rainbow trout and brown trout in isolation, call these values r and b, respectively. The experts tell him that rainbow trout and brown trout have the same food sources and will compete for the oxygen in the water as well as for food for survival. Data from the American Fisheries Society estimates the interaction rates between rainbow trout and brown trout for survival; call these rates m and n respectively. We desire to build a mathematical model to help your uncle determine if the pond can sustain both species of fish. This leads to a *competitive hunter* system of differential equations. We will revisit a similar scenario later in the chapter.

In this section, we will consider a variety of problem-solving methods, models, concepts, and techniques that give the reader some of the basic tools needed in solving and analyzing systems of differential equations.

We start with several models that we will solve and analyze. These models come from a variety of disciplines in science and engineering, including chemistry, physics, biology, fluids mechanics, Newtonian Mechanics, environmental engineering, and financial mathematics. Problem-solving techniques are discussed that help determine the necessary coefficients in the models.

Next we introduce *phase portraits.* Phase portraits provided qualitative viewing of solutions of systems of differential equations. Then we'll cover analytical methods in Maple for solving systems of differential equations with constant coefficients in both the homogeneous and nonhomogeneous cases.

3.2 Applied Systems of Differential Equations

In this section, we introduce many mathematical models from a variety of disciplines. Our emphasis in this section is on building the mathematical model, or expression, that will be solved later in the chapter. Recall previously that we discussed the modeling process. In this section, we will confine ourselves to the first three steps: (1) state the problem, (2) define assumptions and variables, and (3) select or construct the equations.

Example 3.1. Economics: A Basic Supply and Demand Model.
Suppose we are interested in the variation of the price of a specific product. It is observed that a high price for the product attracts more suppliers. However, if we flood the market with the product, the price is driven down. Over time there is an interaction between price and supply. Research the "Tickle Me Elmo" craze from Christmas, 1996.[1]

Problem Identification: Build a model for price and supply for a specific product.
Assumptions and variables: Assume the price is proportional to the quantity supplied. Also assume the change in the quantity supplied is proportional to

[1]See, e.g., https://en.wikipedia.org/wiki/Tickle_Me_Elmo.

the price. We define the following variables.

$$P(t) = \textit{the price of the product at time } t,$$
$$Q(t) = \textit{the quantity supplied at time } t$$

We define two proportionality constants as $a > 0$ and $b > 0$. The constant $(-a)$ is negative and represents a decrease in price as quantity supplied increases; the constant b represents the increase in quantity supplied as price increases.

With our limited assumptions, the initial model would be

$$\frac{dP}{dt} = -aQ$$
$$\frac{dQ}{dt} = bP$$

Example 3.2. An Electrical Network.
Electrical networks with more than one loop give rise to systems of differential equations. Consider the electrical network displayed in Figure 3.1 where there are two resisters and two inductors. We apply Kirchhoff's Voltage Law (the sum of the voltage drops in a closed circuit is equal to the impressed voltage) to each loop. We assume that no other factors interact with the flow of electricity in this circuit.

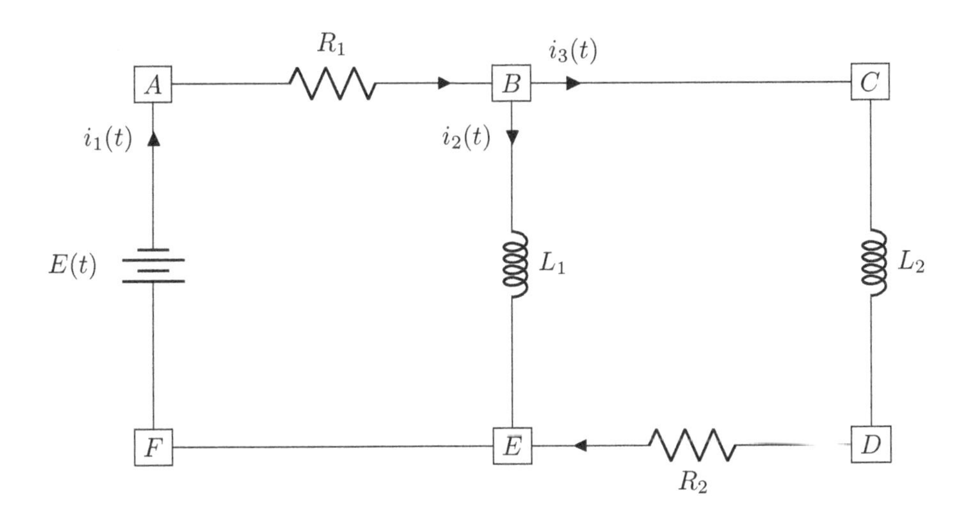

FIGURE 3.1: Electrical Circuit Diagram

Apply Kirchhoff's Voltage Law to the two loops:
Loop ABEF:

$$E(t) = i_1 R_1 + L_1 \frac{di_2}{dt}$$

Loop ABCDEF:

$$E(t) = i_1 R_1 + L_2 \frac{di_3}{dt} + i_3 R_2$$

From Kirchhoff's Current Law (the sum of currents flowing through a node in an electrical circuit is equal to zero), we know that $i_1(t) = i_2(t) + i_3(t)$. We substitute this expression for i_1 into the loop equations to obtain the model:

$$\begin{aligned}\frac{di_2}{dt} &= -\frac{R_1}{L_1} i_2 - \frac{R_1}{L_1} i_3 + \frac{E(t)}{L_1},\\ \frac{di_3}{dt} &= -\frac{R_1}{L_2} i_2 - \frac{R_1 + R_2}{L_2} i_3 + \frac{E(t)}{L_2}.\end{aligned}$$

Example 3.3. Competition between Species.

Imagine a small fish pond supporting both brown trout and rainbow trout. Let $B(t)$ denote the population of brown trout at time t and $R(t)$ denote the population of rainbow trout at time t. We want to know if both can coexist in the pond. Although population growth depends on many factors, we will limit ourselves to considering basic isolated growth and the interaction with the other competing species for scarce life-support resources.

We assume that the species grow in isolation. The level of the population of the rainbow trout or the brown trout, $B(t)$ and $R(t)$, depend on many variables such as their initial numbers, the amount of competition, the existence of predators, their individual species birth and death rates, and so forth. In isolation we assume the following proportionality models (following the same arguments as the basic populations models that we have discussed before) to be true where the environment can support an unlimited number of trout and/or bass. Later, we might refine this model to incorporate the limited growth assumptions of the logistic model:

$$\begin{aligned}\frac{dB}{dt} &= aB\\ \frac{dR}{dt} &= mR\end{aligned}$$

Next, we modify the preceeding differential equations to take into account the competition of the trout and bass for living space, oxygen, and food supply. The effect is that the interaction decreases the growth of the species. The interaction terms for competition led to the decay rate that we call b for brown trout and n for rainbow trout. This leads to following simplified model:

$$\begin{aligned}\frac{dB}{dt} &= aB - bBR\\ \frac{dR}{dt} &= mR - nBR\end{aligned}$$

If we have the initial stockage levels, B_0 and R_0, we can determine how the species coexist over time.

If the model is not reasonable, then we might try logistic growth instead of isolated growth. Logistic growth in isolation was discussed in first-order ODE models as a refinement.

Example 3.4. Predator-Prey Relationships.
We now consider a model of population growth for two species in which one animal is hunted by another animal. An example of this might be wolves and rabbits where the rabbits are the primary food source for the wolves.

Let $R(t)$ be the population of the rabbits at time t and $W(t)$ be the population of the wolves at time t.

We assume that rabbits grow in isolation, but are killed by the interaction with the wolves. We further assume that the constants are proportionality constants. Then

$$\frac{dR}{dt} = aR - bRW.$$

We assume that the wolves will die out without food and grow through their interaction with the rabbits. We further assume that these constants are also proportionality constants. Then

$$\frac{dW}{dt} = -mW + nRW.$$

Example 3.5. Diffusion Models.
Diffusion through a membrane leads to a first-order system of ordinary linear differential equations. For example, consider the situation in which two solutions of substance are separated by a membrane of permeability P. Assume the amount of substance that passes through the membrane at any particular time is proportional to the difference between the concentrations of the two solutions. Therefore, if we let x_1 and x_2 represent the two concentrations, and V_1 and V_2 represent their corresponding volumes, then the system of differential equations is given by

$$\begin{aligned}\frac{dx_1}{dt} &= \frac{P}{V_1}(x_2 - x_1)\\ \frac{dx_2}{dt} &= \frac{P}{V_2}(x_1 - x_2)\end{aligned}$$

where the initial amounts of x_1 and $_2$ are given.

If this model does not yield satisfactory results in terms of realism we might try a refinement of the Diffusion Model as follows.

Diffusion through a double-walled membrane, where the inner wall has permeability P_1 and the outer wall has permeability P_2 with $0 < P_1 < P_2$. Suppose the volume of the solution within the inner wall is V_1 and between the two walls is V_2. We let x represent the concentration of the solution within the inner wall and y, the concentration between the two walls. This leads to the following system

$$\frac{dx}{dt} = \frac{P_1}{V_1}(y - x)$$
$$\frac{dy}{dt} = \frac{1}{V_2}\left(P_2(C - y) + P_1(x - y)\right)$$

with $x(0) = x_0$, $y(0) = y_0$, and $C > 0$.

Example 3.6. Insurgencies Models.
As we look around the world, we see many conflicts involving insurgencies. We have the political faction (usually the status quo or the new regime) battling the insurgents or the rebels that are resisting the change or the political status. This also can be seen from history if we look at our own Revolutionary War.

In insurgency operations we find the following assumptions: They are messy, grass roots fights that are confused and brutally contested. We find the definition of the enemy is loosely defined. We find that positive control of the forces is usually weak. There are few rules of engagement (they are often permissive). There are political divisions that are deep seated that leave little room for compromise.

Further as we consider building a mobilization model, we assume that growth is subject to the same laws as any other natural or man-made population (basic growth or logistical growth as discussed before). Additionally, there are three considerations: pool of potential recruits, number of recruiters, and the transformation rate.

We assume logistical growth. Our systems could look like

$$X(t) = \text{insurgency}$$
$$Y(t) = \text{regime}$$

$$\frac{dX}{dt} = a \cdot (k_1 - X) \cdot X$$
$$\frac{dY}{dt} = b \cdot (k_2 - Y) \cdot Y$$

where

a measures insurgency growth rate,

b measures regime growth rate,

k_1 and k_2 are the respective carrying capacities.

3.3 Phase Portraits and Qualitative Assessment of Autonomous Systems of First-Order Differential Equations

Consider the system of differential equations

$$\frac{dx}{dt} = f(x, y)$$
$$\frac{dy}{dt} = g(x, y)$$

These systems are called *autonomous* because they do not include t, the independent variable.

The solution is a pair of parametric equations, $x = x(t)$, $y = y(t)$. The solution is also a curve that varies over time. We call the solution curve a *trajectory*, *path*, or *orbit*. The $x - y$ plane is called the *phase-plane*. We can also obtain plots of x versus t and y versus t. *Rest Points* or *equilibrium points* are points that satisfy both $f(x, y) = 0$ and $g(x, y) = 0$, simultaneously. Once we have the equilibrium values, we desire information about their stability.

Rules of Stability

We classify equilibrium values as *stable*, *asymptotically stable*, or *unstable*. We define these as follows:

Stable: If a trajectory starts close to a rest point, it remains close for all future time.

Asymptotically Stable: If a trajectory starts close to a rest point, then it tends toward the rest point as t tends to infinity.

Unstable: If a trajectory does not follow the rules for either a stable or an asymptotically stable trajectory.

The following results are useful in investigating solutions to autonomous systems. We offer these without proof:

Proposition.

1. There is at most one trajectory through any point in the phase-plane.
2. A trajectory that starts at a point other than a rest point cannot reach a rest point in a finite amount of time.
3. No trajectory can cross itself unless it is a closed curve. If the trajectory is a closed curve, then it is a periodic solution.

4. Implications and properties of motion from a given starting point (not a rest point):
   (a) A trajectory will follow the same path regardless of starting time.
   (b) A trajectory cannot return to the starting point unless the motion is periodic.
   (c) A trajectory can *never* cross another trajectory.
   (d) A trajectory can only approach, and can never reach, a rest point.

   We illustrate the Maple commands below to obtain phase portraits.

Qualitative Graphical Assessment

Consider the autonomous competitive hunter system of differential equations:

$$\frac{dx}{dt} = y(t) \cdot (4.5 - 0.9 \cdot x(t))$$
$$\frac{dy}{dt} = 0.24 \cdot x(t) - 0.08 \cdot x(t) \cdot y(t)$$

First, we plot dx/dt and dy/dt respectively. We want to see where $dx/dt = 0$ and $dy/dt = 0$ simultaneously. In the equation $dx/dt = 0$, we find that either $y(t) = 0$ (which is the x-axis) or $x = 5$ (a vertical line). In the equation $y(t) = 0$, we find either $x(t) = 0$ (the y-axis) or the line $y(t) = 3$. There are two equilibrium points. They are the points $(0,0)$ where the x-axis and y-axis intersect and the point $(5,3)$ where the vertical line $x = 5$ intersects the horizontal line $y = 3$.

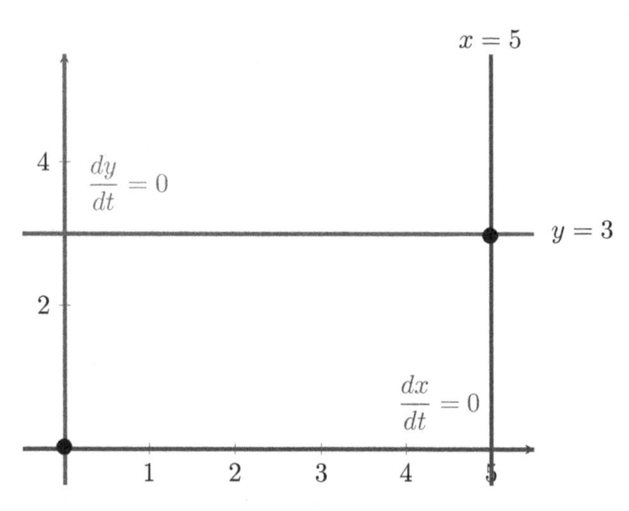

FIGURE 3.2: Graphs of $dx/dt = 0$ and $dy/dt = 0$

Analysis shows that both $(0,0)$ and $(5,3)$ are unstable equilibrium values. See Figure 3.3.

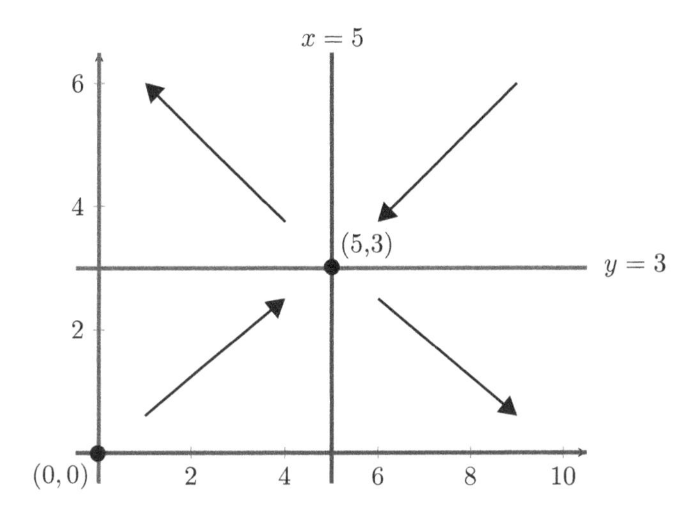

FIGURE 3.3: Equilibrium Analysis

Phase Portrait

We now examine the phase portrait that provides the same information to us, but in a slightly different format. Begin by loading the *DEtools* package to make *DEplot* available.

```
> with(DEtools) :
```

Define the differential equations.

```
> diffeq1 := diff(x(t), t) = 0.24 · x(t) − 0.08 · y(t) · x(t);
  diffeq2 := diff(y(t), t) = 4.5 · y(t) − 0.9 · y(t) · x(t);
```

$$diffeq1 := \frac{d}{dt}\,x(t) = 0.24 \cdot x(t) - 0.08 \cdot y(t) \cdot x(t)$$

$$diffeq2 := \frac{d}{dt}\,y(t) = 4.5 \cdot y(t) - 0.9 \cdot y(t) \cdot x(t)$$

And plot the phase portrait.

```
> inits := [[x(0) = 5.6, y(0) = 4.8], [x(0) = 5.6, y(0) = 4.5], [x(0) =
  4.6, y(0) = 2.2],
  [x(0) = 4.6, y(0) = 2.0]] :
```

> $DEplot(\{diffeq1, diffeq2\}, [x(t), y(t)], t = 0..10, x = 2.5..7.5, y = 0..6, inits,$
> $linecolor = black, title =$ "Competing Species");

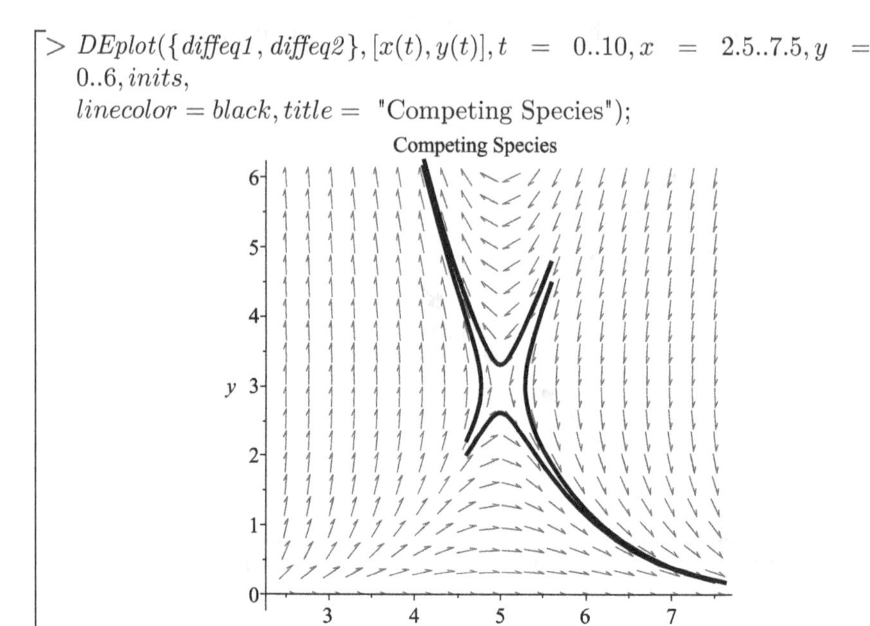

The phase portrait shows, from a given starting point (the initial condition), who survives. The phase portrait shows the traces of possible solution curves. Note how, even though starting very close, trajectories can diverge wildly.

Example 3.7. Competitive Hunter System for the Fish Pond Problem.
Let $B(t)$ represent the number of brown trout at time t, and $R(t)$ represent the number of rainbow trout at time t. Then

$$\textit{Rate of Change} = \textit{Rate in Isolation} + \textit{Rate in Competition}$$

gives us the *competitive hunter system*

$$\frac{dB}{dt} = 0.8B - 0.02B \cdot R$$
$$\frac{dR}{dt} = 0.6R - 0.01B \cdot R$$

First, find the rest points by solving $dB/dt = 0$ and $dR/dt = 0$ simultaneously.

> $solve(\{0.8B - 0.02B \cdot R = 0, 0.6R - 0.01B \cdot R\}, \{B, R\});$

$$\{B = 0., R = 0.\}, \{B = 60., R = 40.\}$$

As in the last example, both equilibrium values, $(0, 0)$ and $(60, 40)$, are not stable. Thus, the phase portrait will be very useful for giving us a sense of the possible solutions.

```
> deqn1 := diff(B(t),t) = 0.8 · B(t) − 0.02 · B(t) · R(t);
```

$$deqn1 := \frac{d}{dt}B(t) = 0.8 \cdot B(t) - 0.02 \cdot B(t) \cdot R(t)$$

```
> deqn2 := diff(R(t),t) = 0.6 · R(t) − 0.01 · B(t) · R(t);
```

$$deqn2 := \frac{d}{dt}R(t) = 0.6 \cdot R(t) - 0.01 \cdot B(t) \cdot R(t)$$

```
> inits := [[B(0) = 10, R(0) = 8], [B(0) = 10, R(0) = 9],
  [B(0) = 100, R(0) = 65], [B(0) = 100, R(0) = 62]] :
```

```
> DEplot({deqn1, deqn2}, [B(t), R(t)], t = 0..20, B = 0..110, R =
  0..75, inits,
  linecolor = black, title = "Fish Pond");
```

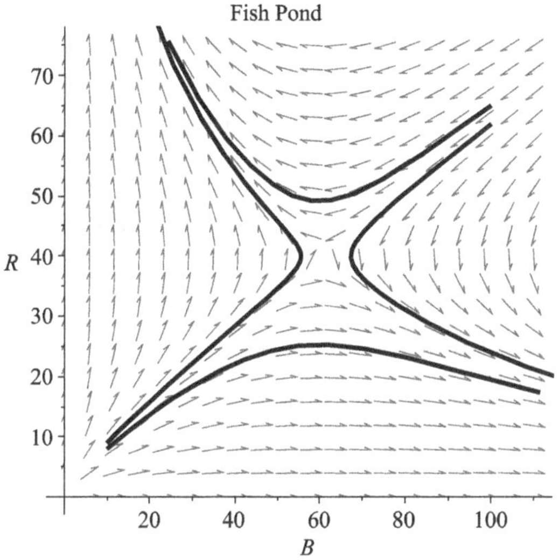

We see that depending on the starting value, one species will dominate over time. As before, the starting values are quite close, but the trajectories go in very different directions. Our competitive hunter system for the fish pond is very sensitive to initial conditions.

Exercises

Find and classify all the rest points or equilibria. Plot the phase portrait, then sketch a few trajectories to indicate the motion.

1. $dx/dt = x + 1$, $dy/dt = y - 2$
2. $dx/dt = x$, $dy/dt = -2y$

3. $dx/dt = -y$, $dy/dt = 2x$

4. $dx/dt = -x + 1$, $dy/dt = -2y + 1$

3.4 Solving Homogeneous and Nonhomogeneous Systems of ODEs in Maple

Maple can solve systems of differential equations of the form

$$\frac{dx}{dt} = ax + by + g(t)$$
$$\frac{dy}{dt} = mx + ny + h(t)$$

where (1) a, b, m, n are constants, and (2) the functions $g(t)$ and $h(t)$ can either be 0 or functions of t alone with real coefficients.

When $g(t)$ and $h(t)$ are both 0 then the system of differential equations is called a *homogeneous system*, otherwise it is nonhomogeneous. We will begin with homogeneous systems.

The method we will use involves eigenvalues and eigenvectors.

Example 3.8. A Homogeneous System with Initial Values.
Consider the following homogeneous system with initial conditions

$$x' = 2x - y + 0$$
$$y' = 3x - 2y + 0$$
$$x(0) = 1, y(0) = 2$$

We rewrite the system of differential equation in matrix form $X' = AX$ where

$$A = \begin{bmatrix} 2 & 1 \\ 3 & -2 \end{bmatrix}, \quad X = \begin{bmatrix} x(t) \\ y(t) \end{bmatrix}, \quad \text{and} \quad X' = \begin{bmatrix} dx/dt \\ dy/dt \end{bmatrix}$$

then we can solve $X' = AX$. This form is highly suggestive of the first-order separable equation that we saw in the previous chapter. We can assume the solution to have a similar form: $X = Ke^{\lambda t}$, where λ is a constant and X and K are vectors. The values of λ are called *eigenvalues* and the components of K are the corresponding *eigenvectors*. We note that a full discussion of the theory and applications of eigenvalues and eigenvectors can be found in linear algebra textbooks as well as many differential equations textbooks.

Since we have a 2×2 nondegenerate system, there are two linearly independent solutions that we will call X_1 and X_2. The *complementary solution* $X = c_1X_1 + c_2X_2$ where c_1 and c_2 are arbitrary constants. We use the initial conditions to find specific values for c_1 and c_2.

The following steps can be used when we have distinct real eigenvalues.

Step 1. Set up the system as a matrix, $X' = AX$, $X(0) = X_0$.

Step 2. Find A's eigenvalues λ_1 and λ_2.

Step 3. Find the corresponding eigenvectors K_1 and K_2.

Step 4. Write the complementary solution $X_c = c_1X_1 + c_2X_2$ where $X_1 = K_1e^{\lambda_1 t}$ and $X_2 = K_2e^{\lambda_2 t}$.

Step 5. Use the initial conditions to solve for c_1 and c_2, and rewrite the solution for X_c.

Maple allows us to easily find the homogeneous solution in matrix form. The following commands illustrate the steps listed above.

```
> with(LinearAlgebra) :
> A := Matrix([[2, -1], [3, -2]]);
```

$$A := \begin{bmatrix} 2 & -1 \\ 3 & -2 \end{bmatrix}$$

Maple's command *Eigenvectors* returns two items: a vector and a matrix. The vector contains the eigenvalues $\begin{bmatrix} \lambda_1 \\ \lambda_2 \end{bmatrix}$; the matrix is made from the respective eigenvectors $[\vec{v}_1, \vec{v}_2]$.

```
> λ, EVs := Eigenvectors(A);
  K1 := Column(EVs, 1);
  K2 := Column(EVs, 2);
```

$$\lambda, EVs := \begin{bmatrix} 1 \\ -1 \end{bmatrix}, \begin{bmatrix} 1 & 1/3 \\ 1 & 1 \end{bmatrix}$$

$$K_1 := \begin{bmatrix} 1 \\ 1 \end{bmatrix}$$

$$K_2 := \begin{bmatrix} 1/3 \\ 1 \end{bmatrix}$$

```
> X := t → c1 · K1 · exp(λ1 · t) + c2 · K2 · exp(λ2 · t);
  X(t);
```

$$X := t \to c_1 \cdot K_1 \cdot e^{\lambda_1 t} + c_2 \cdot K_2 \cdot e^{\lambda_2 t}$$

$$\begin{bmatrix} \dfrac{c_1e^{-t}}{3} + c_2e^{t} \\ c_1e^{-t} + c_2e^{t} \end{bmatrix}$$

Of course, since this is an important technique, Maple has a command *matrixDE* in the *DEtools* package to perform this quickly.

```
> with(DEtools) :
```

> $HomogeneousSoln := matrixDE(A, t):$
> $\phi := HomogeneousSoln[1];$

$$\phi := \begin{bmatrix} e^t & e^{-t} \\ e^t & 3e^{-t} \end{bmatrix}$$

> $Xc := \phi . \langle c_1, c_2 \rangle;$ # *Period = matrix multiply,* $\langle c_1, c_2 \rangle$ *= column vector.*

$$Xc := \begin{bmatrix} c_1 e^t + c_2 e^{(-t)} \\ c_1 e^t + 3c_2 e^{(-t)} \end{bmatrix}$$

Using the initial conditions $x(0) = 1$, $y(0) = 2$, we next find the complementary solution.

> $IC := eval(Xc, t = 0);$

$$IC := \begin{bmatrix} c_1 + c_2 \\ c_1 + 3c_2 \end{bmatrix}$$

> $const := solve(\{IC_1 = 1, IC_2 = 2\}, \{c_1, c_2\});$

$$const := \left\{ c_1 = \frac{1}{2}, c_2 = \frac{1}{2} \right\}$$

> $CompSoln := eval(Xc, const);$

$$CompSoln := \begin{bmatrix} \frac{e^t}{2} + \frac{e^{-t}}{2} \\ \frac{e^t}{2} + \frac{3e^{-t}}{2} \end{bmatrix}$$

We plot the solutions to the X_1 and X_2 components, each a function of t.

> $plot(CompSoln, t = 0..2);$

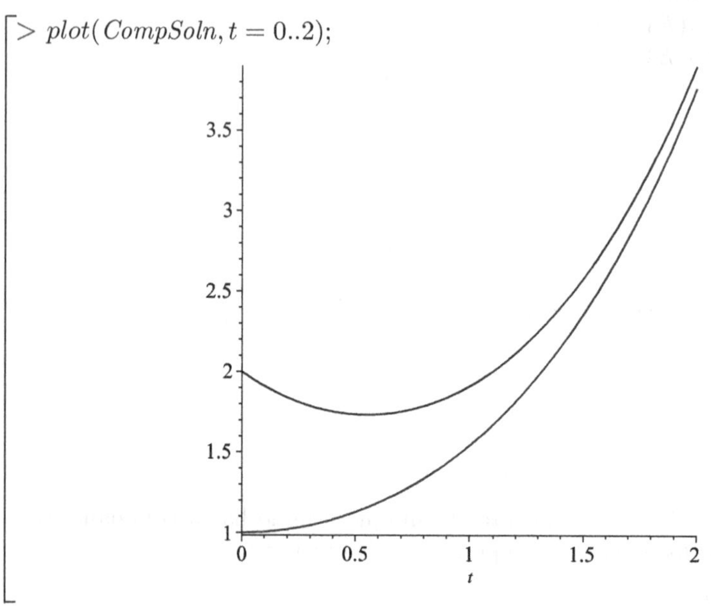

Note that both solutions grow without bound as t goes to infinity as both components are asymptotic to $y = e^t/2$.

Example 3.9. A Homogenous System with Complex Eigenvalues $\lambda = a \pm bi$.
Complex eigenvalues for a real 2×2 matrix must always come in conjugate pairs, so we can use a moderate amount of algebra with a little analysis to eliminate the i terms leaving only real functions in our solution. The key to finding two real linearly independent solutions from complex eigenvalues is Euler's identity

$$e^{it} = \cos(t) + i\sin(t).$$

Using Euler's identity, we can rewrite the solutions for X_1 and X_2. For $\lambda_{1,2} = a \pm bi$,

$$\begin{aligned} K_1 e^{\lambda_1 t} &= K_1 e^{(a+bi)t} = K_1 e^{at}\left(\cos(bt) + i\sin(bt)\right) \\ K_2 e^{\lambda_2 t} &= K_2 e^{(a-bi)t} = K_2 e^{at}\left(\cos(bt) - i\sin(bt)\right) \end{aligned}$$

The following steps are a summary for when we encounter complex eigenvalues.

Step 1. Find the complex eigenvalues $\lambda = a \pm bi$.

Step 2. Find the complex eigenvectors

$$K_1 = \begin{bmatrix} u_1 + iv_1 \\ u_2 + iv_2 \end{bmatrix} \quad \text{and} \quad K_2 = \begin{bmatrix} u_1 - iv_1 \\ u_2 - iv_2 \end{bmatrix}.$$

Step 3. Form the real vectors

$$B_1 = \begin{bmatrix} u_1 \\ u_2 \end{bmatrix} \quad \text{and} \quad B_2 = -\begin{bmatrix} v_1 \\ v_2 \end{bmatrix}.$$

Step 4. Write the complementary solution $X_c = c_1 X_1 + c_2 X_2$ where

$$\begin{aligned} X_1 &= e^{at}\left(B_1\cos(bt) + B_2\sin(bt)\right) \\ X_2 &= e^{at}\left(B_2\cos(bt) - B_2\sin(bt)\right) \end{aligned}$$

Maple will allow for easy manipulation of these steps for us.

```
> with(LinearAlgebra) :
  with(DEtools);
```

```
> A := Matrix([[6, -1], [5, 4]]);
```

$$A := \begin{bmatrix} 6 & -1 \\ 5 & 4 \end{bmatrix}$$

> $\lambda, K := Eigenvectors(A);$

$$\lambda, K := \begin{bmatrix} 5+2i \\ 5-2i \end{bmatrix}, \begin{bmatrix} \frac{1}{5}+\frac{2I}{5} & \frac{1}{5}-\frac{2I}{5} \\ 1 & 1 \end{bmatrix}$$

> $\phi := matrixDE(A, t)_1;$

$$\phi := \begin{bmatrix} e^{5t}\sin(2t) & e^{5t}\cos(2t) \\ e^{5t}\sin(2t) - 2e^{5t}\cos(2t) & e^{5t}\sin(2t) + 2e^{5t}\cos(2t) \end{bmatrix}$$

We next find the complementary solution.

> $Xc := \phi \,.\, \langle c_1, c_2 \rangle;$

$$Xc := \begin{bmatrix} e^{5t}\sin(2t)c_1 + e^{5t}\cos(2t)c_2 \\ \left(e^{5t}\sin(2t) - 2e^{5t}\cos(2t)\right)c_1 + \left(e^{5t}\sin(2t) + 2e^{5t}\cos(2t)\right)c_2 \end{bmatrix}$$

We will solve for c_1 and c_2 using the initial conditions, $x(0) = 1$, $y(0) = 2$.

> $IC := eval(Xc, t = 0);$

$$\begin{bmatrix} c_2 \\ -2c_1 + c_2 \end{bmatrix}$$

> $const := solve(\{IC_1 = 1, IC_2 = 2\}, \{c_1, c_2\});$

$$const := \left\{c_1 = -\frac{1}{2}, c_2 = 1\right\}$$

> $Xg := eval(Xc, const);$

$$Xg := \begin{bmatrix} -\frac{e^{5t}\sin(2t)}{2} + e^{5t}\cos(2t) \\ \frac{3e^{5t}\sin(2t)}{2} + 2e^{5t}\cos(2t) \end{bmatrix}$$

Again, we obtain plots of X_1 and X_2 as functions of t.

> $plot(Xg, t = 0..0.5);$

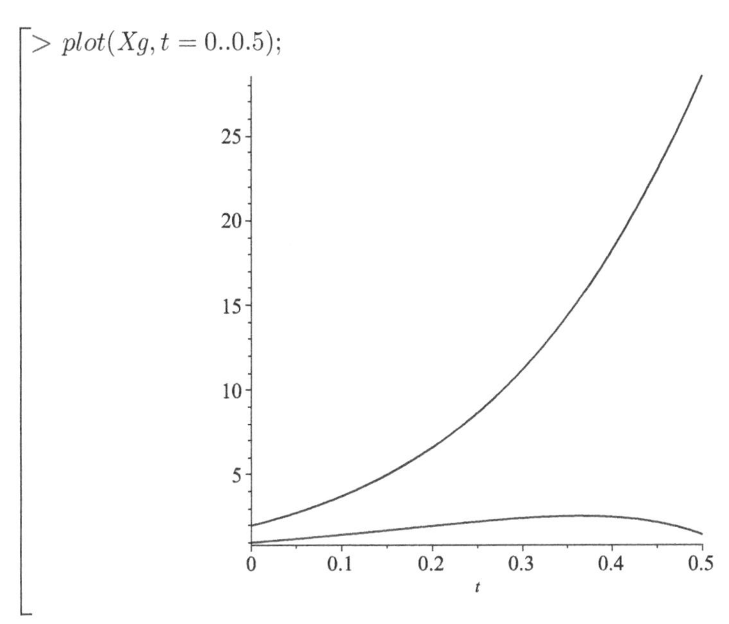

The solutions will oscillate due to the cos and sin terms, and grow to infinity because of the exponentials. Try graphing with $t = n..n+1$ for several values of n.

Example 3.10. A Homogeneous System with Repeated Eigenvalues. When eigenvalues are repeated we must find a method to obtain linearly independent solutions. (A system must be at least 4×4 to have repeated complex eigenvalues since we know that systems with real coefficients have complex eigenvalues that must occur as conjugate pairs of complex numbers.)

The following is a summary for repeated real eigenvalues for the system $X' = AX$ with $A = \begin{bmatrix} a & b \\ c & d \end{bmatrix}$.

Step 1. Find *the repeated eigenvalues* $\lambda_1 = \lambda_2 = \lambda$.

Step 2. One solution is $X_1 = Ke^{\lambda t}$, and the second linearly independent solution is

$$X_2(t) = Kte^{\lambda t} + Pe^{\lambda t}$$

where $P = \begin{bmatrix} p_1 \\ p_2 \end{bmatrix}$ must satisfy the system

$$\begin{aligned} (a-\lambda)p_1 + bp_2 &= k_1 \\ cp_1 + (d-\lambda)p_2 &= k_2 \end{aligned} \quad \text{or, as a matrix equation} \quad (A - \lambda I)\cdot P = K$$

We now use Maple to present an example of this method. Maple's *matrixDE* recognizes repeated eigenvalues and places the solution in the correct form.

> $A := Matrix([[3, -18], [2, -9]]);$

$$A := \begin{bmatrix} 3 & -18 \\ 2 & -9 \end{bmatrix}$$

> $\lambda, K := Eigenvectors(A);$

$$\lambda, K := \begin{bmatrix} -3 \\ -3 \end{bmatrix}, \begin{bmatrix} 3 & 0 \\ 1 & 0 \end{bmatrix}$$

The repeated eigenvalue is $\lambda = -3$. We know that, by definition, $v = \begin{bmatrix} 0 & 0 \end{bmatrix}^T$ cannot be an eigenvector, so only one eigenvector, $\langle 3, 1 \rangle$, is shown above.

> $\phi := matrixDE(A, t)_1;$

$$\phi := \begin{bmatrix} e^{-3t} & e^{-3t}t \\ \frac{e^{-3t}}{3} & \frac{e^{-3t}t}{3} - \frac{e^{-3t}}{18} \end{bmatrix}$$

We next find the complementary solution.

> $Xc := \phi \;.\; \langle c_1, c_2 \rangle;$

$$Xc := \begin{bmatrix} e^{-3t}c_1 + e^{-3t}tc_2 \\ \frac{e^{-3t}c_1}{3} + \left(\frac{e^{-3t}t}{3} - \frac{e^{-3t}}{18}\right)c_2 \end{bmatrix}$$

We again solve for c_1 and c_2 using the initial conditions, $x(0) = 1$, $y(0) = 2$.

> $IC := eval(Xc, t = 0);$

$$IC := \begin{bmatrix} c_1 \\ \frac{c_1}{3} - \frac{c_2}{18} \end{bmatrix}$$

> $const := solve(\{IC_1 = 1, IC_2 = 2\}, \{c_1, c_2\});$

$$const := \{c_1 = 1, c_2 = -30\}$$

> $Xg := eval(Xc, const);$

$$Xg := \begin{bmatrix} e^{-3t} - 30e^{-3t}t \\ 2e^{-3t} - 10e^{-3t}t \end{bmatrix}$$

Generate plots of X_1 and X_2 as functions of t.

```
> plot(Xg, t = 0..2);
```

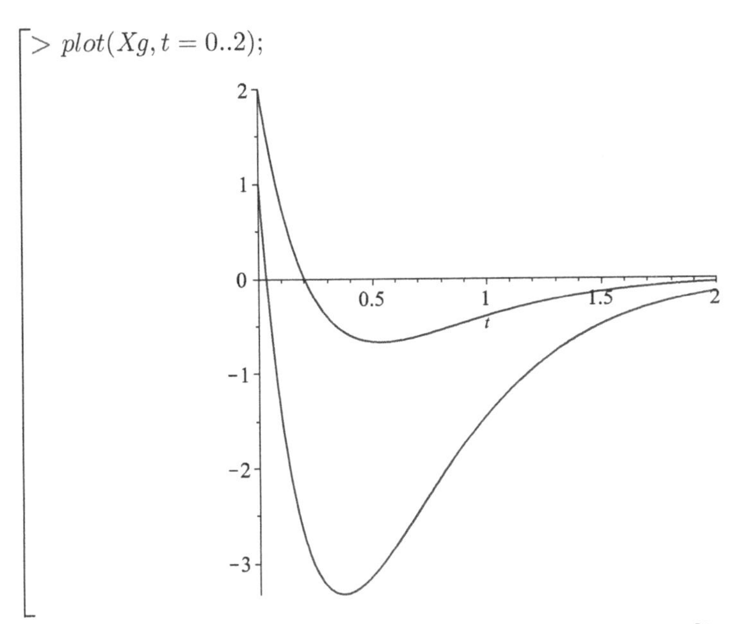

In the graph above, we see the quick damping effect of e^{-3t}.

Nonhomogeneous Systems of Differential Equations and Maple

In the nonhomogeneous form, $X' = AX + F(t)$, we need both a complementary solution X_c and a particular solution X_p to find the general solution $X_g = X_c + X_p$. The complementary solution is found by solving the homogeneous part of the system $X' = AX$. The particular solution is found using the *variation of parameters* technique. The following summary of the solution procedure is provided.

Step 1. Find the complementary solution X_c by solving the homogeneous system $X' = AX$. Form the linear independent set of solutions, put them as columns of the matrix Φ.

Step 2. *Vary the parameters:* write the form of the particular solution

$$X_p = u_1(t)X_1(t) + u_2(t)X_2(t),$$

(u_1 and u_2 are the parameters.)

Step 3. Find the matrix inverse Φ^{-1}.

Step 4. Determine the parameters $u_1(t)$ and $u_2(t)$ from

$$U = \begin{bmatrix} u_1 \\ u_2 \end{bmatrix} = \int \Phi^{-1} \cdot F(t)\, dt.$$

(*When the integrand* $\Phi^{-1} \cdot F$ *is difficult to integrate, Laplace transform techniques may be easier to use.*)

Step 5. Calculate the particular solution

$$X_p = \Phi \cdot U = \Phi \cdot \int \Phi^{-1} \cdot F(t)\, dt$$

Step 6. Form the general solution for $X = X_c + X_p$.

Consult a differential equations textbook for a more detailed explanation and for proofs of the procedure.

Now we turn to an example using Maple. Given the following system of nonhomogeneous differential equations:

$$x' = 2x - y + 0$$
$$y' = 3x - 2y + 4t$$
$$x(0) = 1, y(0) = 2$$

Note that the nonhomogeneous term F is $g(t) = 0$ and $h(t) = 4t$.

Remember to load the *LinearAlgebra* and *DEtools* packages before beginning the procedure.

Part 1: Find the Homogeneous Solution X_c

```
> A := Matrix([[2, -1], [3, -2]]);
```

$$A := \begin{bmatrix} 2 & -1 \\ 3 & -2 \end{bmatrix}$$

```
> λ, K := Eigenvectors(A);
```

$$\lambda, K := \begin{bmatrix} 1 \\ -1 \end{bmatrix}, \begin{bmatrix} 1 & 1/3 \\ 1 & 1 \end{bmatrix}$$

```
> Φ := matrixDE(A, t)[1];
```

$$\Phi := \begin{bmatrix} e^t & e^{-t} \\ e^t & 3e^{-t} \end{bmatrix}$$

The complementary solution is:

```
> Xc := Φ . <c1, c2>;
```

$$Xc := \begin{bmatrix} e^t c_1 + e^{-t} c_2 \\ e^t c_1 + 3e^{-t} c_2 \end{bmatrix}$$

If the original system was homogeneous, we would substitute the initial conditions to determine c_1 and c_2, but since the original system is nonhomgeneous, we must first find the particular solution X_p.

Part 2: Find the Particular Solution X_p

```
> Φi := evalm(Φ^(-1));   # evalm mean "evaluate as a matrix"
```

$$\Phi i := \begin{bmatrix} \frac{3}{2e^t} & -\frac{1}{2e^t} \\ -\frac{1}{2e^{-t}} & \frac{1}{2e^{-t}} \end{bmatrix}$$

```
> F := <0, 4t>;
```

$$F := \begin{bmatrix} 0 \\ 4t \end{bmatrix}$$

```
> U := int~(Φi . F, t);   # 'int~' = 'apply int to each entry of the matrix.'
```

$$U := \begin{bmatrix} \frac{2(t+1)}{e^t} \\ \frac{2(t-1)}{e^{-t}} \end{bmatrix}$$

The particular solution is

```
> Xp := Φ . U;
```

$$Xp := \begin{bmatrix} 4t \\ 8t-4 \end{bmatrix}$$

Combining the results above gives the general solution as

```
> X := Xc + Xp;
```

$$X := \begin{bmatrix} e^t c_1 + e^{-t} c_2 + 4t \\ e^t c_1 + 3e^{-t} c_2 + 8t - 4 \end{bmatrix}$$

We finish the problem by determining c_1 and c_2 by using the initial conditions $x(0) = 1$ and $y(0) = 2$.

```
> IC := eval(X, t = 0);
```

$$IC := \begin{bmatrix} c_1 + c_2 \\ c_1 + 3c_2 - 4 \end{bmatrix}$$

```
> const := solve({IC[1] = 1, IC[2] = 2}, {c1, c2});
```

$$const := \left\{ c_1 = -\frac{3}{2}, c_2 = \frac{5}{2} \right\}$$

> $Xg := eval(X, const);$

$$Xg := \begin{bmatrix} -\frac{3e^t}{2} + \frac{5e^{-t}}{2} + 4t \\ -\frac{3e^t}{2} + \frac{15e^{-t}}{2} + 8t - 4 \end{bmatrix}$$

As always, graph the solution.

> $plot(Xg, t = 0..1);$

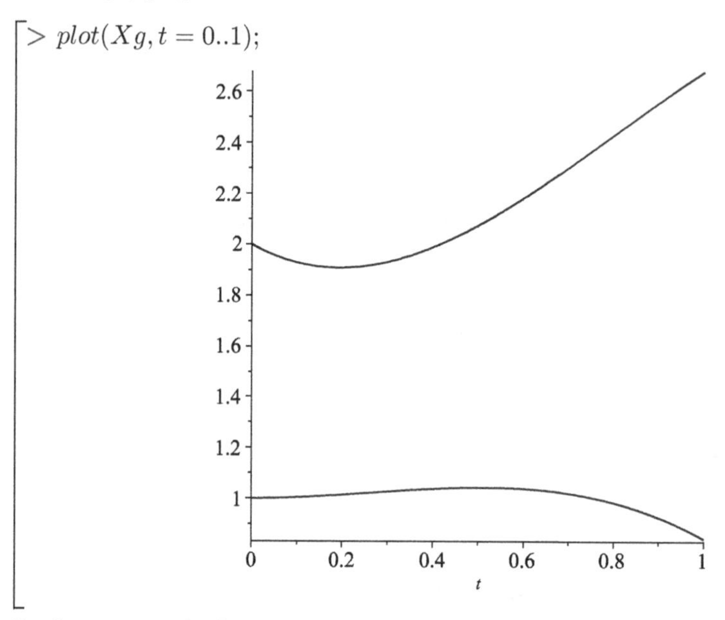

Look at several plots over larger domains to understand the nature of the solution.

Applied Problem Solving of Systems of ODEs with Maple

Recall the diffusion model from Example 3.5. We will use Maple's *dsolve* to assist us in the solving the diffusion model.

Example 3.11. Chemical Diffusion through a Single Membrane. Consider the diffusion initial value problem:

$$\frac{dx_1}{dt} = \frac{P}{V_1}(x_2 - x_1) = -x_1 + x_2$$
$$\frac{dx_2}{dt} = \frac{P}{V_2}(x_1 - x_2) = x_1 - x_2$$

with $x_1(0) = 2$ and $x_2(0) = 10$, and assuming that $P = V_1 = V_2$.

First, we set up the differential equation.

$> Chem1 := Matrix([[1,1],[-1,1]]);$

$$Chem1 := \begin{bmatrix} 1 & -1 \\ -1 & 1 \end{bmatrix}$$

$> inits := x_1(0) = 2, x_2(0) = 10;$

$$inits := x_1(0) = 2, x_2(0) = 10$$

$> X := \langle x_1(t), x_2(t) \rangle;$

$$X := \begin{bmatrix} x_1(t) \\ x_2(t) \end{bmatrix}$$

$> Chem1DE := diff\sim(X,t) = Chem1 \;.\; X;$

$$Chem1DE := \begin{bmatrix} \frac{d}{dt} x_1(t) \\ \frac{d}{dt} x_2(t) \end{bmatrix} = \begin{bmatrix} -x_1(t) + x_2(t) \\ x_1(t) - x_2(t) \end{bmatrix}$$

We'll use *dsolve* to obtain the solution.

$> dsolve(\{Chem1DE, inits\}, \{x_1(t), x_2(t)\});$
$Chem1\_soln := rhs\sim(\%);$

$$\{x_1(t) = 6 - 4e^{-2t}, x_2(t) = 4e^{-2t} + 6\}$$

$$Chem1\_soln := \{6 - 4e^{-2t}, 4e^{-2t} + 6\}$$

The usual graph:

$> plot(Chem1\_soln, t = 0..10, thickness = 2);$

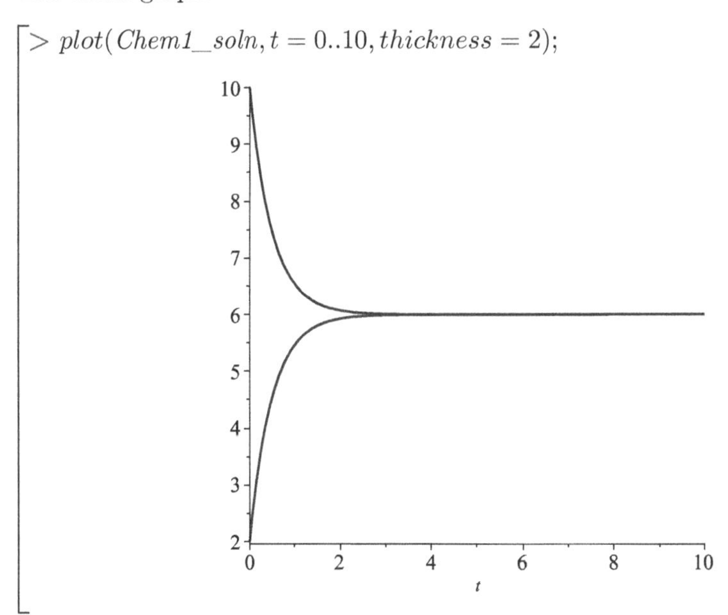

The solution's plot shows that we have a stable equilibrium value at 6 units.

Our next application is solving a nonhomogeneous system. We'll use the steps outlined above (on pg. 102). Remember to load the *LinearAlgebra* and *DEtools* packages before starting on the solution.

Example 3.12. Chemical Diffusion through a Double-Walled Membrane.
Consider the system of nonhomogeneous differential equations

$$\frac{dx}{dt} = 1.5(y - x)$$
$$\frac{dy}{dt} = 0.8(10 - y) + 0.3(x - y)$$

with

$$x(0) = 2, y(0) = 1$$

$> A := Matrix([[-1.5, 1.5], [0.3, -1.1]]);$

$$A := \begin{bmatrix} -1.5 & 1.5 \\ 0.3 & -1.1 \end{bmatrix}$$

$> \lambda, K := Eigenvectors(A);$

$$\lambda, K := \begin{bmatrix} -2.0 + 0.0\,i \\ -0.59999999999999978 + 0.0\,i \end{bmatrix},$$
$$\begin{bmatrix} -0.948683298050513768 + 0.0\,i & -0.857492925712544318 + 0.0\,i \\ 0.316227766016837997 + 0.0\,i & -0.514495755427526458 + 0.0\,i \end{bmatrix}$$

The results above illustrate two aspects of Maple:

- Maple does computations in the complex domain even though the inputs and outputs are real, and
- *Eigenvectors* increases the number of digits used in calculations to improve accuracy.

Let's *simplify* both λ and K.

$> \lambda, K := simplify(\lambda), simplify(K);$

$$\Lambda, K := \begin{bmatrix} -2.0 \\ -0.6000000000 \end{bmatrix}, \begin{bmatrix} -0.9486832981 & -0.8574929257 \\ 0.3162277660 & -0.5144957554 \end{bmatrix}$$

$> \phi := matrixDE(A, t)_1;$

$$\phi := \begin{bmatrix} 1.e^{-2t} & 1.e^{-\frac{3t}{5}} \\ -0.333333333e^{-2t} & 0.6000000000e^{-\frac{3t}{5}} \end{bmatrix}$$

```
> φi := evalm(φ^(-1));
```

$$\phi i := \begin{bmatrix} \frac{0.6428571431}{e^{-2t}} & -\frac{1.071428572}{e^{-2t}} \\ \frac{0.3571428569}{e^{-\frac{3t}{5}}} & \frac{1.071428572}{e^{-\frac{3t}{5}}} \end{bmatrix}$$

```
> F := <0,8>;
```

$$F := \begin{bmatrix} 0 \\ 8 \end{bmatrix}$$

```
> U := int~(φi . F,t);
  # Remember: '~' means apply the operation to each element of the array.
```

$$U := \begin{bmatrix} -\frac{4.285714288}{e^{-2.t}} \\ \frac{14.28571429}{e^{-.6000000000t}} \end{bmatrix}$$

```
> φ . U;
  Xp := simplify(%);
```

$$\begin{bmatrix} -\frac{4.285714288\, e^{-2\,t}}{e^{-2.0\,t}} + \frac{14.28571429\, e^{-3t/5}}{e^{-0.6000000000\,t}} \\ \frac{1.428571428\, e^{-2\,t}}{e^{-2.0\,t}} + \frac{8.571428574\, e^{-3t/5}}{e^{-0.6000000000\,t}} \end{bmatrix}$$

$$Xp := \begin{bmatrix} 10.00000000 \\ 10.00000000 \end{bmatrix}$$

```
> X := φ . <c1,c2> + Xp;
```

$$X := \begin{bmatrix} 1.\, e^{-2t} c_1 + 1.\, e^{-\frac{3t}{5}} c_2 + 10.00000000 \\ -0.333333333\, e^{-2t} c_1 + 0.6000000000\, e^{-\frac{3t}{5}} c_2 + 10.00000000 \end{bmatrix}$$

```
> IC := eval(X,t = 0);
```

$$IC := \begin{bmatrix} 1.\, c_1 + 1.\, c_2 + 10.00000000 \\ -0.333333333\, c_1 + 0.6000000000\, c_2 + 10.00000000 \end{bmatrix}$$

```
> const := solve({IC[1] = 2, IC[2] = 1}, {c1, c2});
```

$$const := \{c_1 = 4.500000002, c_2 = -12.50000000\}$$

```
> Xg := eval(X, const);
```

$$Xg := \begin{bmatrix} 4.500000002\, e^{-2\,t} - 12.50000000\, e^{-(3/5)\,t} + 10.0 \\ -1.499999999\, e^{-2\,t} - 7.500000000\, e^{-(3/5)\,t} + 10.0 \end{bmatrix}$$

> $plot(Xg, t = 0..15, 0..10.5, thickness = 2);$

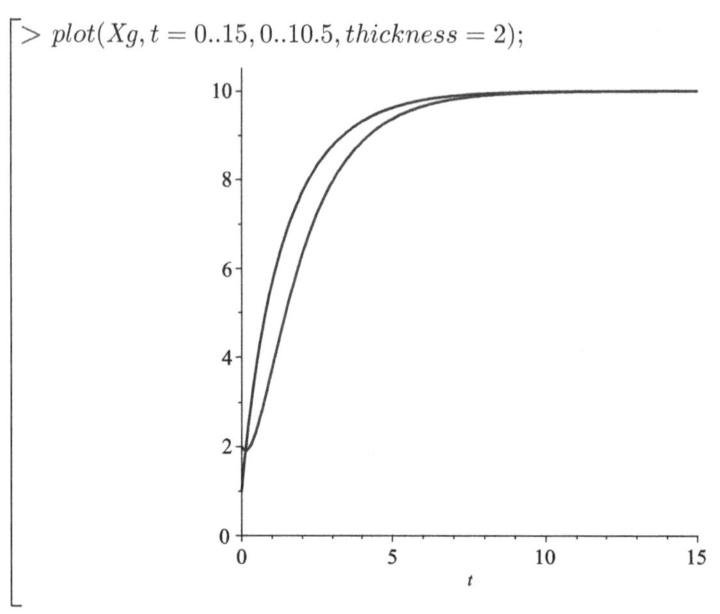

The graph of the solutions shows a stable equilibrium value at 10.0 units.

In an electric circuit, Ohm's Law (*voltage drop* $\propto$ *current* for a resistor), Faraday's Law (*voltage drop* $\propto$ *current'* for an inductor), and Coulomb's Law (*charge'* $\propto$ *current* for a capacitor) describe the voltage and current relationships across resistors, inductors, and capacitors, respectively. Kirchhoff's Voltage Law (*conservation of voltage around a circuit*) and Current Law (*conservation of current through any node of a circuit*) complete the picture.

Example 3.13. Electrical Circuit: A Homogeneous System.
Analyze the behavior of the electric circuit described by the system

$$\frac{dQ}{dt} = -Q - i_2$$
$$\frac{di_2}{dt} = Q - i_2$$

and

$$Q(0) = 3, i_2(0) = 1$$

As usual, remember to load the *LinearAlgebra* and *DEtools* packages.

> $A := Matrix([[-1, -1], [1, -1]]);$

$$A := \begin{bmatrix} -1 & -1 \\ 1 & -1 \end{bmatrix}$$

> $X := \langle Q(t), i_2(t) \rangle;$

$$X := \begin{bmatrix} Q(t) \\ i_2(t) \end{bmatrix}$$

```
> IVP := [diff~(X,t) = A . X, Q(0) = 3, i2(0) = 1];
```

$$IVP := \left[\left[\begin{array}{c}\frac{d}{dt}Q(t)\\ \frac{d}{dt}i_2(t)\end{array}\right] = \left[\begin{array}{c}-Q(t)-i_2(t)\\ Q(t)-i_2(t)\end{array}\right], Q(0)=3, i_2(0)=1\right]$$

```
> dsolve(IVP, X);
  Xc := rhs(%);
```

$$\left[\begin{array}{c}Q(t)\\ i_2(t)\end{array}\right] = \left[\begin{array}{c}-e^{-t}\left(\sin(t)-3\cos(t)\right)\\ e^{-t}\left(3\sin(t)+\cos(t)\right)\end{array}\right]$$

$$Xc := \left[\begin{array}{c}-e^{-t}\left(\sin(t)-3\cos(t)\right)\\ e^{-t}\left(3\sin(t)+\cos(t)\right)\end{array}\right]$$

```
> plot(Xc, t = 0..10, thickness = 2);
```

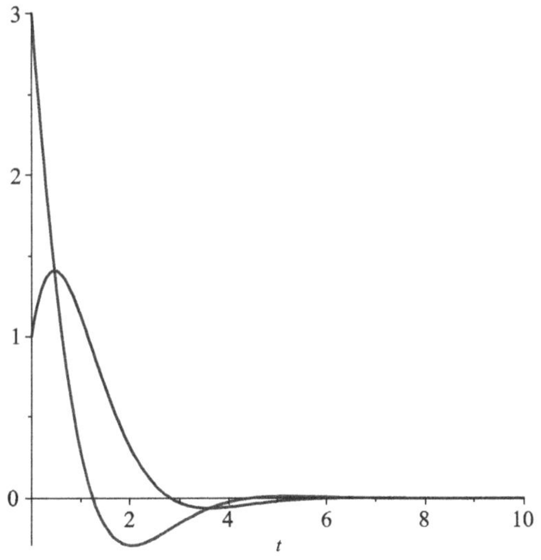

The plot tells us that both Q and i_2 quickly go to zero.

We add an external source, often called a *forcing function*, to the previous system thus changing it from homogeneous to nonhomogeneous.

Example 3.14. Electrical Circuit: A Nonhomogeneous System.
Analyze the behavior of the electric circuit described by the system

$$\frac{dQ}{dt} = -Q - i_2 + 10\,e^{-t}$$

$$\frac{di_2}{dt} = Q - i_2$$

and

$$Q(0) = 3, i_2(0) = 1$$

```
> A := Matrix([[-1, -1], [1, -1]]);
```

$$A := \begin{bmatrix} -1 & -1 \\ 1 & -1 \end{bmatrix}$$

```
> φ := matrixDE(A, t)[1];
```

$$\phi := \begin{bmatrix} e^{-t}\sin(t) & e^{-t}\cos(t) \\ -e^{-t}\cos(t) & e^{-t}\sin(t) \end{bmatrix}$$

```
> φi := simplify(evalm(φ^(-1)));
```

$$\phi i := \begin{bmatrix} e^{t}\sin(t) & -e^{t}\cos(t) \\ e^{t}\cos(t) & e^{t}\sin(t) \end{bmatrix}$$

```
> F := <10 exp(-t), 0>;
```

$$F := \begin{bmatrix} 10\,e^{-t} \\ 0 \end{bmatrix}$$

```
> U := int~(φi . F, t);   # '~=' apply the operation to each element
```

$$U := \begin{bmatrix} 10\cos(t) \\ 10\sin(t) \end{bmatrix}$$

```
> φ . U;
  Xp := simplify(%);
```

$$\begin{bmatrix} 0 \\ 10\,e^{-t}\cos(t)^2 + 10\,e^{-t}\sin(t)^2 \end{bmatrix}$$

$$\begin{bmatrix} 0 \\ 10\,e^{-t} \end{bmatrix}$$

```
> X := φ . <c[1], c[2]> + Xp;
```

$$X := \begin{bmatrix} e^{-t}\sin(t)c_1 + e^{-t}\cos(t)c_2 \\ -e^{-t}\cos(t)c_1 + e^{-t}\sin(t)c_2 + 10\,e^{-t} \end{bmatrix}$$

```
> IC := eval(X, t = 0);
```

$$IC := \begin{bmatrix} c_2 \\ -c_1 + 10 \end{bmatrix}$$

```
> const := solve({IC[1] = 3, IC[2] = 1}, {c[1], c[2]});
```

$$const := \{c_1 = 9, c_2 = 3\}$$

```
> Xg := eval(X, const);
```

$$Xg := \begin{bmatrix} 9\,e^{-t}\sin(t) + 3\,e^{-t}\cos(t) \\ -9\,e^{-t}\cos(t) + 3\,e^{-t}\sin(t) + 10\,e^{-t} \end{bmatrix}$$

> $plot(Xg, t = 0..15, 0..10.5, thickness = 2);$

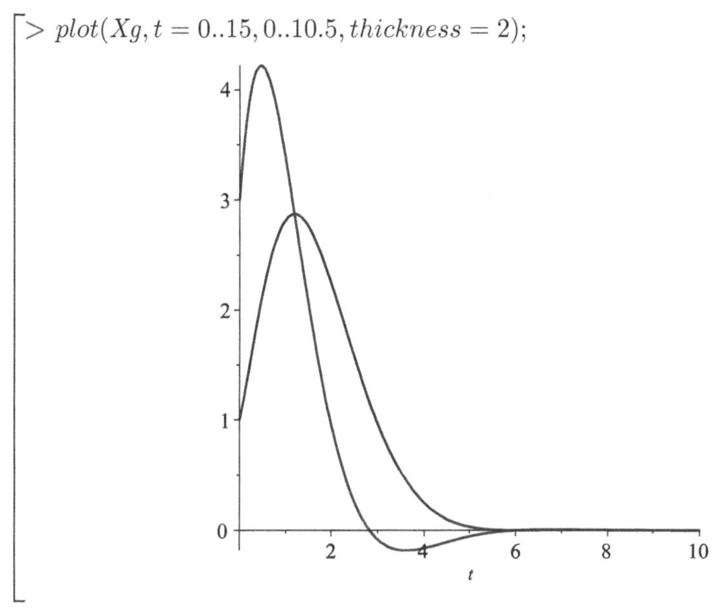

The graph tells us that both Q and i_2 go to zero, but less quickly than the unforced circuit. Adjusting the plot range will show that $i_2(t)$ no longer goes negative.

3.5 Numerical Solutions to Systems of Ordinary Differential Equations with Maple

In the previous chapter, we discussed the use of numerical solutions — Euler, Improved Euler, and Runge-Kutta methods — to first-order differential equations. In this chapter, we extend the use of numerical solutions to systems of differential equations. We show the Euler and Runge-Kutta methods. Our goal here is to provide a solution method for many models of systems of ODEs that do not have closed-form analytical solutions.

Throughout most of this chapter, we have investigated and modeled autonomous systems of first-order differential equations. A more general form of systems of two ordinary first-order differential equations is given by

$$\begin{aligned} \frac{dx}{dt} &= f(t, x, y) \\ \frac{dy}{dt} &= g(t, x, y) \end{aligned} \tag{3.1}$$

If the variable t appears explicitly in one of the functions f or g, the system is not autonomous. In this section, we present numerical techniques for approximating solutions for $x(t)$ and $y(t)$ subject to initial conditions $x(t_0) = x_0$ and $y(t_0) = y_0$.

We will give the algorithm for each method, and show the Maple commands to execute a numerical solution. We also show how to obtain both the phase portraits of the system and the plots of approximate numerical solutions.

Euler's Method

The iterative formula for Euler's Method for systems is

$$x_n = x_{n-1} + f(t_{n-1}, x_{n-1}, y_{n-1}) \cdot \Delta t$$
$$y_n = y_{n-1} + f(t_{n-1}, x_{n-1}, y_{n-1}) \cdot \Delta t$$

We illustrate a few iterations of the following autonomous initial value problem with a step size of $\Delta t = 0.1$.

$$x' = 3x - 2y, \quad x_0 = 3,$$
$$y' = 5x - 4y, \quad y_0 = 6.$$

We compute

$$x_1 = 3 + (3 \cdot 3 - 2 \cdot 6) \cdot 0.1 = 2.7$$
$$y_1 = 6 + (5 \cdot 3 - 4 \cdot 6) \cdot 0.1 = 5.1$$

and then

$$x_2 = 2.7 + (3 \cdot 2.7 - 2 \cdot 5.1) \cdot 0.1 = 2.49$$
$$y_2 = 5.1 + (5 \cdot 2.7 - 4 \cdot 5.1) \cdot 0.1 = 4.41$$

and so forth.

In Maple, we enter the system of ODEs and initial conditions, and then use *dsolve* specifying the optional classical numerical methods. The command sequence to obtain the Euler estimates to our example follows.

First, we set the size of arrays that can be displayed to 100 so that we can view our output directly.

```
> interface(rtablesize = 100) :
```

Define the IVP and list the domain points desired to be shown in the output array.

```
> ODEs := (diff(x(t), t) = 3·x(t)−2·y(t), diff(y(t), t) = 5·x(t)−4·y(t));
  inits := (x(0) = 3, y(0) = 6);
```

$$ODEs := \frac{d}{dt}\,x(t) = 3x(t) - 2y(t),\ \frac{d}{dt}\,y(t) = 5x(t) - 4y(t)$$

$$inits := x(0) = 3, y(0) = 6$$

> $Pts := Array([seq(0.1 \cdot k, k = 0..22)]);$

$$Pts := [0., 0.1, 0.2, 0.3, 0.4, 0.5, 0.6, 0.7, 0.8, 0.9, 1.0, 1.1, 1.2, 1.3, 1.4, 1.5, 1.6, 1.7, 1.8, 1.9, 2.0, 2.1, 2.2]$$

To use Euler's method with *dsolve*, set *method=classical[foreuler]*.

> $EulerSoln := dsolve(\{ODEs, inits\}, [x(t), y(t)],$
> $numeric, method = classical[foreuler], output = Pts, stepsize = 0.1);$

t	$x(t)$	$y(t)$
0.0	3.0	6.0
0.1	2.70000000000000018	5.09999999999999964
0.2	2.49000000000000021	4.41000000000000014
0.3	2.35500000000000043	3.89100000000000046
0.4	2.28330000000000055	3.51210000000000022
0.5	2.26587000000000049	3.24891000000000041
0.6	2.29584900000000047	3.08228100000000049
0.7	2.36814750000000052	2.99729310000000071
0.8	2.47913313000000057	2.98244961000000064
0.9	2.62638314700000075	3.02903633100000080
1.0	2.80849082490000113	3.13061337210000090
1.1	3.02491539795000097	3.28261343571000141
1.2	3.27586733019300080	3.48202576040100098
1.3	3.56222237717070067	3.72714912133710108
1.4	3.88545926605448955	4.01740066138760987
1.5	4.24761691359331550	4.35317002985981105
1.6	4.65126798169934741	4.73571047471254403
1.7	5.09950628126664274	5.16706027567719950
1.8	5.59594611051119450	5.64998930603964045
1.9	6.14473208245662317	6.18796663887938081
2.0	6.75055837941773440	6.78514602455594051
2.1	7.41869668833186857	7.44636680444243293
2.2	8.15503233394294114	8.17716842683139333

The power of Euler's method is two-fold. First, it is very easy to use, and second, as a numerical method, it can be used to estimate a solution to a system of differential equations that does not have a closed-form solution.

Let's investigate the following predator-prey system that does not have a closed-form analytical solution

$$\frac{dx}{dt} = 3x - xy$$
$$\frac{dy}{dt} = xy - 2y$$
$$x(0) = 1, y(0) = 2, \Delta t = 0.1$$

We will obtain an estimate of the solution using Euler's method. Start by loading the *plots* package to make *odeplot* available.

```
> with(plots) :
```

Next define the problem and the array of t values we wish to generate.

```
> PredPrey := diff(x(t), t) = 3 · x(t) − x(t) · y(t), diff(y(t), t) = x(t) · y(t) − 2 · y(t);
```

$$PredPrey := \frac{d}{dt}x(t) = 3 \cdot x(t) - x(t) \cdot y(t), \frac{d}{dt}y(t) = x(t) \cdot y(t) - 2 \cdot y(t)$$

```
> inits := x(0) = 1, y(0) = 2;
```

$$inits := x(0) = 1, y(0) = 2;$$

```
> Pts := Array([seq(0.1 · k, k = 0..26.5)]) :
```

Now generate the solution using Euler's method.

```
> EulerSoln := dsolve({PredPrey, inits}, [x(t), y(t)], output = Pts,
      numeric, stepsize = 0.1, method = classical[foreuler])
```

The *odeplot* command from the *plots* package is very versatile. We can easily graph x versus t, y versus t, and y versus x by using simple options.

```
> odeplot(EulerSoln, [[t, x(t)], [t, y(t)]]);
    # [t, x(t)] is plot x(t), and [t, y(t)] is plot y(t).
```

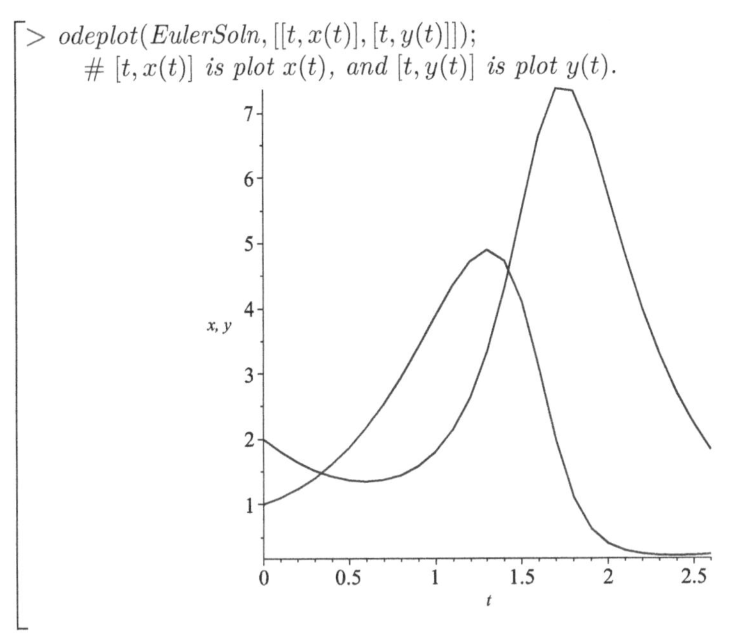

Plotting solutions y versus x gives us a good deal more information about the relation of the two functions. In an ODE system like this, the xy-plane is called the *phase plane*. We experiment and find that when we plot $y(t)$ versus $x(t)$ in the phase plane, we have an approximately closed loop.

```
> odeplot(EulerSoln, [x(t), y(t)]);
```

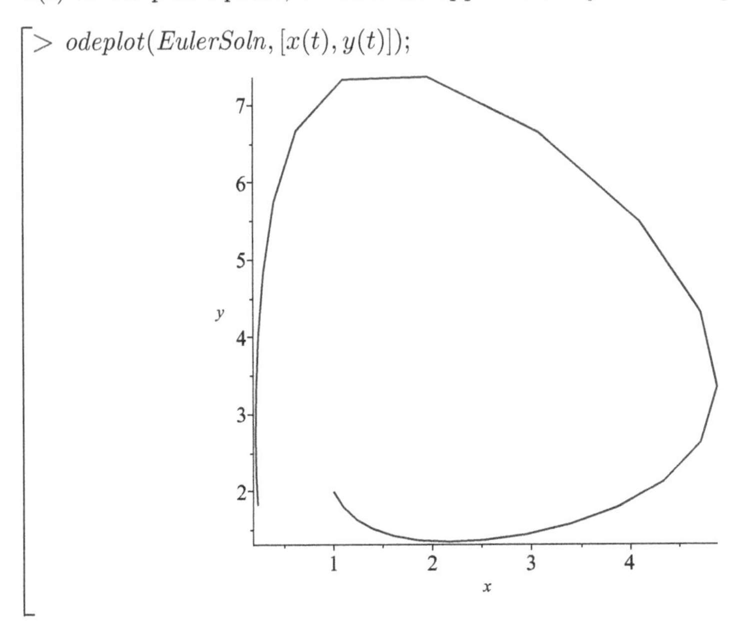

The Runge-Kutta Method

Another method we use for numerical estimates is the Runge-Kutta,[2] RK4. The '4' indicates that 4 slope estimates are combined for each step. This method can also be applied to systems in the same fashion as Euler's method. We illustrate with the same predator-prey example. (*The Maple output has been "snipped" to fit the page.*)

> $RK4Soln := dsolve(\{PredPrey, inits\}, [x(t), y(t)], output = Pts, numeric, stepsize = 0.1, method = classical[rk4])$

$[t, x(t), y(t)]$

0.0	1.0	2.0
0.1	1.11554071453956705	1.81968188493959970
0.2	1.26463746535620780	1.67761981669985927
0.3	1.45146389024689193	1.57278931221810914
0.4	1.68031037053259169	1.50542579227745321
0.5	1.95456986108324959	1.47762563817635395
0.6	2.27500034092209802	1.49412980191491540
0.7	2.63687047094120208	1.56336399876409615
0.8	3.02561346371586382	1.69866644464148697
0.9	3.41107833928229720	1.91916110241906601
⋮	⋮	⋮
1.7	1.71962027608752543	5.27274016506954890
1.8	1.38441109626295589	5.03717315896791806
1.9	1.14891557150367585	4.67751425643222785
2.0	0.991726731312949417	4.25984946993614155
2.1	0.893335577155144112	3.83076393278872684
2.2	0.839403883501402825	3.41909584046305914
2.3	0.820581249716352268	3.04076921363085662
2.4	0.831441511759657415	2.70331493424722602
2.5	0.869478412981861348	2.40921992688754827
2.6	0.934370815459076520	2.15818803799247361

[2] See Giordano and Weir [GW1991], pg 118, for RK4's formulas and a geometric interpretation.

Generate the same set of plots for comparison with Euler's method.

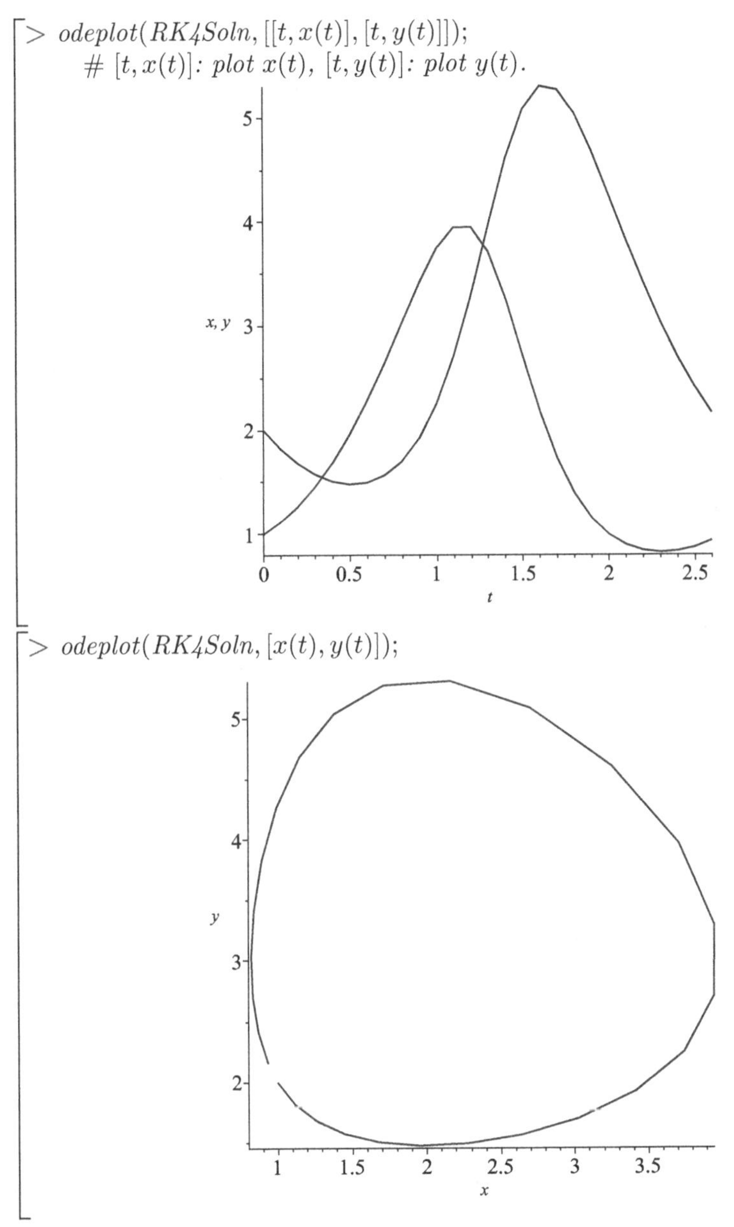

> odeplot(RK4Soln, [[t, x(t)], [t, y(t)]]);
[t, x(t)]: plot x(t), [t, y(t)]: plot y(t).
x, y
t
> odeplot(RK4Soln, [x(t), y(t)]);
y
x

Compare the two plots above with the plots from Euler's method. Remember, these plots are all of the same system, but we used different methods to estimate the functions' values.

Exercises

1. Given the following system of linear first-order ODE's of species cooperation (symbiosis):

$$\begin{aligned}\frac{dx_1}{dt} &= -0.5x_1 + x_2\\ \frac{dx_2}{dt} &= 0.25x_1 - 0.5x_2\\ x_1(0) &= 200, x_2(0) = 500\end{aligned}$$

   (a) Perform Euler's Method with step size $h = 0.1$ to obtain graphs of numerical solutions for $x_1(t)$ and $x_2(t)$ versus t, and for x_2 versus x_1. You can put both $x_1(t)$ and $x_2(t)$ versus t on one graph, if you want.

   (b) From the graphs, discuss the long-term behavior of the system (discuss stability).

   (c) Analytically, using eigenvalues and eigenvectors, solve the system of DEs to determine the population of each species for $t \geq 0$.

   (d) Determine if there is a steady-state solution for this system.

   (e) Obtain real plots of $x_1(t)$ and $x_2(t)$ versus t and for $x_2(t)$ versus $x_1(t)$. Compare to the numerical plots. Briefly discuss.

2. Given a competitive hunter model defined by the system:

$$\begin{aligned}\frac{dx_1}{dt} &= 15x - x^2 - 2xy = x(15 - x - 2y)\\ \frac{dx_2}{dt} &= 12y - y^2 - 1.5xy = y(12 - y - 1.5x)\\ x_1(0) &= 200, x_2(0) = 500\end{aligned}$$

   (a) Perform a graphical analysis of this competitive hunter model in the x-y plane.

   (b) Identify all equilibrium points and classify their stability.

   (c) Find the numerical solutions using Euler's method with step size $h = 0.05$. Try two separate initial conditions: first, use $x(0) = 5$ and $y(0) = 4$, then use $x(0) = 3$, $y(0) = 9$. Obtain graphs of $x(t)$, $y(t)$ individually (or on the same axis) and then a plot of y versus x using your numerical approximations. Compare to your phase portrait analysis.

3. Since brown trout (B) and rainbow trout (T) both live in the same lake and eat the same food sources, they are competing for survival. The rates of growth for brown trout (dB/dt) and for rainbow trout (dT/dt) are

estimated by the following equations (*coefficients and values are in thousands*):

$$\frac{dB}{dt} = (10 - B - T)B$$
$$\frac{dT}{dt} = (15 - B - 3T)T$$

(a) Obtain a qualitative graphical solution of this system. Find all equilibrium points of the system and classify each as unstable, stable, or asymptotically stable.

(b) If the initial conditions are $B(0) = 5$ and $T(0) = 2$, determine the long-term behavior of the system from your graph in part (a). Sketch the behavior.

(c) Using Euler's method with $h = 0.1$ and the same initial conditions as above, obtain estimates for B and T. Using these estimates determine a more accurate graph by plotting B versus T for the solution from $t = 0$ to $t = 7$. Recall Euler's method for autonomous systems can be written as

$$x_{n+1} = x_n + h \cdot f(x_n, y_n)$$
$$y_{n+1} = y_n + h \cdot g(x_n, y_n)$$

(d) Compare the graph in part (c) to the possible solutions found in (a) and (b). Briefly comment.

Chapter 3 Projects

Project 3.1. Diffusion.

Diffusion through a membrane leads to a first-order system of ordinary linear differential equations. For example, consider the situation in which two solutions of substance are separated by a membrane of permeability P. Assume the amount of substance that passes through the membrane at any particular time is proportional to the difference between the concentrations of the two solutions. Therefore, if we let x_1 and x_2 represent the two concentrations, and V_1 and V_2 represent their corresponding volumes, then the system of differential equations is given by:

$$\frac{dx_1}{dt} = \frac{P}{V_1}(x_2 - x_1)$$
$$\frac{dx_2}{dt} = \frac{P}{V_2}(x_1 - x_2)$$

where the initial amounts of x_1 and x_2 are given.

Consider two salt concentrations of equal volume V separated by a membrane of permeability P. Given that $P = V$, determine the amount of salt in each concentration at time t if $x_1(0) = 2$ and $x_2(0) = 10$.

(a) Write out the system of differential equations that models this behavior.

(b) Using the methods described in this chapter, solve the system. Clearly indicate the system's eigenvalues and eigenvectors.

(c) Plot the solutions for x_1 and x_2 on the same axis, labeling each. Comment about the plots.

(d) Use a numerical method (Euler or Runga-Kutta) to iterate a numerical solution to predict $x_i(4)$ using a step size of 0.5. Obtain a plot of your numerical approach. Compare it to the analytical plot. Comment about the plots.

Consider the diffusion through a double-walled membrane, where the inner wall has permeability P_1 and the outer wall has permeability P_2 with $0 < P_1 < P_2$. Suppose the volume of the solution within the inner wall is V_1 and between the two walls is V_2. Let x represent the concentration of the solution within the inner wall and y, the concentration between the two walls which leads to the following system

$$\frac{dx}{dt} = \frac{P_1}{V_1}(y - x)$$
$$\frac{dy}{dt} = \frac{1}{V_2}\left(P_2(C - y) + P_1(x - y)\right)$$
$$x(0) = 2, y(0) = 1, C = 10$$

Also assume the following parameter values: $P_1 = 3$, $P_2 = 8$, $V_1 = 2$, and $V_2 = 10$.

(a) Set up the system of ODEs with all coefficients inserted.

(b) Use the method of *variation of parameters* for systems,

$$X = X_c + \phi(t) \int \phi^{-1}(t) \cdot F(t)\, dt$$

to find both X_c and X_p.

(c) Use the initial conditions to find the particular solution, and find the coefficients for X_c in the solution $X_c + X_p$.

(d) Plot the solutions for $x(t)$ and $y(t)$ on the same axis. Comment about the solution.

Project 3.2. An Electrical Network.

An electrical network containing more than one loop also gives rise to a system of differential equations. For instance, in the electrical network displayed in Figure 3.1 (pg. 85), there are two resistors and two inductors. At branch point B in the network, the current $i_1(t)$ splits in two directions. Thus,

$$i_1(t) = i_2(t) + i_3(t).$$

Kirchhoff's law applies to each loop in the network. For loop $ABEF$, we find that

$$E(t) = i_1 R_1 + L_1 \frac{d\,i_2}{dt}.$$

The sum of the voltage drops across the loop $ABCDEF$ is

$$E(t) = i_1 R_1 + L_2 \frac{d\,i_3}{dt} + i_3 R_3.$$

Substituting, we obtain the following systems for equations

$$\frac{d\,i_1}{dt} = -\frac{R_1 + R_2}{L_1} + \frac{R_2}{L_2} i_2 + 0$$

$$\frac{d\,i_2}{dt} = \left[\frac{R_2}{L_2} - \frac{1}{R_2 C}\right] i_2 - \frac{R_1 + R_2}{L_2} i_1 + \frac{E(t)}{L_2}$$

$$i_1(0) = 0, i_2(0) = 1.$$

Initially, let $E(t) = 0$ volts, $L_1 = 1$ henry, $L_2 = 1$ henry, $R_1 = 1$ ohm, $R2 = 1$ ohm, and $C = 3$ farads.

(a) Write the system of differential equations that models this circuit.

(b) Using the methods described in this chapter, solve this system. Clearly indicate the system's eigenvalues and eigenvectors.

(c) Plot the solutions for $i_1(t)$ and $i_2(t)$ on the same axis and label each. Comment about the plots.

(d) Use a numerical method (Euler or Runge-Kutta) and iterate a numerical solution to predict $i_1(4)$ and $i_2(4)$ using a step size of $h = 0.5$. Obtain a plot of your numerical approach. Compare it to a graph of the analytical solution. Comment about the plots.

Now, let $E(t) = 100 \cdot \sin(t)$ volts.

(e) Set up the system of ODEs with all coefficients.

(f) Use the method of *variation of parameter* for systems,

$$X = X_c + \phi(t) \int \phi^{-1}(t) \cdot F(t)\, dt$$

to find X_c and X_p.

(g) Use the initial conditions to find the particular solution, and find the coefficients for X_c in the solution $X = X_c + X_p$.

(h) Plot the solutions for $i_1(t)$ and $i_2(t)$ on the same axis and label each. Comment about the plots.

(i) Plot i_2 versus i_3. Comment on the plot.

Project 3.3. Interacting Species.

Suppose $x(t)$ and $y(t)$ represent respective populations of two species over time t. One model might be

$$\frac{dx}{dt} = R_1 x, x(0) = x_0$$
$$\frac{dy}{dt} = R_2 y, y(0) = y_0$$

where R_1 and R_2 are intrinsic coefficients. Models involving competition between species or predator-prey models most often include interaction terms between the variables. These interactions terms, if included, will preclude any analytical solution attempts, so we will simplify these models for this project.

Let's model brown trout (B) and rainbow trout (T) attempting to coexist in a small pond in South Carolina.

$$B' = -0.5B + T + H$$
$$T' = 0.25B - 0.5T + K$$
$$B(0) = 2000, T(0) = 5000$$

Initially, we set $H = K = 0$.

(a) Write the system of differential equations that models fish population in the pond.

(b) Using the methods described in this chapter, solve the system. Clearly indicate your eigenvalues and eigenvectors.

(c) Plot the solutions for $B(t)$ and $T(t)$ on the same axis and label each. Comment about the plots.

(d) Use a numerical method (Euler or Runge-Kutta) and iterate a numerical solution to predict $B(10)$ and $T(10)$ using a step size of $h = 0.5$. Obtain a plot of your numerical approach. Compare it to a graph of the analytical solution. Comment about the plots.

Now, let $H = 1500$ and $K = 1000$.

(e) Set up the system of ODEs with all coefficients.

(f) Use the method of *variation of parameter* for systems,

$$X = X_c + \phi(t) \int \phi^{-1}(t) \cdot F(t)\, dt$$

to find X_c and X_p.

(g) Use the initial conditions to find the particular solution, find the coefficients for X_c in the solution $X = X_c + X_p$.

(h) Plot the solutions for $B(t)$ and $T(t)$ on the same axis and label each. Comment about the plots.

(i) Plot the $B(t)$ versus $T(t)$. Comment about the plot.

(j) Do the species coexist? Briefly explain. If any die out, determine when this happens.

Project 3.4. Trapezoidal Method.

The trapezoidal method is a more stable numerical method that is shown in numerical analysis textbooks. (See, e.g., Burden and Faires, *Numerical Analysis.* Brooks-Cole Pub., pg. 344–346.)

(a) Find the trapezoidal algorithm and modify it for systems of ODEs.

(b) Write a Maple program to obtain the trapezoidal estimates and compare the results to both Euler and Runge-Kutta estimates.

Case Studies with Systems of Differential Equations

Case 3.1. Predator-Prey Revisited.

Problem: We want to determine the winner of predators (foxes) versus a prey (hares) in a constrained region.

We make the following assumptions.

(1) The predator species is totally dependent on the prey species as its only food supply.

(2) The prey species has an unlimited food supply and no threat to its growth other than the specific predator.

If there were no predators, the second assumption would imply that the prey species grows exponentially; i.e., if $x = x(t)$ is the size of the prey population at time t, then we would have $dx/dt = ax$. This represents exponential growth when $a > 0$ and exponential decay when $a < 0$.

But there *are* predators, which must engender a negative component in the prey growth rate. Suppose we write $y = y(t)$ for the size of the predator population at time t. The crucial assumptions for completing the model are as follows:

- The rate at which predators encounter prey is jointly proportional to the sizes of the two populations.
- A fixed proportion of encounters between predator and prey lead to the prey's death.

These assumptions lead to the conclusion that the negative component of the prey growth rate is proportional to the product $x \cdot y$ of the population sizes, i.e.,

$$\frac{dx}{dt} = ax - bxy.$$

Now we consider the predator population. If there were no food supply, the population would die out at a rate proportional to its size, i.e. we would find $dy/dt = -cy$. (Keep in mind that the *natural growth rate* is a composite of birth and death rates, both presumably proportional to population size.) In the absence of food, there is no energy supply to support the birth rate. But there is a food supply: the prey. And what's bad for hares is good for lynx. That is, the energy to support growth of the predator population is proportional to deaths of prey, so

$$\frac{dy}{dt} = -cy + pxy.$$

This analysis leads to the Lotka-Volterra Predator-Prey model:

$$\frac{dx}{dt} = ax - bxy$$
$$\frac{dy}{dt} = -cy + pxy,$$

where a, b, c, and p are positive constants. (See Volterra, "Fluctuations in the Abundance of a Species Considered Mathematically," *Nature* 118 (1926), pg. 558–560.)

For our problem, we take the constants to be $\{a, b, c, p\} = \{0.2, 0.008/60, 0.03, 0.00003\}$. Load the *DEtools* library to have *DEplot*, and the *plots* package for *display*. Then define the predator-prey system.

```
> with(DEtools) :
  with(plots) :
```

```
> a, b, c, p := 0.2, 0.008/60, 0.03, 0.00003;
```

```
> hare := diff(h(t), t) = a · h(t) − b · h(t) · f(t);
  fox := diff(f(t), t) = −c · h(t) + p · h(t) · f(t);
```

$$hare := \frac{d}{dt}h(t) = 0.2h(t) - 0.0001333333333 \cdot h(t) \cdot f(t);$$
$$fox := \frac{d}{dt}f(t) = -0.03f(t) + 0.00003 \cdot h(t) \cdot f(t);$$

```
> init1 := [h(0) = 2000, f(0) = 600] :
```

```
> domain := t = 0..320 :
```

Next, draw the graphs. First, build the fox and hare component plots. Combine the two plots on the same axes. Then graph foxes versus hares in the phase plane.

> $HarePlot := DEplot(\{hare, fox\}, [h(t), f(t)], domain, stepsize = 0.5,$
> $[init1], scene = [t, h(t)], arrows = none, linecolor = blue):$
> $FoxPlot := DEplot(\{hare, fox\}, [h(t), f(t)], domain, stepsize = 0.5,$
> $[init1], scene = [t, f(t)], arrows = none, linecolor = blue):$

> $display(HarePlot, FoxPlot, title =$ "Predator-Prey Model");

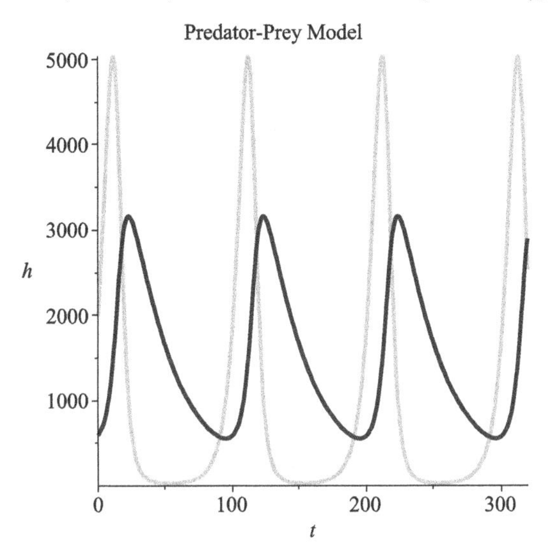

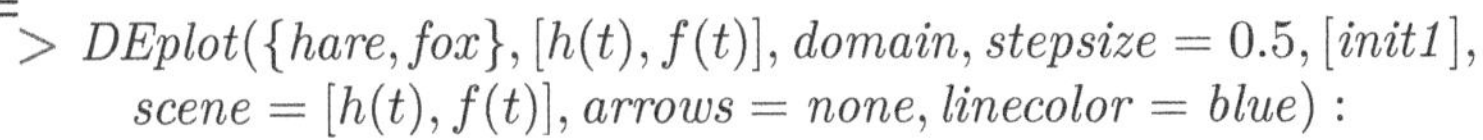

> $DEplot(\{hare, fox\}, [h(t), f(t)], domain, stepsize = 0.5, [init1],$
> $scene = [h(t), f(t)], arrows = none, linecolor = blue):$

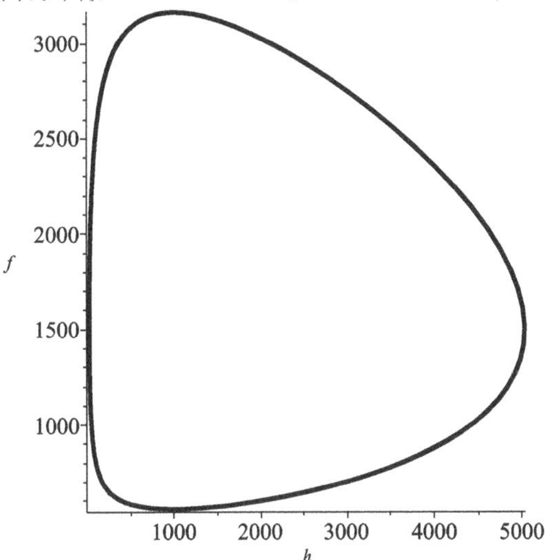

Now for a phase portrait showing several trajectories.

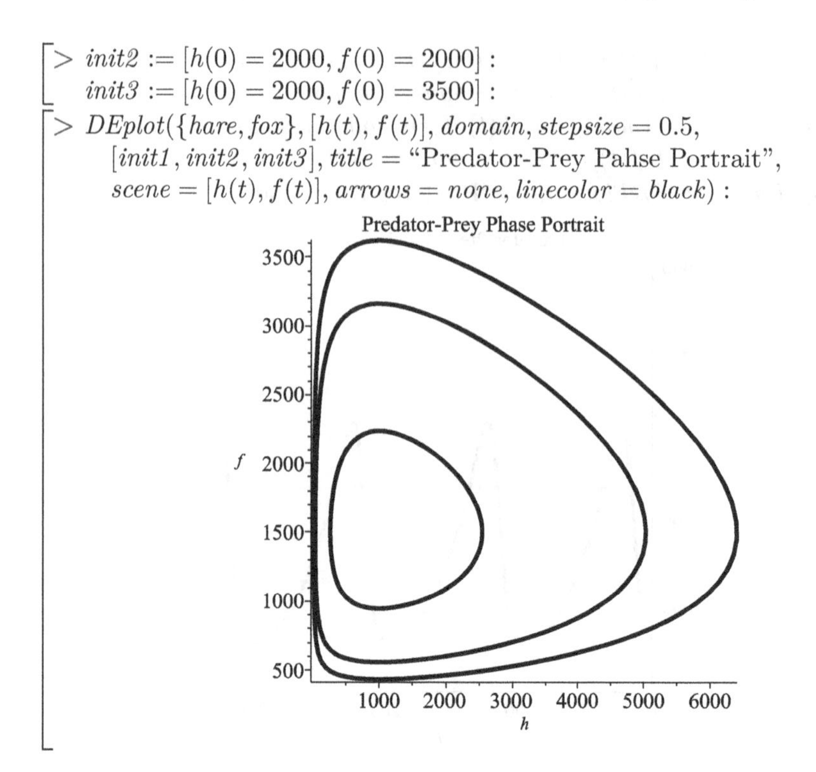

$> init2 := [h(0) = 2000, f(0) = 2000]:$
$init3 := [h(0) = 2000, f(0) = 3500]:$
$> DEplot(\{hare, fox\}, [h(t), f(t)], domain, stepsize = 0.5,$
$[init1, init2, init3], title =$ "Predator-Prey Pahse Portrait",
$scene = [h(t), f(t)], arrows = none, linecolor = black):$

Predator-Prey Model Interpretation:
We see that our predator-prey model is in equilibrium as we move around the equilibrium value, $(hares, foxes) = (1000, 1500)$. The point $(0, 0)$ is not stable. The ecological system appears stable, and does need human intervention at this time. Compare this situation to the wolves and moose of Isle Royale National Park in Michigan where human intervention in the wolf-moose ecological system is a fiercely debated controversy.

Case 3.2. Continuous SIR Models of Epidemics.
Consider a new viral disease that is spreading throughout a region. For example, recently at our college, we had a breakout of mumps. This disease is making a resurgence. The CDC is interested in knowing and experimenting with a model for the new disease prior to it actually becoming a "real" epidemic. Let us consider the population being divided into three categories: susceptible, infected, and removed. We make the following assumptions for our model:

- No one enters or leaves the community and there is no contact outside the community.
- Each person is either susceptible, s (able to catch this new virus); infected, i (currently has the virus and can spread it); or removed, r (al-

ready had the virus and has immunity or has died, i.e., will not get it again).

- Initially every person is either s or i.
- Once someone gets the virus this year, they cannot contract it again.
- The average length of the disease is 2 weeks over which the person is deemed infected and can spread the disease.
- Our time period for the model will be per week.

The model we will consider is the SIR model.[3]

We define our variables:

$$\begin{aligned} s(n) &= \text{number in the population susceptible in period } n \\ i(n) &= \text{number infected in period } n \\ r(n) &= \text{number removed in period } n \end{aligned}$$

Let's start our modeling process with $r(n)$. Our assumption for the length of time someone has the disease is 2 weeks. Thus, half the infected people will be removed each week,

$$\frac{dr}{dt} = 0.5 \cdot i(t)$$

The value, 0.5, is called the *removal rate per week*. It represents the proportion of the infected individuals who are removed from the set of "infecteds" each week. If real data is available, we could perform "data analysis" in order to obtain a very good estimate of the removal rate.

The function $i(t)$ will have terms that both increase and decrease its amount over time. It is decreased by the number that are removed each week, $0.5 \cdot i(n)$. The function is increased by the numbers of susceptible that come into contact with an infected person and catch the disease, $a \cdot s(n) \cdot i(n)$. We define the rate, a, as the rate at which the disease is spread, *the transmission coefficient*. We realize this is a probabilistic coefficient. We will assume, initially, that this rate is a constant value that can be determined with the initial conditions as follows.

Assume we have a population of 1,000 students in the dorms. Our nurse found 2 students reporting to the infirmary initially with the new virus. The next week, 3 students came in to the infirmary with symptoms like the new disease: $i(0) = 2$, $s(0) = 998$. In week 1, the number of newly infected is 3. Thus

$$\begin{aligned} 3 &= a\, i(1)\, s(1) = a \cdot 2 \cdot 997 \\ a &\approx 0.0015 \end{aligned}$$

[3] Elizabeth S. Allman, John A. Rhodes, *Mathematical Models in Biology: An Introduction*, Cambridge University Press, 2004.

Therefore

$$\frac{di}{dt} = 0.0015 \cdot s(n) \cdot i(n) - 0.5i(n).$$

Last, consider $s(n)$. This value is decreased only by the number that become infected. We may use the same rate, a, as before to obtain the model.

$$\frac{ds}{dt} = -0.0015 \cdot s(n) \cdot i(n).$$

Our coupled SIR model is the system of differential equations

$$\begin{aligned} \frac{ds}{dt} &= -0.0015 \cdot s(n) \cdot i(n) \\ \frac{di}{dt} &= 0.0015 \cdot s(n) \cdot i(n) - 0.5i(n) \\ \frac{dr}{dt} &= 0.5 \cdot i(t) \\ s(0) &= 998, i(0) = 2, r(0) = 0 \end{aligned} \tag{3.2}$$

The SIR Model in Eq (3.2) can be solved iteratively and viewed graphically. Let's use Maple's *DEplot* to obtain graphs and to observe the behavior of the solution so that we gain some insight into the model's dynamics.

```
> with(DEtools) :
  with(plots) :
> a, b := 0.5, 0.0015 :
> susceptible := diff(s(t), t) = −b · s(t) · i(t);
  infected := diff(i(t), t) = −a · i(t) + b · s(t) · i(t);
  removed := diff(r(t), t) = a · i(t);
  SIRmodel := [susceptible, infected, removed] :
```

$$susceptible := \frac{d}{dt} s(t) = -b \cdot s(t) \cdot i(t)$$

$$infected := \frac{d}{dt} i(t) = -a \cdot i(t) + b \cdot s(t) \cdot i(t)$$

$$removed := \frac{d}{dt} r(t) = a \cdot i(t)$$

```
> vars := [s(t), i(t), r(t)] :
> init1 := [s(0) = 998, i(0) = 2, r(0) = 0] :
> domain := t = 0..25 :
> sPlot := DEplot(SIRmodel, vars, domain, [init1], scene = [t, s],
    linecolor = RoyalBlue) :
> iPlot := DEplot(SIRmodel, vars, domain, [init1], scene = [t, i],
    linecolor = Maroon) :
```

```
> rPlot := DEplot(SIRmodel, vars, domain, [init1], scene = [t, r],
    linecolor = ForestGreen) :
> display(sPlot, iPlot, rPlot, legend = [s, i, r], title = "SIR Model");
```

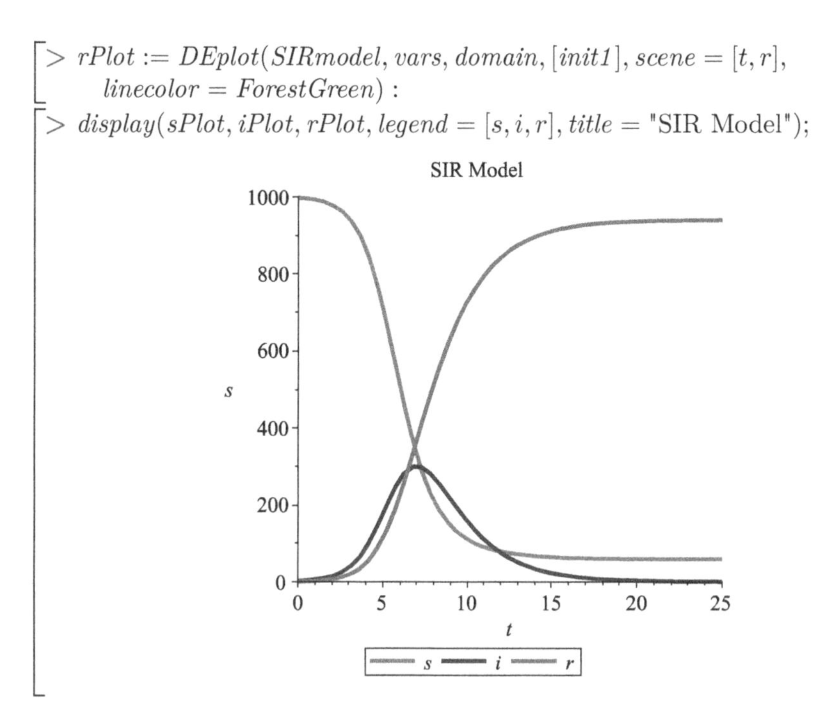

In this example, we see that the maximum number of inflected persons occurs at about day 7. Everyone survives, and not everyone gets the new disease.

Let's see what happens in another case example.

Case Example: Analysis of the Hong Kong Flu.[4]
During the winters of 1968-69 and 1970-71, a pandemic was caused by a virulent new strain, H3N2, of influenza A. This new strain was named the *Hong Kong flu* after the site it was first isolated in July, 1968. Travelers brought the Hong Kong flu to California in August, 1968, but it didn't reach epidemic levels until late November. Soon after arriving on the west coast, the virus traveled to New York City. At the time, few people were immune to the new Hong Kong flu. Dr's Jonas Salk, of polio vaccine fame, and Thomas Francis had developed the first flu vaccine in 1938; however, it wasn't until the 1980s that wide-spread public vaccination programs were initiated. Since the flu is not a "reportable disease" and many people do not seek medical attention, proxy measurements such as school or work absenteeism are used to monitor the infection's progression. *Excess mortalities*, the number of flu-related deaths beyond normal expectations, is one of the best proxies for tracking the disease.

We will investigate the course of Hong Kong flu through one of the first urban centers infected in the United States. Table 3.1 gives weekly data of excess pneumonia-influenza deaths in New York City. Figure 3.4 displays the data as a column graph.

[4] Adapted from: Keith Stroyan, *Calculus Using Mathematica: Scientific Projects*

TABLE 3.1: Excess Pneumonia-Influenza Deaths, New York City (Data source: Centers for Disease Control.)

Week	Excess deaths	Week	Excess deaths
1	14	8	108
2	31	9	68
3	50	10	77
4	66	11	33
5	156	12	65
6	190	13	24
7	156		

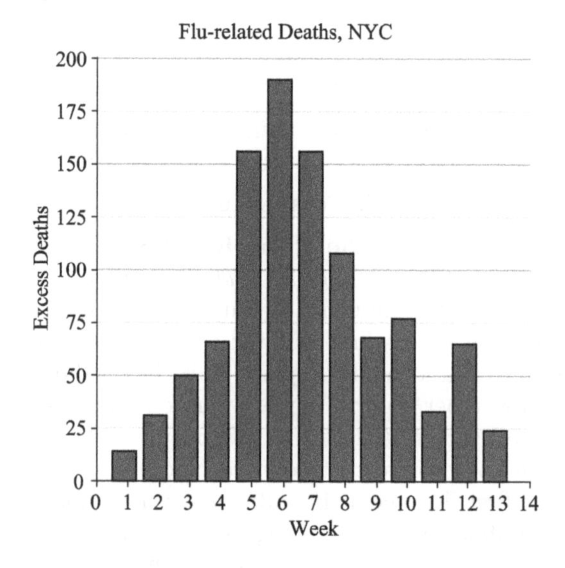

FIGURE 3.4: Flu-Related Deaths, Frequency v Week. (Data source: Centers for Disease Control.)

Even though healthy adults normally recover from the flu and its complica-

and Mathematical Background, Academic Press, 1993, and David Smith & Lang Moore, "The SIR Model for Spread of Disease," *Convergence*, Dec., 2004. https://www.maa.org/press/periodicals/loci/joma/the-sir-model-for-spread-of-disease.

tions, we can successfully view excess mortality as proportional to the number of cases of flu in a previous week. That is, given the standard course of the disease, excess deaths are proportional to the number of infections three weeks earlier. Table 3.1's data represents the growth and decline in the number of new cases of the flu three weeks earlier. We model the spread of the flu in order to be able to predict the course of future outbreaks of similar epidemics.

We will investigate whether we can use a SIR model for the Hong Kong flu pandemic as it affected New York City.

Several assumptions are necessary. The population of New York City in 1968 was 7,900,000. We know that population is not static, but as the time period is relatively short—just a few months, we can reasonably assume a constant population. Several of the computations will be easier, and be more readily ported to other situations, if we use proportions of the whole. Immunity to this strain of the flu comes upon recovery or death. Letting t be the time period, we have the *scaled* variables

$$s(t) = \text{proportion susceptible} = \frac{\text{number susceptible}}{7900000}$$

$$i(t) = \text{proportion infected} = \frac{\text{number infected}}{7900000}$$

$$r(t) = \text{proportion recovered} = \frac{\text{number recovered}}{7900000}$$

The initial conditions we choose are

$$s(0) = 1.0; \text{ there was no initial immunity, everyone is susceptible,}$$

$$i(0) = \frac{\text{10 initial cases, the infected people arriving in NYC}}{7900000} = 1.27 \cdot 10^{-6},$$

$$r(0) = 0.$$

Now, we need to determine values for the parameters a, the *transmission coefficient*, and b, the *removal rate*, for our SIR model.

Most people become infectious one day before symptoms appear and stay infectious up to 5 days, with an average of 3 days being infectious (according to the CDC). This average suggests we set $a = 1/3$. Assuming each infected person makes a possibly infecting contact every two days suggests we set $b = 1/2$. We must stress that these are initial estimates for the parameters, estimates that need to be refined during the modeling cycle. The graph in Figure 3.5 shows the solution curves for these parameter choices.

Let's interpret the results shown in the graph. The model shows the number of susceptible decreasing as a percent from 100% to just over 40%. We also see the number recovered/removed goes from 0% to just under 60%. Finally, we see the number infected peaking around $t = 75$ days to a value just over 6%.

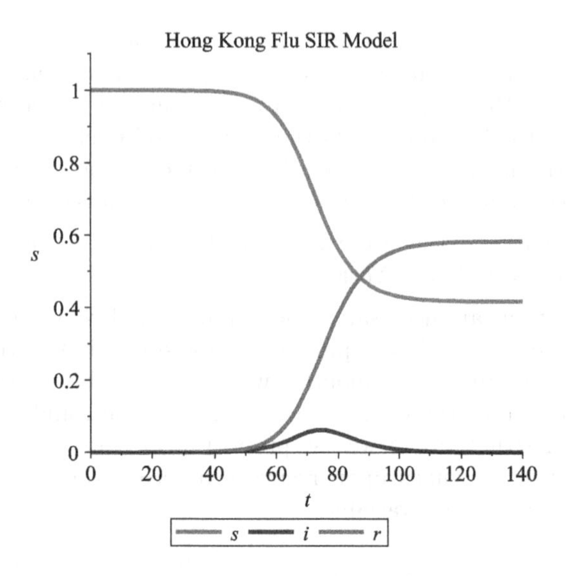

FIGURE 3.5: Initial Hong Kong Flu SIR Model

We started with 7,900,000 people, so about 475,000 have the flu at the time period of maximum infection. The area under the curve would represent the total number that had been infected in the epidemic. What about our losses due to deaths? The average deaths per week was 79.6 or about 0.001% of the population.

Let's compare our SIR model with the data. Recall the data gave the excess number of deaths each week that could be attributed to the flu epidemic. If we assume that the fraction of deaths among infected individuals is constant, then the number of deaths per week should be roughly proportional to the number of infected people in an earlier week. We repeat the column graph of the data, along with a graph of $i(t)$ versus t, the percent infected over time in Figure 3.6. The shape of the model appears to be a reasonable match for the shape of the data.

Case 3.3. Models of Combat: Iwo Jima with Lanchester Equations.

At Iwo Jima in WWII, the Japanese had 21,500 soldiers and the US had 73,000 soldiers. They engaged in conventional warfare, but the Japanese were fighting from reinforced entrenchments. The *kill rate* for the Japanese forces against the US was 0.0544 while that of the US side against the Japanese forces was 0.0106 (based on data after the battle). If these rates are correct, which side should win? How many should remain on the winning side when the other side has only 1,500 remaining? Give a brief explanation on the kill rates.

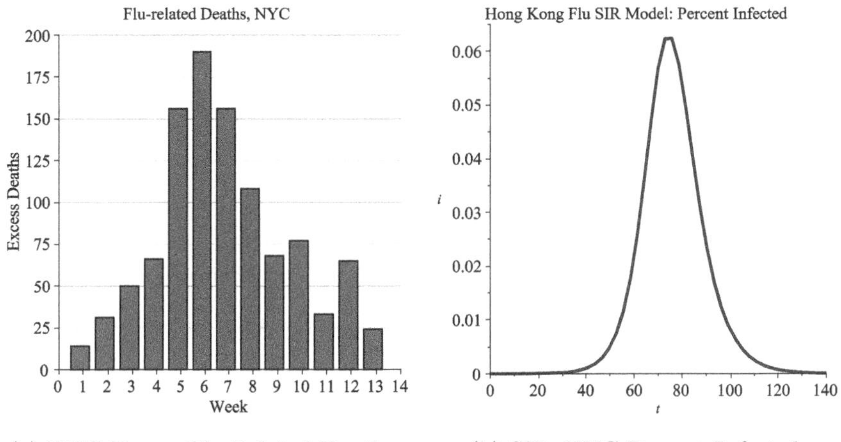

(a) NYC Excess Flu-Related Deaths. (b) SIR: NYC Percent Infected

FIGURE 3.6: Comparison of the Data and SIR Model

(Historical note: The battle ended with 1,500 Japanese survivors and 44,314 US survivors and took approximately 33–34 days.).

Fredrick W. Lanchester, an English engineer, devised equations in his 1916 book *Aircraft in Warfare* that have been used to model combat for 100 years. He developed the *Square Law Model for Modern Combat*:

$$\frac{dx}{dt} = -a \cdot y(t)$$

$$\frac{dy}{dt} = -b \cdot x(t)$$

where a and b represent the kill rates against the x and y force, respectively, by their opponents.

Investigate Lanchester's model with the Iwo Jima data. Remember, before you begin, load the *DEtools* and *plots* packages.

$> a, b := 0.0106, 0.0544 :$

$> IwoJimaModel := \{diff(x(t), t) = -a \cdot y(t), diff(y(t), t) = -b \cdot x(t), \};$

$$IwoJimaModel := \left\{ \frac{d}{dt}x(t) = -a \cdot y(t), \frac{d}{dt}y(t) = -b \cdot x(t), \right\};$$

$> vars, domain := \{x(t), y(t)\}, t = 0..35 :$

$> inits := [x(0) = 21500, y(0) = 73500] :$

$> Xp := DEplot(IwoJimaModel, vars, domain, [inits], stepsize = 0.5,$
$\quad scene = [t, x], linecolor = blue) :$

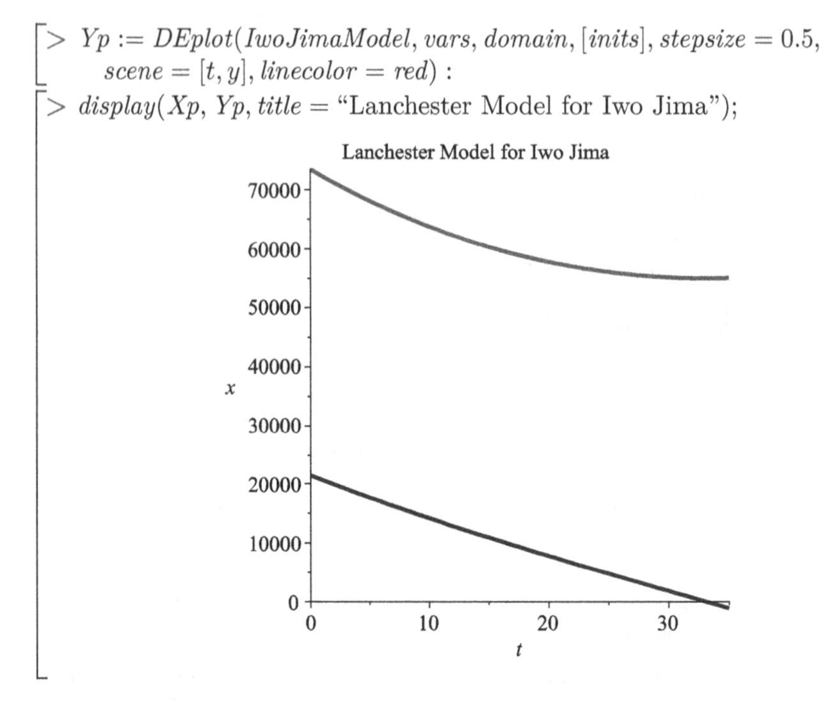
> Yp := DEplot(IwoJimaModel, vars, domain, [inits], stepsize = 0.5, scene = [t, y], linecolor = red) :
> display(Xp, Yp, title = "Lanchester Model for Iwo Jima");
Lanchester Model for Iwo Jima
70000
60000
50000
40000
30000
20000
10000
0
x
0
10
20
30
t

How did our model do? Actually, not very well. We had a 91% error for the Japanese troops, and a 24% error for the US force. What could account for this?

History shows several facts that were not modeled correctly by the square law. First, the Japanese were imbedded in a hillside as in guerrilla warfare. They saw the US forces attacking, whereas the US troops probably could not see the Japanese soldiers well. Additionally, the US force landed amphibiously over a two-week period—they were not there all at once. You will be asked to consider these aspects in the following exercise set to see if you can improve the model of this historical event.

You might want to consider *Brackney's Mixed Law* model, also called the *Parabolic Law*, which was developed in 1959. Brackney's Mixed Law is used to represent guerrilla warfare. Let x represent the conventional force and y, the guerilla force. Then

$$\frac{dx}{dt} = -a \cdot y(t)$$

$$\frac{dy}{dt} = -b \cdot x(t) \cdot y(t)$$

where a and b represent the respective kill rates.

Project Exercises

1. *The Battle of Iwo Jima.*

   The validity of Lanchester's equation can be demonstrated in an actual situation where US forces captured the island of Iwo Jima. Information required for the verification and "what if" analysis is the number of friendly troops put ashore each day and the number of friendly causalities for each day's engagement, knowledge that the enemy troops were not reinforced or withdrawn, the number of enemy troops at the start of the battle and the number at the end of the engagement, and the length of the engagement. The enemy was well entrenched into the rocks on the island. The US forces were attacking into the enemy's prepared defenses. In an idealized situation the US forces would be considered to follow a modified Lanchester's square law with replacement troops landing each day while the enemy could be considered to follow the standard square law.

   Part 1. Since the enemy is entrenched and looking down onto the US troops attacking, it is easier for them to hit and kill US forces. The probability

   $$P(\text{hit US troops with an enemy weapon}) = 0.54,$$

   and the probability

   $$P(\text{kill US troops} \mid \text{a hit}) = 0.1.$$

   We assume these events are independent and that their product represents the kill coefficient of the aggregated Japanese forces against US forces. The probability

   $$P(\text{hit Japanese troops with a US weapon}) = 0.12,$$

   and we will assume the probability

   $$P(\text{Kill Japanese troops} \mid \text{a hit}) = 0.1$$

   also. We assume these events are independent and that their product represents the kill coefficient of the aggregated US forces against the Japanese forces.

   1) Determine the kill rates for the US and Japanese forces. Explain from your knowledge of probability and statistics why they might be reasonable.
   2) Determine who wins a "fight to the finish."
   3) Parity: Is parity possible in this problem? Can we find it easily? Is it possible or easier to find parity after all the U.S. troops have landed, and then assume that the battle is new? Under this scenario at what kill ratio could the enemy have reached parity? Is that value feasible? Explain.

4) The real battle ended with 1,500 Japanese survivors, 44,314 US survivors, and took approximately 33–34 days. Relate your result with these real results. If different, why do you think these results are different.
5) Reflect on using the Lanchester model to adequately explain the results of the Battle of Iwo Jima.

Part 2. Enemy initial strength was 21,000 troops in fortified positions on the island. Friendly troop strength was modified by landing as given in Table 3.3:

TABLE 3.3: Iwo Jima U.S. Troop Landings

Day	Troops Landed
Day 1	30,000
Day 2	1,200
Day 3	6,735
Day 4	3,626
Day 5	5,158
Day 6	13,227
Day 7	3,054
Day 8	3,359
Day 9	3,180
Day 10	1,456
Day 11	250
Thereafter	0
Total troops	71,245

1) Modify Lanchester's model to account for reinforcements. How does the modified model do compared to the original?
2) Modify Brackney's Mixed Law to incorporate reinforcements. How does the modified model do compared to the original?
3) Which model best describes the Battle of Iwo Jima? Explain.

2. Find the equilibrium values for:

   a) The Predator-Prey model presented (pg 124).
   b) The SIR model presented in Eq 3.2 (pg 128).
   c) The combat model of the Battle of Iwo Jima.

3. In the Predator-Prey model determine the outcomes with the following sets of parameters.

a) Initial foxes are 200 and initial rabbits are 400.
b) Initial foxes are 2,000 and initial rabbits are 10,000.
c) The birth rate of rabbits increases to 0.1.

4. In the SIR model determine the outcome with the following parameters changed.
   a) Initially 5 are sick, and then 10 are sick the next week.
   b) The flu lasts 1 week.
   c) The flu lasts 4 weeks.
   d) There are 4,000 students in the dorm, and 5 are initially infected. There are 30 more infected the next week.

Further Reading

[AB2014] Martha L. Abell and James P. Braselton, *Differential Equations with Maple V®*, Academic Press, 2014.

[Barrow1997] David Barrow, Art Belmonte, Jack Bryant, Kirby Smith, Mike Stecher, Al Boggess, Jeff Morgan, Maury Rahe, and Tom Kiffe, *Solving Differential Equations with Maple V Release 4*, Brooks/Cole Publishing Co., 1997.

[GW1991] Frank R. Giordano and Maurice D. Weir, *Differential Equations: A Modeling Approach*, Addison-Wesley, 1991.

[GFH2013] Frank Giordano, William P. Fox, and Steven Horton, *A First Course in Mathematical Modeling*, Nelson Education, 2013.

4

Problem Solving with Linear, Integer, and Mixed Integer Programming

Objectives

(1) Formulate a linear programming problem.

(2) Solve and interpret a linear programming problem.

(3) Perform and interpret sensitivity analysis.

(4) Formulate and solve integer and mixed integer programming problems.

Introduction

Consider planning the shipment of needed items from the warehouses where they are manufactured and stored to the distribution centers where they are needed. There are three warehouses at different cities: Detroit, Pittsburgh and Buffalo. They have 260, 140 and 245 tons of paper, respectively. There are four publishers in Boston, New York, Chicago and Indianapolis. The publishers ordered 85, 240, 250 and 80 tons of paper to publish new books.
The costs in dollars for transportation of one ton of paper are given in Table 4.1.

TABLE 4.1: Transportation Costs

From: \ To:	Boston (BS)	New York (NY)	Chicago (CH)	Indianapolis (IN)
Detroit (DT)	15	20	16	21
Pittsburgh (PT)	25	13	5	11
Buffalo (BF)	15	15	7	17

Your boss wants you to minimize the shipping costs while meeting demand. This problem deals with the allocation of resources and can be modeled as a *linear programming problem* as we will discuss. (Here ‘programming’ does not refer to computers, even though computers are required for solving large linear programming problems.)

Consider starting a new diet which needs to be healthy. You go to a nutritionist who gives you lots of information on foods. The nutritionist recommends sticking to six different foods: Bread, Milk, Cheese, Fish, Potato and Yogurt, and provides you with Table 4.2 of nutritional information which also includes the average cost of the items.

TABLE 4.2: Nutritional Data

	Bread	Milk	Cheese	Potato	Fish	Yogurt
Cost, ($)	2.0	3.5	8.0	1.5	11.0	1.0
Protein, (g)	4.0	8.0	7.0	1.3	8.0	9.2
Fat, (g)	1.0	5.0	9.0	0.1	7.0	1.0
Carbohydrates, (g)	15.0	11.7	0.4	22.6	0.0	17.0
Calories, (cal)	90	120	106	97	130	180

The nutritionist recommends that our diet contains not less than 150 calories, not more than 10 g of protein, not less than 10 g of carbohydrates and not less than 8 g of fat. Also, we decide that our diet should have *minimal cost.* In addition, we conclude that our diet should include at least 0.5 g of fish and not more than 1 cup of milk. This is another *allocation of recourses* problem where we want the optimal diet at minimum cost. We have six unknown variables that define the weight of the food. There is a lower bound for fish of 0.5 g. There is an upper bound for milk of 1 cup. To model and solve this problem, we can use linear programming.

Modern linear programming was the result of a research project undertaken by the US Department of the Air Force under the title of Project SCOOP for *Scientific Computation of Optimum Programs.* As the number of fronts in the Second World War increased, it became more and more difficult to coordinate troop supplies effectively. Mathematicians looked for ways to use the new computers being developed to perform calculations quickly. One of the Project SCOOP team members, George Dantzig, developed the *simplex algorithm* in 1947 for solving simultaneous linear programming problems. (See Dantzig's "Maximization of a Linear Form Whose Variables Are Subject to a System of Linear Inequalities," *USAF Comptroller*, Nov. 1949.) The simplex method has several advantageous properties along with being very efficient, allowing its use for solving problems with a large number of variables. Linear programming uses very accessible methods from linear algebra. (Later, we'll encounter another Project SCOOP member, Thomas Saaty, inventor of the *Analytic Hierarchy Process.*)

In January 1952, the first successful solution to a linear programming (LP) problem was found using a high-speed electronic computer on the National Bureau of Standards SEAC machine. Today, most LPs are solved via high-speed computers. Computer-specific software packages, such as LINDO, Excel's Solver, GAMS, etc., have been developed to help in the solving and analysis of LP problems.[1] We will use the power of Maple to solve our linear programming problems.

To provide a framework for our discussions, we offer the following basic linear programming model

$$\text{Maximize (or minimize) } Z = f(X)$$
$$\text{subject to}$$
$$g_i(X) \left\{ \begin{array}{c} \geq \\ = \\ \leq \end{array} \right\} b_i \text{ for all } i.$$

Now let's explain this notation. The various components x_i of the vector X are called the *decision variables* of the model. These are the variables that can be controlled or manipulated. The linear function $f(X)$ is called the *objective function.* By "subject to," we connote that there are certain resource requirements, resource limitations, or side conditions that must be met. These conditions are called *functional constraints*. The constant b_i represents the limiting level of the associated linear constraint $g_i(X)$, and is called the right-hand side of the model. Restrictions requiring decision variables to be greater than or equal to zero are called *nonnegativity constraints.*

Linear programming is a method for solving linear programs, which occur very frequently in almost every modern industry. In fact, areas using linear programming are as diverse as agriculture, defense, health, transportation, manufacturing, advertising, and telecommunications. The reason for this is that in most situations, the classic economic problem exists — maximize an objective while competing for limited resources. The 'Linear' in Linear Programming means that in the case of production, the quantity produced is proportional to the resources used and also the revenue generated. The coefficients are constants. No products of variables are allowed.

In order to use this technique, the company must identify a number of constraints that will limit the production or transportation of their goods; these may include factors such as labor hours, energy, and raw materials. Each constraint must be quantified in terms of one unit of output, as the problem-solving method uses the constraints in a significant way.

An optimization problem that satisfies the following five properties is said to be a linear programming problem.

- There is a unique objective function $f(X)$.
- A decision variable component of X in either the objective function or a constraint function must appear with an exponent of 1, and may be multiplied by a constant.

[1] See "2017 Linear Programming: Software Survey," at http://www.orms-today.org/surveys/LP/LP-survey.html from INFORMS.

- No terms contain products of decision variables.
- All coefficients of decision variables are constants.
- Decision variables are permitted to assume real values; i.e., fractional as well as integer values.

Linear programming problems, by the nature of the many unknowns, are very hard to solve by human inspection, but methods have been developed to use the power of computers to do the hard work quickly. We will graphically illustrate the solution methods with two-variable linear programs in Section 4.2.

4.1 Formulating Linear Programming Problems

A linear programming problem is a problem that requires an objective function to be maximized or minimized subject to resource constraints. The key to formulating a linear programming problem is recognizing the decision variables. The objective function and all constraints are written in terms of these decision variables.

The conditions we stated for a mathematical model to be a linear program (LP) were:

- All variables are continuous (i.e., can take fractional values).
- There is a single objective (to maximize or minimize).
- The objective and constraints are linear functions; i.e., any term is either a constant or a constant multiplied by a single variable.
- The decision variables must be non-negative.

LPs form a very important class of problems because many practical situations can be formulated as LPs, and there exists an algorithm — the *simplex algorithm* — that enables us to solve even very large LPs numerically relatively easily and quickly.

We will return later to the simplex algorithm for solving LPs, but for the moment we will concentrate upon formulating LPs.

We'll consider several specific examples of the types of problem that can be formulated as LPs. Note here that the key to formulating an LP is *practice.* However a useful hint is that a very common objective for an LP is *minimize cost* or *maximize profit.*

Some of the major application areas modeled by LPs are:

- Blending
- Production planning
- Oil refinery management
- Distribution
- Financial and economic planning
- Manpower planning
- Blast furnace burdening
- Farm planning
- Animal feed composition
- Airline route planning

Example 4.1. Production Mix of New Drinks.
Consider the following problem statement:
A company wants to can two new different drinks for the holiday season. It takes 2 hours to can one gross of Drink A, and it takes 1 hour to label the cans. It takes 3 hours to can one gross of Drink B, and it takes 4 hours to label the cans. The company makes $10 profit on one gross of Drink A, and $20 profit of one gross of Drink B. Given that there is 20 hours to devote to canning the drinks and 15 hours to devote to labeling cans per week, how many cans of each type drink should the company package to maximize profits?

Production Mix LP Formulation

Problem: Maximize the profit of selling these new drinks.
Define variables:

$$x_1 = \text{the number of gross cans produced for Drink A per week}$$
$$x_2 = \text{the number of gross cans produced for Drink B per week}$$

Objective Function:

$$Z = 10x_1 + 20x_2$$

Constraints:

(1) Canning has only 20 hours available per week

$$2x_1 + 3x_2 \leq 20$$

(2) Labeling has only 15 hours available per week

$$x_1 + 4x_2 \leq 15$$

(3) Non-negativity restrictions

$$x_1 \geq 0 \text{ (non-negativity of Drink A production)}$$
$$x_2 \geq 0 \text{ (non-negativity of Drink B production)}$$

The Complete LP Formulation:

Maximize $Z = 10x_1 + 20x_2$
subject to:

$$2x_1 + 3x_2 \leq 20$$
$$x_1 + 4x_2 \leq 15$$
$$x_1 \geq 0$$
$$x_2 \geq 0$$

We will see how to solve these two-variable problems graphically in the next section.

Example 4.2. Financial Planning.
A bank makes four kinds of loans to its personal customers. The loans yield the following annual interest rates to the bank:

- First mortgage: 14%
- Second mortgage: 20%
- Home improvement: 20%
- Personal overdraft: 10%

The bank has a maximum foreseeable lending capability of $250 million, and is further constrained by the policies:

- First mortgages must be at least 55% of all mortgages issued and at least 25% of all loans issued (in $ terms).
- Second mortgages cannot exceed 25% of all loans issued (in $ terms).
- To avoid public displeasure and the introduction of a new windfall tax, the average interest rate on all loans must not exceed 15%.

Formulate the bank's loan problem as an LP so as to maximize interest income while satisfying the policy limitations.

Note here that the policy conditions, while potentially limiting the profit that the bank can make, also limit its exposure to risk in a particular area. It is a fundamental principle of risk reduction that risk is reduced by spreading money (appropriately) across different areas.

Here, as in *all* formulation exercises, we are translating a verbal description of the problem into an *equivalent* mathematical description. A useful tip when formulating LPs is to express the variables, objective, and constraints in words before attempting to express them in mathematics.

Financial Planning LP Formulation

Define variables
Essentially we are interested in the amount (in dollars) the bank loans to customers in each of the four different areas (not specifically in the actual number of such loans). Hence, let x_i = amount loaned in area i in millions of dollars (where $i = 1$ corresponds to first mortgages, $i = 2$ to second mortgages, etc.) and note that each $x_i \geq 0$ for $(i = 1, 2, 3, 4)$.

It is conventional in LPs to have all variables greater than or equal 0. Any variable (X, say) which can be positive *or* negative can be written as $X = X_1 - X_2$ (the difference of two new variables) where $X_1 \geq 0$ and $X_2 \geq 0$.
Objective Function
To maximize total interest income:

$$\text{Maximize } Z = 0.14x_1 + 0.20x_2 + 0.20x_3 + 0.10x_4$$

Constraints

(a) Limit on amount lent

$$x_1 + x_2 + x_3 + x_4 \leq 250$$

(b) Policy condition 1: First mortgages must be at least 0.55 · (total mortgage lending) and first mortgages must be at least 0.25 · (total loans)

$$x_1 \geq 0.55(x_1 + x_2)$$
$$x_1 \geq 0.25(x_1 + x_2 + x_3 + x_4)$$

(c) Policy condition 2: Second mortgages are not more than 0.25 · (total loans)

$$x_2 \leq 0.25(x_1 + x_2 + x_3 + x_4)$$

(d) Policy condition 3: The average annual interest is not more than 0.15. This constraint can be written as

$$0.14x_1 + 0.20x_2 + 0.20x_3 + 0.10x_4 \leq 0.15(x_1 + x_2 + x_3 + x_4)$$

(e) Nonnegativity: All loan amounts are nonnegative

$$x_i \geq 0, \text{ for } i = 1, 2, 3, 4$$

The Complete LP Formulation:

$$\begin{aligned}
&\text{Maximize } Z = 0.14x_1 + 0.20x_2 + 0.20x_3 + 0.10x_4 \\
&\text{subject to} \\
&\quad 0.45x_1 - 0.55x_2 \geq 0 \\
&\quad 0.25x_1 - 0.75x_2 + 0.25x_3 + 0.25x_4 \geq 0 \\
&\quad 0.01x_1 - 0.05x_2 - 0.05x_3 + 0.05x_4 \geq 0 \\
&\quad x_i \geq 0 \text{ for } i = 1, 2, 3, 4
\end{aligned}$$

Example 4.3. Blending Problem.
Consider the example of a manufacturer of animal feed who is producing feed mix for dairy cattle. In our simple example, the feed mix contains two active ingredients. One kg of the feed mix must contain a minimum quantity of each of four nutrients as shown in Table 4.3 (columns). The ingredients have the nutrient values and cost in Table 4.3 (rows).

TABLE 4.3: Feed Nutritional Data

Nutrient	A	B	C	D	Cost/kg
Ingredient 1 (g/kg)	100	80	40	10	40
Ingredient 2 (g/kg)	200	150	20	0	60
Min requirement (g/kg)	90	50	20	2	

What should be the amounts of active ingredients in one kg of feed mix that minimizes cost?

Feed Blending LP Formulation

Problem: Minimize the cost of blending the feed.
Define variables:
In this problem, it is best to think in terms of one kilogram of feed mix. That kilogram is made up of two parts: ingredient 1 and ingredient 2,

$$x_1 = \text{amount (kg) of ingredient 1 in one kg of feed mix}$$
$$x_2 = \text{amount (kg) of ingredient 2 in one kg of feed mix}$$

Essentially, the variables, x_1 and x_2, can be thought of as the recipe telling us how to make up one kilogram of feed mix.
Objective Function:
Minimize the cost of the ingredients in the mix.

$$Z = 40x_1 + 60x_2$$

Constraints:

- Nutrient constraints.

$$\begin{aligned} 100x_1 + 200x_2 &\geq 90 \quad \text{(Nutrient A)} \\ 80x_1 + 150x_2 &\geq 50 \quad \text{(Nutrient B)} \\ 40x_1 + 20x_2 &\geq 20 \quad \text{(Nutrient C)} \\ 10x_1 &\geq 2 \quad \text{(Nutrient D)} \end{aligned}$$

- Balancing constraint (an *implicit* constraint due to the definition of the variables).

$$x_1 + x_2 = 1$$

- Nonnegativity constraint: All ingredient amounts are nonnegative.

$$x_i \geq 0, \text{ for } i = 1, 2, 3, 4$$

This gives us our complete LP model for the blending problem.
The Complete LP Formulation:

$$\begin{aligned} &\text{Maximize } Z = 40x_1 + 60x_2 \\ &\text{subject to} \\ &\qquad 100x_1 + 200x_2 \geq 90 \\ &\qquad 80x_1 + 150x_2 \geq 50 \\ &\qquad 40x_1 + 20x_2 \geq 20 \\ &\qquad 10x_1 \geq 2 \\ &\qquad x_1 + x_2 = 1 \\ &\qquad x_1 \geq 0 \text{ and } x_2 \geq 0 \end{aligned}$$

Blending animal feed was one of the very early applications of linear programming. See F. Waugh's "The Minimum Cost Dairy Feed," *J. of Farm Econ.*, Vol. 33, June 1951, p. 299, and I. Katzman's "Solving Feed Problems through Linear Programming," *J. of Farm Econ.*, Vol. 38, No. 2 (May, 1956), pp. 420–429, for examples.

Example 4.4. Production Planning Problem.
A company manufactures four styles of the same table. In the final part of the manufacturing process, there are assembly, polishing and packing operations. For each style, the time required for these operations is shown in Table 4.4 (in minutes) as is the profit per unit sold.

TABLE 4.4: Table Production Data

	Assembly	Polish	Pack	Profit
Style 1	2	3	2	1.50
Style 2	4	2	3	2.50
Style 3	3	3	2	3.00
Style 4	7	4	5	4.50

Given the current state of the labor force, the company estimates that, each year, they have 100,000 minutes of assembly time, 50,000 minutes of polishing time and 60,000 minutes of packing time available. How many of each style should the company make per year, and what is the associated profit?

Production Planning LP Formulation

Problem: Maximize the profit from producing the 4 styles of tables.
Define variables:
Let x_i be the number of units of style i for $i = 1$, 2, 3, and 4 made per year.
Objective Function:
Maximize the profit.

$$Z = 1.50x_1 + 2.50x_2 + 3.00x_3 + 4.50x_4$$

Constraints:
Available resources for the operations of assembly, polishing, and packing are:

$$\begin{aligned} 2x_1 + 4x_2 + 3x_3 + 7x_4 &\leq 100{,}000 \quad \text{(assembly)} \\ 3x_1 + 2x_2 + 3x_3 + 4x_4 &\leq 50{,}000 \quad \text{(polish)} \\ 2x_1 + 3x_2 + 2x_3 + 5x_4 &\leq 60{,}000 \quad \text{(pack)} \end{aligned}$$

The Complete LP Formulation:

Maximize $Z = 1.5x_1 + 2.5x_2 + 3.0x_3 + 4.5x_4$
subject to:

$$\begin{aligned} 2x_1 + 4x_2 + 3x_3 + 7x_4 &\leq 100000 \\ 3x_1 + 2x_2 + 3x_3 + 4x_4 &\leq 50000 \\ 2x_1 + 3x_2 + 2x_3 + 5x_4 &\leq 60000 \\ x_i \geq 0 \text{ for } i &= 1,2,3,4 \end{aligned}$$

Exercises

Give the complete LP formulation for the following problems.

1. A company wants to can two different drinks for the holiday season. It takes 4 hours to can one gross of Drink A, and it takes 3 hours to label the cans. It takes 3.5 hours to can one gross of Drink B, and it takes 3.5 hours to label the cans. The company makes $18 profit on one gross of Drink A and $21 profit on one gross of Drink B. Given that we have 48 hours to devote to canning the drinks and 38 hours to devote to labeling cans per week, how many cans of each type drink should the company package to maximize profits?

2. The Mariners Toy Company wishes to make three models of ships to maximize their profits. They found that a model tugboat takes the cutter 1.5 hours, the painter 2.5 hours, and the assembler 4.5 hours of work; the tugboat produces a profit of \$6.50. The sailboat takes the cutter 3.5 hours, the painter 3.5 hours, and the assembler 2.5 hours. The sailboat produces a \$4.00 profit. The destroyer takes the cutter one hour, the painter 3 hours, and the assembler one hour. The destroyer produces a profit of \$2.50. The cutter is only available for 48 hours per week, the painter for 52 hours, and the assembler for 60 hours. Assume that they will sell all the ships that they make. Formulate this LP to determine how many ships of each type the Mariners Toy Company should produce.

3. In order to produce 1000 tons of non-oxidizing steel for BMW engine valves, at least the following units of manganese, chromium, and molybdenum, will be needed weekly: 11 units of manganese, 13 units of chromium, and 15 units of molybdenum (1 unit is 10 lb). These materials are obtained from a dealer who markets these metals in three sizes small (S), medium (M), and large (L). One S case costs \$11 and contains 2.5 units of manganese, 2.5 units of chromium, and one unit of molybdenum. One M case costs \$11.50 and contains 2.5 units of manganese, 3.5 units of chromium, and one unit of molybdenum. One L case costs \$14.50 and contains 1.5 units of manganese, 1.5 units of chromium, and 5 units of molybdenum. How many cases of each kind (S, M, L) should be purchased weekly so that we have enough manganese, chromium, and molybdenum at the lowest cost?

4. The Super Bowl Advertising Agency wishes to plan an advertising campaign in three different media—television, radio, and magazines. The purpose or goal is to reach as many potential customers as possible. Results of a marketing study are given in Table 4.5.

TABLE 4.5: Super Bowl Advertising Data

	Day Time TV	Night Time TV	Radio	Magazine
Advertising Cost per Unit	$\$50,000$	$\$85,000$	$\$40,000$	$\$25,000$
Potential Customers Reached per Unit	450,000	900,000	550,000	250,000
Female Customers Reached per Unit	300,000	400,000	200,000	100,000

The company does not want to spend more than $\$900,000$ total on advertising. They further require (1) at least 2 million exposures take place among women; (2) TV advertising be limited to $500,000; (3) at least 3 advertising units be bought on daytime TV and 2 units on primetime TV, and (4) the number of radio and magazine advertisement units should each be between 5 and 10 units.

5. A tomato cannery has 6000 pounds of Grade A tomatoes and 12,000 pounds of Grade B tomatoes from which they will make whole canned tomatoes and tomato paste. Whole canned tomatoes must be composed of at least 80% Grade A tomatoes, whereas tomato paste must be made with at least 10% Grade A tomatoes. Whole Grade A tomatoes sell for $0.08 per pound, and Grade B tomatoes sell for $0.05 per pound. Maximize revenue.

   HINT: Let x_{wa} = pounds of Grade A tomatoes used for canned whole tomatoes, and x_{wb} = pounds of Grade B tomatoes used for canned whole tomatoes. The total number of whole tomato cans produced is the sum $x_{\text{wa}} + x_{\text{wb}}$ after each is found. Also remember, a percent is a fraction of the whole times 100%.

6. Acme Butchers is a large-scale distributor of dressed meats for Virginia Beach restaurants and hotels. The Texas Steakhouse orders meat for meatloaf (mixed ground beef, pork, and veal) for 1,500 pounds according to the following specifications:

   (a) Ground beef is to be no less than 400 pounds and no more than 600 pounds.
   (b) The ground pork is to be between 200 and 300 pounds.
   (c) The ground veal must weigh between 100 and 400 pounds.
   (d) The weight of the ground pork must be no more than one and one half (3/2) times the weight of the ground veal.

   The contract calls for the steakhouse to pay $1,200 for the meat. The cost per pound for the meat is: $0.70 for hamburger, $0.60 for pork, and $0.80 for the veal. How can this be modeled?

7. A portfolio manager in charge of a bank wants to invest $10 million. The securities available for purchase, as well as their respective quality ratings, maturate, and yields, are shown in Table 4.6. The bank places certain policy limitations on the portfolio manager's actions:

   (1) Government and Agency Bonds must total at least $4 million.
   (2) The average quality of the portfolios cannot exceed 1.4 on the bank's quality scale. Note a lower number means higher quality.
   (3) The average years to maturity must not exceed 5 years.

   Assume the objective is to maximize after-tax earnings on the total investment.

TABLE 4.6: Portfolio Investments Data

Bond Name	Bond Type	Moody's Quality Scale	Bank's Quality Scale	Years to Maturity	Yield at Maturity	After-Tax Yield
A	Municipal	Aa	2	9	4.3%	4.3%
B	Agency	Aa	2	15	5.4%	2.7%
C	Govt 1	Aaa	1	4	5%	2.5%
D	Govt 2	Aaa	1	3	4.4%	2.2%
E	Local	Ba	5	2	4.5%	4.5%

4.2 Understanding Two-Variable Linear Programming: A Graphical Simplex

Many applications in business and economics involve a process called *optimization.* In optimization problems, you are asked to find the minimum or the maximum result. This section graphically illustrates the strategy used in the simplex method of linear programming. We will restrict ourselves in this graphical context to two dimensions. Variables in the simplex method are restricted to nonnegative variables.

A two-dimensional linear programming problem consists of a linear *objective function* and a system of linear inequalities called *constraints.* The objective function gives the linear quantity that is to be maximized (or minimized). The constraints determine the *set of feasible solutions*, the set of values of the variables that satisfy all constraints.

Example 4.5. Memory Chips for CPUs (modified from Fox, 2012). Let's start with a manufacturing example. Suppose a small business wants to know how many of two types of high-speed computer chips to manufacture weekly to maximize their profits.

First, we need to define our decision variables. Let

$$x_1 = \text{number of high-speed chip type A to produce weekly}$$
$$x_2 = \text{number of high-speed chip type B to produce weekly}$$

The company reports a profit of \$150 for each type A chip and \$130 for each type B chip sold. The production line reports the information in Table 4.7.

The profit equation is:

$$\text{Profit} = 150x_1 + 130x_2$$

TABLE 4.7: Memory Chip Data

	Chip A	Chip B	Quantity available
Assembly time (hours)	2	4	1500
Installation time (hours)	4	3	1600
Profit (per unit)	150	130	

The constraint information from Table 4.7 becomes inequalities written mathematically as

$$2x_1 + 4x_2 \leq 1500 \text{ (assembly)}$$
$$4x_1 + 3x_2 \leq 1600 \text{ (installation)}$$
$$x_1 \geq 0, x_2 \geq 0 \text{ (nonnegativity)}$$

The Feasible Region

The constraints of a linear program, which include any bounds on the decision variables, essentially shape the region in the x_1-x_2 plane that will be the domain for the objective function prior to any optimization being performed. Every linear inequality constraint that is part of the LP formulation divides the entire space defined by the decision variables into 2 parts:

- the portion of the space containing points that violate the constraint, and
- the portion of the space containing points that satisfy the constraint.

The intersection of all the regions that satisfy the constraints results in a convex set.

It is very easy to determine which portion will contribute to shaping the domain. We can simply substitute the value of some point in either *half-space* into the constraint. Any point will do, but the origin is particularly appealing. Since there's only one origin, if it satisfies the constraint, then the *half-space* containing the origin will contribute to the domain of the objective function, otherwise, the other *half-space* not containing the origin will contribute to the domain. We do this for each of the constraints in the problem. The intersection of all the *half-spaces* that satisfied the constraints *individually* forms the domain for the objective function for the optimization. Because it contains the points that satisfy all the constraints *simultaneously*, this set is considered the points feasible to the problem. Naturally, the common name for this domain is the *feasible region.*

Consider our example's constraints:

$$2x_1 + 4x_2 \leq 1500$$
$$4x_1 + 3x_2 \leq 1600$$
$$x_1 \geq 0, x_2 \geq 0$$

For our graphical work, we begin with the nonnegativity constraints: $x_1 \geq 0$ and $x_2 \geq 0$ to begin identifying the region. We are strictly in the first quadrant of the x_1-x_2 plane.

The Maple commands to enter the LP and obtain and graph the feasible region are:

```
> with(plots) :
> ObjectiveFcn := (x1, x2) → 150 · x1 + 130 · x2;
          ObjectiveFcn := (x1, x2) ↦ 150 · x1 + 130 · x2;
> Constraints := {2 · x1 + 4 · x2 ≤ 1500, 4 · x1 + 3 · x2 ≤ 1600, x1 ≥
  0, x2 ≥ 0}
  Constraints := {0 ≤ x1, 0 ≤ x2, 2 · x1 + 4 · x2 ≤ 1500, 4 · x1 + 3 · x2 ≤ 1600}
> theRange := (x1 = −100..800, x2 = −75..600) :
> FeasibleRegion := inequal(Constraints, theRange,
  optionsfeasible = (color = grey), optionsclosed = (color =
  black), thickness = 3) :
> Objective := contourplot(ObjectiveFcn(x1, x2), theRange, contours =
  9,
  color = red) :
> display(Objective, FeasibleRegion, title =
  text"MemoryChipProblemGraph")
```

Memory Chip Problem Graph

The graph shows a plot of (1) the assembly hour's constraint, and (2) the installation hour's constraint, both in the first quadrant, along with the non-negativity restrictions on the decision variables. The intersection of the four half-spaces defined by these constraints forms the feasible region, the shaded quadrilateral in the plot. This area represents the domain for the objective function's optimization.

Example 4.6. Finding a Feasible Region.
Shade the feasible region defined by the set of constraints

$$\begin{aligned} x_1 + 4x_2 &\leq 40 \\ 4x_1 + x_2 &\leq 40 \\ 2x_1 + 3x_2 &\leq 35 \\ x_1 \geq 0, x_2 &\geq 0 \end{aligned}$$

Solution: The feasible region is the set of ordered pairs (x_1, x_2) that satisfy all four constraints simultaneously. They are points that lie below $x_1 + 4x_2 = 40$, below $4x_1 + x_2 = 40$, below $2x_1 + 3x_2 = 35$, to the right of $x_1 = 0$, and above $x_2 = 0$. Note that the non-negativity constraints, $x_1 \geq 0$ and $x_2 \geq 0$, restrict the feasible region to the first quadrant. We see this in Figure 4.1. We note the region will be a *convex set*. For more information on convex sets and their important relationship to linear programming, please see the references.

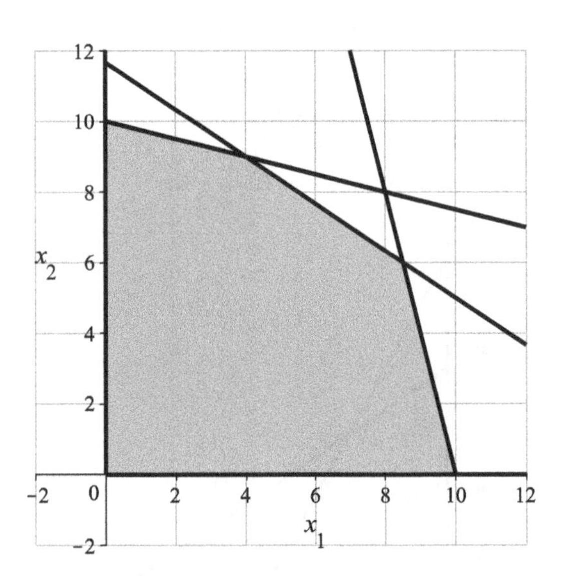

FIGURE 4.1: Shaded Feasible Region

If the problem is *well-behaved*, the feasible region will be a closed and bounded polygonal shape, called a *polyhedron*, such as the one shaded in Figure 4.1. It does not have to be so—not all problems are well-behaved. Sometimes the orientation and location of the constraints fail to hold back the objective function in the direction of the optimization. When this happens, the problem is *unbounded*; the objective function's value goes off to positive or negative infinity. Can you draw a sketch of a situation in which this will happen?

Other times, the intersection of the half-spaces is an empty set. In this case, the problem is *infeasible*; there are no possible solutions that will simultaneously satisfy the requirements of all the constraints. Can you draw a sketch of a situation in which this will happen?

Solving a Linear Programming Problem Graphically

Recall that we have decision variables defined and an objective function that is to be maximized or minimized. Although all points inside the feasible region, if the region exists, provide feasible solutions to the problem, the location of the optimal solution is given by the following theorem.

The Fundamental Theorem of Linear Programming.
If the optimal solution to a linear program exists, then it occurs at a corner point of the feasible region.

Notice the various corner points formed by the intersections of constraints in Example 4.6. The corner points are of great importance to us. There is a cool theorem (*Didn't know there were any of those, huh?*) in linear optimization that states, "If an optimal solution exists, then an optimal corner point exists." Therefore, any algorithm searching for the optimal solution to a linear program should have some mechanism of heading toward the corner point where the optimum solution will occur. If the search procedure stays on the outside border of the feasible region while pursuing the optimal solution, it is called an *exterior point* method. If the search procedure cuts through the interior of the feasible region, it is called an *interior point* method.

Thus, in a linear programming problem, if there exists an optimal solution, it must occur at a corner point of the set of feasible solutions (the vertices of the polyhedral region). Note that in Figure 4.1, the corner points of the feasible region are: $(0,0)$, $(0,10)$, $(4,9)$, $(8.5,6)$, and $(10,0)$.

How did we find the point $(4,9)$?

This point is the intersection of the lines $x_1 + 4x_2 = 40$ and $2x_1 + 3x_2 = 35$. You have solved systems of two equations many times before. Use your favorite method: substitution, elimination, or augmented matrices. All the corner points come from the intersection of boundary lines (replace the $\leq$ with $=$) of pairs of constraints. Be careful though, the intersection point $(8,8)$ from the intersection of $x_1 + 4x_2 = 40$ and $4x_1 + x_2 = 40$ lies outside the feasible region; this point violates the constraint $2x_1 + 3x_2 \leq 35$.

Now that we have all the possible solution coordinates, we need to know which is the *optimal solution*. It's easy to determine the answer: Evaluate the objective function at each corner point, and simply choose the best solution.

Assume our objective is to maximize $Z = 2x_1 + 2x_2$. We can set up a table of coordinates and corresponding Z-values. From Table 4.8, we see the best

TABLE 4.8: Evaluating Z at the Corner Points

Corner Point	$Z = 2x_1 + 2x_2$
$(0, 0)$	$Z = 0$
$(0, 10)$	$Z = 20$
$(4, 9)$	$Z = 26$
$(8.5, 6)$	$Z = 29$
$(10, 0)$	$Z = 20$

solution, the maximum value of Z occurs at the corner point $(8.5, 6)$ where $Z = 29$.

Graphically, we see the result by plotting the objective function line, $Z = 2x + 2y$, for a specific value of Z on top of the feasible region. Determine the parallel direction for the line to maximize (in this case) Z. Remember that the slope of a line in standard form $z = ax + by$ is $m = -b/a$. Move the line parallel until it crosses the last point in the feasible set. That point, which must be a corner point, is the solution. The line that goes through the origin at a slope of, in this example, $-2/2$ is called the *iso-profit line*, 'iso' because any point on that line gives the same value for profit. See Figure 4.2 below. (Note: Use the option *contourlabels=true* in Maple's *contourplot* to see the value of the objective function when the cursor is held over the contour line.) Experiment with animating the iso-profit lines in Exercise 6.

The steps for graphically solving a linear programming problem involving only two variables are:

1. Sketch the region corresponding to the system of constraints. The points satisfying all constraints make up the feasible region.
2. Find all the corner points (or intersection points) in the feasible region.
3. Evaluate the objective function at each corner point. Select the corner point that optimizes the objective function. For bounded regions, both a maximum and a minimum will exist. If a solution for an unbounded region exists, it will exist at a corner point.

If two corner points tie for the maximum of the objective function, then, by linearity, all points on the line segment connecting those two corner points are also maximum points.

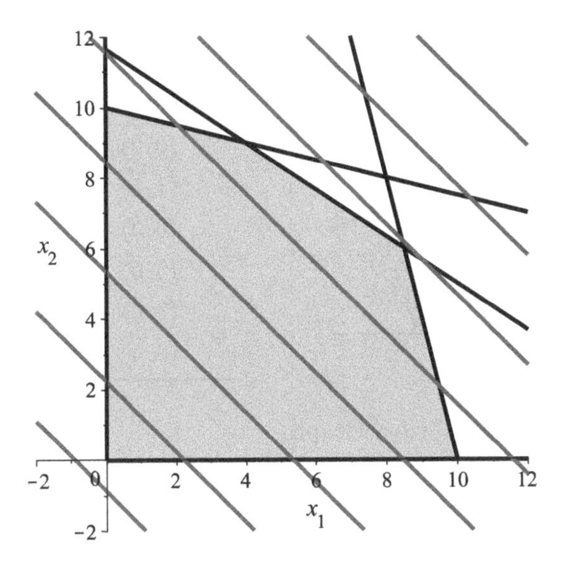

FIGURE 4.2: Iso-Profit Lines on the Feasible Region

Example 4.7. A Minimization Problem.
The procedure for minimizing is the same as for maximizing: we check the corner points, choosing the least value of the objective function when it exists.

Minimize $Z = 5x + 7y$
subject to:

$$
\begin{aligned}
2x + 3y &\geq 6 \\
3x - y &\leq 15 \\
-x + y &\leq 4 \\
2x + 5y &\leq 27 \\
x \geq 0 \text{ and } y &\geq 0
\end{aligned}
$$

The corner points in Figure 4.3 are $(0, 2)$, $(0, 4)$, $(1, 5)$, $(6, 3)$, $(5, 0)$, and $(3, 0)$. See if you can find all these corner points. Examining the values of Z in Table 4.9 shows us that the minimum value occurs at $(0, 2)$ with a Z value of 14. Notice in our graph that an iso-objective function line will last cross the point $(0, 2)$ as it leaves the feasible region in the direction ("southeast") that minimizes Z.

Example 4.8. An Unbounded Case.
To examine the concept of an unbounded feasible region, consider the con-

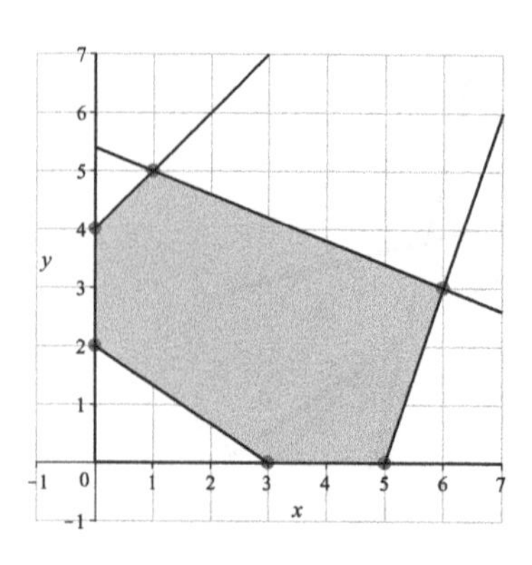

FIGURE 4.3: Feasible Region Graph

TABLE 4.9: Objective Evaluation

Corner Point	$Z = 5x + 7y$
$(0, 2)$	$Z = 14$
$(1, 5)$	$Z = 40$
$(6, 3)$	$Z = 51$
$(5, 0)$	$Z = 25$
$(3, 0)$	$Z = 15$
$(0, 4)$	$Z = 28$

straints

$$x + 2y \geq 7$$
$$3x + y \geq 6$$
$$x \geq 0 \text{ and } y \geq 0$$

Unbounded feasible region

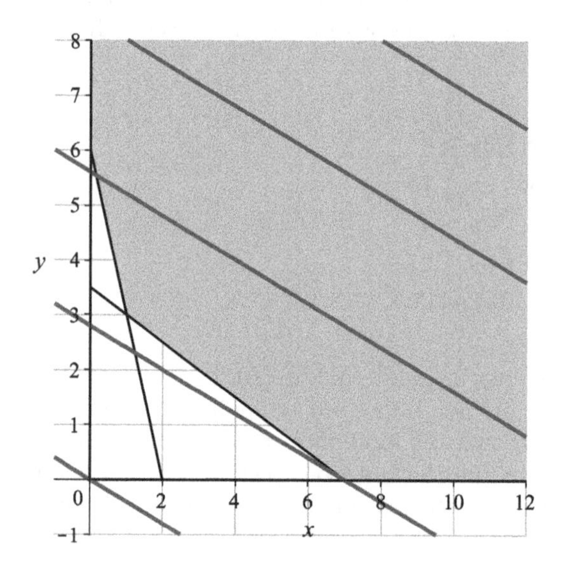

FIGURE 4.4: An Unbounded Feasible Region

Note that the corner points are $(0, 6)$, $(1, 3)$ and $(7, 0)$; the region is unbounded. If our objective is to minimize $Z = 2x + 5y$, then our solution is $(7, 0)$ with $Z = 14$.

Determine why there is no solution to the LP to maximize $Z = 2x + 5y$.

Exercises

In each problem: Find both the maximum and minimum solutions when possible. Assume nonnegativity constraints.

1. Optimize $Z = 2x + 3y$
   subject to:

$$
\begin{aligned}
2x + 3y &\geq 6 \\
3x - y &\leq 15 \\
-x + y &\leq 4 \\
2x + 5y &\leq 27
\end{aligned}
$$

2. Optimize $Z = 6x + 4y$
   subject to:

$$
\begin{aligned}
-x + y &\leq 12 \\
x + y &\leq 24 \\
2x + 5y &\leq 80
\end{aligned}
$$

3. Optimize $Z = 6x + 5y$
   subject to:

$$
\begin{aligned}
x + y &\geq 6 \\
2x + y &\geq 9
\end{aligned}
$$

4. Optimize $Z = x - y$
   subject to:

$$
\begin{aligned}
x + y &\geq 6 \\
2x + y &\geq 9
\end{aligned}
$$

5. Optimize $Z = 5x + 3y$
   subject to:

$$
\begin{aligned}
1.2x + 0.6y &\leq 24 \\
2.0x + 1.5y &\leq 80
\end{aligned}
$$

6. Investigate iso-profit lines with an animated version of Figure 4.2 by experimenting with the following Maple code.

> $with(plots):$

> $Constraints := [x_1 + 4x_2 \leq 40, 4x_1 + x_2 \leq 40, 2x_1 + 3x_2 \leq 35, x_1 \geq 0, x_2 \geq 0]:$

> $ObjFcn := (a, b) \rightarrow 2a + 2b:$

> $theRange := (x_1 = -2..12, x_2 = -2..12):$

> $FeasibleRegion := inequal(Constraints, theRange, optionsfeasible = (color = grey)):$

> $animate(implicitplot, [z = ObjFcn(x_1, x_2), theRange, thickness = 3], z = 0..35, frames = 36, background = FeasibleRegion);$

Projects

Project 4.1. Fuel Costs.

With the rising cost of gasoline and increasing prices to consumers, the use of additives to enhance performance of gasoline is being considered. Consider two additives, Additive 1 and Additive 2. The following conditions must hold for the use of additives:

- Harmful carburetor deposits must not exceed 1/2 lb per car's gasoline tank.
- The quantity of Additive 2 plus twice the quantity of Additive 1 must be at least 1/2 lb per car's gasoline tank.
- 1 lb of Additive 1 will add 10 octane units per tank, and 1 lb of Additive 2 will add 20 octane units per tank. The total number of octane units added must not be less than six (6).
- Additives are expensive and cost \$1.53 per lb for Additive 1 and \$4.00 per lb for Additive 2.

We want to determine the quantity of each additive that will minimize their cost while meeting the above restrictions.

Required:

1. List the decision variables and define them.
2. List the objective function.
3. List the resources that constrain this problem.
4. Graph the feasible region.

5. Label all intersection points of the feasible region.
6. Plot the objective function in a different color (highlight the objective function line, if necessary) and label it the iso-cost line.
7. Clearly indicate on the graph the point that is the optimal solution.
8. List the coordinates of the optimal solution and the value of the objective function.
9. Assume now that the manufacturer of additives has the opportunity to sell you a nice "TV special deal" to deliver at least 0.5 lb. of Additive 1 and at least 0.3 lb. of Additive 2. Use graphical LP methods to help recommend whether you should buy this TV offer. Support your recommendation.
10. Write a one-page cover letter to the CEO of the company that summarizes the results you found.

Project 4.2. Crop Planning.

A farmer has 30 acres on which to grow tomatoes and corn. Each 100 bushels of tomatoes require 1000 gallons of water and 5 acres of land. Each 100 bushels of corn require 6000 gallons of water and 2.5 acres of land. Labor costs are $1 per bushel for both corn and tomatoes. The farmer has available 30,000 gallons of water and $750 in capital. He knows that he cannot sell more than 500 bushels of tomatoes or 475 bushels of corn. He estimates a profit of $2 on each bushel of tomatoes and $3 of each bushel of corn. How many bushels of each should he raise to maximize profits?

Required:

1. List the decision variables and define them.
2. List the objective function.
3. List the resources that constrain this problem.
4. Graph the feasible region.
5. Label all intersection points of the feasible region.
6. Plot the Objective function in a different color (highlight the objective function line, if necessary) and label it the iso-profit line.
7. Clearly indicate on the graph the point that is the optimal solution.
8. List the coordinates of the optimal solution and the value of the objective function.
9. Assume now that farmer has the opportunity to sign a nice contract with a grocery store to grow and deliver at least 300 bushels of tomatoes and at least 500 bushels of corn. Use graphical LP methods to help recommend a decision to the farmer. Support your recommendation.
10. If the farmer can obtain an additional 10,000 gallons of water for a total cost of $50, is it worth the cost to obtain the additional water? Determine the new optimal solution caused by adding this level of resource.

11. Write a one-page cover letter to the farmer that summarizes the results you found.

Project 4.3. Tire Production.

Firestone Tires headquartered in Nashville, TN, has a plant in Florence, SC that manufactures two types of tires: SUV 225 radials and SUV 205 radials. Demand is high because of the recent recall of tires. Each batch of 100 SUV 225 radials requires 100 gallons of synthetic plastic and 5 lb. of rubber. Each batch of 100 SUV 205 radials requires 60 gallons of synthetic plastic and 2 1/2 lb. of rubber. Labor costs are $1 per tire for each type tire. The manufacturer has weekly quantities available of 660 gallons of synthetic plastic, $750 in capital, and 300 lb. of rubber. The company estimates a profit of $3 on each SUV 225 radial and $2 on each SUV 205 radial. How many of each type tire should the company manufacture in order to maximize its profits?

Required:

1. List the decision variables and define them.
2. List the objective function.
3. List the resources that constrain this problem.
4. Graph the feasible region.
5. Label all intersection points of the feasible region.
6. Plot the objective function in a different color (highlight the objective function line, if necessary) and label it the iso-profit line.
7. Clearly indicate on the graph the point that is the optimal solution.
8. List the coordinates of the optimal solution and the value of the objective function.
9. Assume now that the manufacturer has the opportunity to sign a favorable contract with a tire outlet store to deliver at least 500 SUV 225 radial tires and at least 300 SUV 205 radial tires. Use graphical LP methods to help recommend a decision to the manufacturer. Support your recommendation.
10. If the manufacturer can obtain an additional 1,000 gallons of synthetic plastic for a total cost of $50, is it worth the extra expense to obtain this amount? Determine the new optimal solution caused by adding this level of resource.
11. If the manufacturer can obtain an additional 20 lb. of rubber for $50, should it obtain the rubber? Determine the new solution after adding this amount.
12. Write a one-page cover letter to your boss that summarizes the results you found.

Project 4.4. Toy Production.
Consider a toy maker that carves wooden soldiers. The company specializes in two types: Confederate soldiers and Union soldiers. The estimated profit for each is \$28 and \$30, respectively. A Confederate soldier requires 2 units of lumber, 4 hours of carpentry, and 2 hours of finishing to complete. A Union soldier requires 3 units of lumber, 3.5 hours of carpentry, and 3 hours of finishing to complete. Each week the company has 100 units of lumber delivered. The workers can provide at most 120 hours of carpentry and 90 hours of finishing. Determine the number of each type of wooden soldier to produce to maximize weekly profits. Formulate and then solve this linear programming graphically.

References

[Fox2012] William P. Fox, *Mathematical Modeling with Maple®*, Cengage Pub., 2012.

4.3 Solving the Linear Program: The Simplex Method and Maple

In the previous sections we discussed formulating linear programming problems and solving two-dimensional linear programming problems by graphical methods. The graphical method illustrates some key concepts, but is only practical for problems with two variables. As you'll see, linear programming problems often have more than two variables; large-scale LP problems can have thousands of variables and constraints. With problems that have more than two variables, an algebraic method will be used. This method is called the *Simplex Method.* The Simplex Method, developed by George Dantzig in 1947 in Project SCOOP, incorporates both *optimality* and *feasibility* tests to find the optimal solution or solutions to a linear program (if any exist).

An **optimality test** shows whether or not a corner point of the feasible region corresponds to a value of the objective function better than the best value found so far. A **feasibility test** determines whether the proposed corner point is feasible; i.e., the proposed corner point does not violate any of the constraints. The simplex method starts with the selection of a corner point—usually the origin, if it is a feasible point—and then, in a systematic method, moves to an adjacent corner points of the feasible region until the optimal solution is found or it can be shown that no optimal solution exists.

The steps of the method are important because you might need more information than the standard Maple linear programming output provides. For example, knowing whether or not alternate optimal solutions exist requires an understanding of the steps, and is not readily available in Maple's output.

Steps of the Simplex Method

We'll outline the method's steps by re-solving Example 4.5, the Memory Chip Problem (pg. 151).

STEP 1. TABLEAU FORMAT:
Place the linear program in standard Tableau Format, as follows.

The LP formulation of the Memory Chip problem was

Maximize $Z = 150x_1 + 140x_2$
subject to:

$$2x_1 + 4x_2 \leq 1500$$
$$4x_1 + 3x_2 \leq 1600$$
$$x_1, x_2 \geq 0$$

To begin the simplex method, convert each inequality constraint (of the form $\leq$) to an equation. The conversion is accomplished by adding a unique, non-negative variable, called a *slack variable*, to each constraint. For example, the inequality constraint $2x_1 + 4x_2 \leq 1500$ becomes $2x_1 + 4x_2 + s_1 = 1500$ where $s_1 \geq 0$. The inequality $2x_1+4x_2 \leq 1500$ states that the sum $2x_1+4x_2$ is less than or equal to 1500. The slack variable s_1 "takes up the slack" between the sum x_1+4x_2 and the upper limit value 1500. For example, if $x_1 = x_2 = 0$, then $s_1 = 1500$; if $x_1 = 240$ and $x_2 = 0$, then $2x_1 + 4x_2 = 480$, so $s_1 = 1020$. A unique slack variable must be added to each inequality constraint. Our LP becomes

Maximize $Z = 150x_1 + 140x_2$
subject to:

$$2x_1 + 4x_2 + s_1 = 1500$$
$$4x_1 + 3x_2 + s_2 = 1600$$
$$x_1, x_2, s_1, s_2 \geq 0$$

Adding slack variables makes the constraint set a system of linear equations. For standard form, we write these with all variables on the left side of the equation and all constants on the right side. We will also rewrite the objective function by moving all variables to the left side: $Z - 150x_1 - 140x_2 = 0$. Our LP is now ready for the coefficients to be put in a matrix in the Tableau Format. This matrix is called the *simplex tableau.*

Z	x_1	x_2	s_1	s_2		*RHS*
1	−150	−130	0	0	=	0
0	2	4	1	0	=	1500
0	4	3	0	1	=	1600

STEP 2. INITIAL EXTREME POINT:
The Simplex Method begins with a known corner point as an extreme point, usually the origin $(0, 0)$ for many of our examples, called a *basic feasible solution.* The requirement for having a basic feasible solution gives rises to special simplex methods such as the "Big M method" and the "Two-Phase Simplex method" which are studied in a linear programming course.

The tableau previously shown contains the corner point $(x_1, x_2) = (0, 0)$. This corner point is our initial solution.

We substitute $(x_1, x_2) = (0, 0)$, and read the solution from the simplex tableau:

$$\begin{aligned} x_1 &= 0 \\ x_2 &= 0 \\ s_1 &= 1500 \\ s_2 &= 1600 \end{aligned}$$

and

$$Z = 0$$

Let's continue to analyze these variables further. We have 5 variables $\{Z, x_1, x_2, s_1, s_2\}$ and 3 equations. With three equations, we can solve for at most 3 of the 5 variables. The variable Z, although actually a dependent variable, will always be a solution by convention in our tableau. Therefore, we can solve for two non-zero independent variables from among $\{x_1, x_2, s_1, s_2\}$. The non-zero variables are called the **basic variables.** The remaining variables are called the **non-basic variables**. The corresponding solutions are called the **basic feasible solutions** (BFS) and correspond to corner points. A complete step of the simplex method produces a solution that corresponds to a corner point of the feasible region. These solutions are read directly from the tableau matrix.

We also note the basic variables have a column consisting of exactly a single 1 and the rest zeros. We will add an initial column to label the basic variables as shown below.

Basic Variable	Z	x_1	x_2	$\boxed{s_1}$	$\boxed{s_2}$		*RHS*
Z	1	−150	−130	0	0	=	0
s_1	0	2	4	1	0	=	1500
s_2	0	4	3	0	1	=	1600

STEP 3. OPTIMALITY TEST:
We need to determine if moving to an adjacent corner point improves the

value of the objective function. If not, the current extreme point is optimal. If an improvement is possible, the optimality test determines which variable currently in the independent set (having value zero) should *enter* the dependent set as a basic variable and become nonzero. For our maximization problem, we look at the Z-row (The row marked by the basic variable Z). If any coefficients in that row are negative, then we select the variable whose coefficient is the most negative as the entering variable. Think of the coefficients in the original objective function $f(X)$. The variable with the largest partial derivative indicates the direction of maximum improvement for f. Transferring this idea to the Z-row, where we wrote $Z - f(X) = 0$, selects the variable with the most negative value as the direction of maximum improvement.

The Z-row of our tableau gives:

Basic Variable	Z	$\boxed{x_1}$	x_2	s_1	s_2		RHS
Z	1	$\boxed{-150}$	-130	0	0	=	0

The variable with the most negative coefficient is x_1 with value -150. Thus, x_1 wants to become a basic variable; x_1 is called the *entering basic variable.* We can only have two basic variables besides Z, in this example (because we have three equations), so one of the current basic variables $\{s_1, s_2\}$ must be replaced by x_1. Let's proceed to see how we determine which variable exits from the set of basic variables.

STEP 4. FEASIBILITY TEST:

To find a new corner point, or intersection point of the constraints, one of the variables in the basic variable set must *exit* to allow the entering variable from STEP 3 to become basic. The feasibility test determines which current dependent variable to choose for exiting, ensuring we stay inside the feasible region. We will use the *Minimum Positive Ratio Test* as our feasibility test. (Other methods are studied in linear programming courses.) The minimum positive ratio test is: select the current basic variable where the RHS coefficient divided by the entering variable's coefficient in that row gives the minimum positive ratio. In our tableau, compare the ratios $s_1 : 1500/2 = 750$ to $s_2 : 1600/4 = 400$ to choose s_2 as the *departing basic variable.*

Basic Variable	Z	$\boxed{x_1}$	x_2	s_1	s_2		RHS	Min Positive Ratio Test
Z	1	-150	-130	0	0	=	0	–
s_1	0	2	4	1	0	=	1500	$1500/2 = 750$
$\boxed{s_2}$	0	④	3	0	1	=	1600	$1600/4 = \boxed{400}$

We will always disregard all quotients with either 0 or negative values in the denominator. In our example we compared 750 and 400 to select the smallest non-negative value. We now have the location, the x_1 column in the s_2 row, where we should perform a pivot operation. Here, pivot on the ④.

STEP 5. PIVOT:
We form a new equivalent system by using elementary row operations to scale the pivot element to a 1, and all other numbers in the pivot column to zero. We do the row operations by adding a suitable multiple of the pivot row to a multiple of each row in the tableau, thus eliminating the new basic variable. The row operations on the matrix of the tableau we need in this case are:

$$s_2 : \mathit{Row}\ 3 \leftarrow (1/4) \times \mathit{Row}\ 3,$$

then

$$Z : \mathit{Row}\ 1 \leftarrow \mathit{Row}\ 1 + 150 \times \mathit{Row}\ 3,$$
$$s_1 : \mathit{Row}\ 2 \leftarrow \mathit{Row}\ 2 - 2 \times \mathit{Row}\ 3.$$

After pivoting, the tableau becomes

Basic Variable	Z	x_1	x_2	s_1	s_2		*RHS*
Z	1	0	-17.5	0	37.5	=	60000
s_1	0	0	2.50	1	-0.5	=	700
x_1	0	1	0.75	0	0.25	=	400

Then set the new non-basic variables to zero in the new system to find the values of the new basic variables, thereby determining a new intersection point or corner point basic feasible solution.

Let's interpret our current basic feasible solution.
Basic Variables:

$$x_1 = 400$$
$$s_1 = 700$$
$$Z = 60000$$

Non-Basic Variables

$$x_2 = 0$$
$$s_2 = 0$$

STEP 6. REPEAT UNTIL DONE:
STEPS 3-5 are repeated until an **optimal extreme point** is found.

We see that x_2 has a coefficient of -17.5 in the Z-row, therefore we are not yet at an optimal solution; x_2 is the entering variable. The ratios are

$$s_1 : 700/2.5 = 280$$
$$x_1 : 400/0.75 = 533.33$$

Thus, our new departing variable is s_1. We pivot on the s_2 row, x_2 column.

Basic Variable	Z	x_1	x_2	s_1	s_2		*RHS*
Z	1	0	0	7.2	34	=	65000
x_2	0	0	1	0.4	−0.2	=	280
x_1	0	1	0	−0.3	0.4	=	190

The current solution is:
Basic Variables:

$$x_1 = 190$$
$$s_1 = 280$$
$$Z = 65000$$

Non-Basic Variables

$$s_1 = 0$$
$$s_2 = 0$$

Since there are no negative coefficients in the Z-row, the Optimality Test indicates we are at the optimal solution, the maximum value of $Z = 65{,}000$.

Maple has several commands that we will illustrate later in this chapter which directly solve linear programming problems. However, we'll start with a modification of Paul Fishback's "Revised Simplex Method" (see [Fishback2009]), applying the simplex procedure using matrices, which makes it easy to find more information about the solution.

Revised Simplex

Write the LP in matrix form.

$$\left\{\begin{array}{rl} \text{Maximize } Z & = f(X) \\ \text{subject to} & \\ g_i(X) & \leq b_i \text{ for all } i. \\ x_i & \geq 0 \text{ for all } i. \end{array}\right\} \iff \left\{\begin{array}{rl} \text{Maximize } Z & = C{\cdot}X \\ \text{subject to} & \\ AX & \leq B \\ X & \geq \vec{0} \end{array}\right\}$$

The steps of the Revised Simplex Method are:

1. Initialize. Set up the simplex tableau the same as before as the partitioned

block matrix

$$\begin{bmatrix} 1 & -C & \vec{0}^{\mathrm{T}} & 0 \\ \vec{0} & A & I_m & B \end{bmatrix}$$

where I_m is an $m \times m$ identity matrix, and the two zero vectors have appropriate lengths.

2. Optimality Test. Let B be the matrix formed from the columns of the tableau corresponding to the current set of basic variables. Calculate B^{-1} and

$$U = B^{-1} \cdot \begin{bmatrix} -c & 0 \\ A & b \end{bmatrix}.$$

If the solution is optimal; i.e., the Z-row, the first row of U, has no negative entries, stop. Otherwise, continue to the next step: find the new entering variable.

3. Minimum Ratio Test to Determine the Leaving Variable. Calculate only the column of the entering basic variable, say it is x_j, compute b_i/a_{ij} for positive entries of a_{ij} from the entering variable's column. The column of x_j can be calculated by left-multiplying B_{-1} to the x_j column in A. To form the new B, replace the leaving variable's column with the entering variable's column from the original tableau. Compute the new B^{-1}.

4. Go back to Step 2, the Optimality Test.

We will use the matrix approach, using commands from Maple's *LinearAlgebra* package to perform our steps. Several advantages appear in this technique. First, we reduce round-off errors because we are always using the original tableau to update and to compute the results. The approach also helps in doing sensitivity analysis later.

Return again to Example 4.5, the Memory Chip Problem (pg. 151).

Maximize $Z = 150x_1 + 140x_2$
subject to:

$$2x_1 + 4x_2 \leq 1500$$
$$4x_1 + 3x_2 \leq 1600$$
$$x_1, x_2 \geq 0$$

Step 1. Enter the problem's matrices in Maple.

```
> with(LinearAlgebra) :
> c := ⟨150 | 140⟩ :
```

$> A := Matrix([[2,4],[4,3]]);$

$$A := \begin{bmatrix} 2 & 4 \\ 4 & 3 \end{bmatrix}$$

$> b := \langle 1500, 1600 \rangle$

$> b1 := \langle 0, b \rangle;$

$$b1 := \begin{bmatrix} 0 \\ 1500 \\ 1600 \end{bmatrix}$$

$> n, m := 2, 2 :$

$> Labels := \langle z \mid (x_i \$ i = 1..n) \mid (s_i \$ i = 1..m) \mid RHS \rangle;$

$$Labels := \begin{bmatrix} z & x_1 & x_2 & s_1 & s_2 & RHS \end{bmatrix}$$

$> LPMatrix := \langle UnitVector(1, m+1) \mid \langle -c, A \rangle \mid ZeroVector[row](m),$
$IdentityMatrix(m) \mid b1 \rangle;$

$$LPMatrix := \begin{bmatrix} 1 & -150 & -130 & 0 & 0 & 0 \\ 0 & 2 & 4 & 1 & 0 & 1500 \\ 0 & 4 & 3 & 0 & 1 & 1600 \end{bmatrix}$$

A convenient display is given by stacking *Labels* on *LPMatrix*.

$> \langle Labels, LPMatrix \rangle;$

$$\begin{bmatrix} z & x_1 & x_2 & s_1 & s_2 & RHS \\ 1 & -150 & -130 & 0 & 0 & 0 \\ 0 & 2 & 4 & 1 & 0 & 1500 \\ 0 & 4 & 3 & 0 & 1 & 1600 \end{bmatrix}$$

Now we extract B, the matrix formed from the columns of *LPMatrix* corresponding to the current set of basic variables $\{z, s_1, s_2\}$, and then compute B^{-1}. In this case, B is made from the 1st, 4th, and 5th columns of *LPMatrix*. This time, B^{-1} will be obvious.

$> B := LPMatrix[.., [1,4,5]];$ *# select all rows & columns* 1, 4, *and* 5

$$B := \begin{bmatrix} 1 & 0 & 0 \\ 0 & 1 & 0 \\ 0 & 0 & 1 \end{bmatrix}$$

We look at our optimality test and see we are not optimal as $x_1 = -150$ is the most negative, so x_1 will enter the set of basic variables.

Define a Maple function $MRT(k)$ to calculate the RHS ratios b_k/a_k to apply the minimum ratio test to the kth column. Apply the function for x_1 in column 2 of LPMatrix.

$> MRT := k \rightarrow zip((x,y) \rightarrow \frac{x}{y}, A([[2..-1], -1], A([2..-1], k]) :$

> $MRT(LPMatrix, 2);$

$$\begin{bmatrix} 750 \\ 400 \end{bmatrix}$$

The departing basic variable chosen by the minimum positive ratio test is s_2.

We replace the s_2 column of B with the x_1 column of the original tableau to form the new B, and recompute B^{-1}. Then, we left-multiply the original tableau by this new B^{-1} to obtain an updated tableau.

> $BV := Labels[[1, 4, 2]];$
> $B := LPMatrix[.., [1, 4, 2]];$ # *select all rows & columns* 1, 4, *and* 2

$$bv := \begin{bmatrix} z & s_1 & x_1 \end{bmatrix}$$

$$B := \begin{bmatrix} 1 & 0 & -150 \\ 0 & 1 & 2 \\ 0 & 0 & 4 \end{bmatrix}$$

> $Binv := B^{-1};$

$$Binv := \begin{bmatrix} 1 & 0 & \frac{75}{2} \\ 0 & 1 & -\frac{1}{2} \\ 0 & 0 & \frac{1}{4} \end{bmatrix}$$

> $LPUpdate := Binv \,.\, LPMatrix :$
> $\langle Labels, LPUpdate \rangle;$

$$\begin{bmatrix} z & x_1 & x_2 & s_1 & s_2 & RHS \\ 1 & 0 & -\frac{35}{2} & 0 & \frac{75}{2} & 60000 \\ 0 & 0 & \frac{5}{2} & 1 & -\frac{1}{2} & 700 \\ 0 & 1 & \frac{3}{4} & 0 & \frac{1}{4} & 400 \end{bmatrix}$$

> $MRT(LPUpdate, 3);$

$$\begin{bmatrix} 280 \\ 530. \end{bmatrix}$$

```
> BV[2] := Labels[3] :
  BV;
  B[.., 2] := LPMatrix[.., 3] :   # replace col 2 of B with col 3 of LPMatrix
  B;
```

$$[z \quad x_2 \quad x_1]$$

$$\begin{bmatrix} 1 & -130 & -150 \\ 0 & 4 & 2 \\ 0 & 3 & 4 \end{bmatrix}$$

```
> Binv := B^-1;
```

$$Binv := \begin{bmatrix} 1 & 7 & 34 \\ 0 & \frac{2}{5} & -\frac{1}{5} \\ 0 & -\frac{3}{10} & \frac{2}{5} \end{bmatrix}$$

```
> LPUpdate2 := Binv . LPMatrix :
  ⟨Labels, LPUpdate2⟩;
```

$$\begin{bmatrix} z & x_1 & x_2 & s_1 & s_2 & RHS \\ 1 & 0 & 0 & 7 & 34 & 64900 \\ 0 & 0 & 1 & \frac{2}{5} & -\frac{1}{5} & 280 \\ 0 & 1 & 0 & -\frac{3}{10} & \frac{2}{5} & 190 \end{bmatrix}$$

We are now at an optimal solution as no values in the Z-row, the first row, are negative. Read the solution as

$$x_1 = 190, \quad x_2 = 280 \quad Z = 64900.$$

The tableau gives more information. From the Z-row, we have the *shadow prices* $s_1 = 7$ and $s_2 = 34$, respectively. Obtaining these shadow prices is extremely important in our analysis of the problem. The shadow price for a constraint indicates how much the objective function would increase by increasing the amount of that resource by one unit. In the Memory Chip problem, Constraint 1 was "available assembly time (hours)" and Constraint 2 was "available installation time (hours)." The shadow price $s_1 = 7$ tells us that increasing "available assembly time" by 1 hour should increase profit by \$7. The shadow price $s_2 = 34$ tells us that increasing "available installation time" by 1 hour should increase profit by \$34. If the cost of an additional resource unit for Constraint 1 and Constraint 2 are the same, we should increase Constraint 2 by one unit as that will increase the objective function by about 34 units. We will see more on *sensitivity analysis* later in this chapter.

Example 4.9. Simple Simplex.

Maximize $Z = 3x_1 + x_2$
subject to:

$$\begin{aligned} 2x_1 + x_2 &\leq 6 \\ x_1 + 3x_2 &\leq 9 \\ x_1, x_2 &\geq 0 \end{aligned}$$

The Tableau Format with slack variables s_1 and s_2 is

Basic Variable	Z	x_1	x_2	s_1	s_2	*RHS*
Z	1	−3	−1	0	0	0
s_1	0	2	1	1	0	6
s_2	0	1	3	0	1	9

We have

- Initial Basic Variables $\{Z, s_1, s_2\}$
- Initial Non-Basic Variables $\{x_1, x_2\}$
- Initial Extreme Point $(x_1, x_2) = (0, 0)$
- Value of objective function: $Z = 0$

Optimality Test: There are negative values in the Z-row, therefore we are not optimal. The entering variable is x_1 (corresponding to the -3 in the Z-row.)

Feasibility Test: Compute the ratios of the "RHS" divided by the x_1 column entries to determine the minimum positive ratio.

Basic Variable	Z	x_1	x_2	s_1	s_2	*RHS*	Min Positive Ratio Test
Z	1	−3	−1	0	0	0	–
s_1	0	2	1	1	0	6	$6/2 = 3$
s_2	0	1	3	0	1	9	$9/1 = 9$

Choose s_1 to leave since it corresponds to the minimum positive ratio test value of 3.

Pivot: Divide the row containing the exiting variable (the s_1 row in this case) by the coefficient of the entering variable in that row (the coefficient of x_1, in this case, 2), giving a coefficient of 1 for the entering variable in this row. Then eliminate the entering variable x_1 from the remaining rows (which do not contain the exiting variable s_1 and have a zero coefficient for it). The results are summarized in the next tableau.

Basic Variable	Z	x_1	x_2	s_1	s_2	*RHS*
Z	1	0	0.5	1.5	0	9
x_1	0	1	0.5	0.5	0	3
s_2	0	0	2.5	−0.5	1	6

We now have

- Basic Variables $\{Z, x_1, s_2\}$
- Non-Basic Variables $\{s_1, x_2\}$
- Extreme Point $(x_1, x_2) = (3, 0)$
- Value of objective function: $Z = 9$

The basic variable s_2 has value 6 indicating there is unused resource from Constraint 2.

Optimality Test: There are no negative coefficients in the Z-row. Thus $x_1 = 3$ (a basic variable) and $x_2 = 0$ (a non-basic variable) is an extreme point giving the optimal objective function value $Z = 9$.

The following Maple statements show the computations above. Again define *MRT* to do a minimum ratio test with a specified column. Also, define *FullPivot* in Maple to scale and pivot a matrix representing a tableau in a single step. We'll need the *LinearAlgebra* package.

> $with(LinearAlgebra):$

> $MRT := k \to zip\left((x, y) \to \frac{x}{y}, A([[2..-1], -1], A([2..-1], k]\right):$

> $FullPivot := Pivot(RowOperation(A, r, A[r, c]^{-1}), r, c):$

Now, the computations.

> $Tableau := Matrix([[1, -3, -1, 0, 0, 0], [0, 2, 1, 1, 0, 6], [0, 1, 3, 0, 1, 9]]);$

$$Tableau := \begin{bmatrix} 1 & -3 & -1 & 0 & 0 & 0 \\ 0 & 2 & 1 & 1 & 0 & 6 \\ 0 & 1 & 3 & 0 & 1 & 9 \end{bmatrix}$$

> *MRT*(*Tableau*, 2, 2);

$$\begin{bmatrix} 3 \\ 9 \end{bmatrix}$$

> *FullPivot*(*Tableau*, 2, 2);

$$\begin{bmatrix} 1 & 0 & \frac{1}{2} & \frac{3}{2} & 0 & 9 \\ 0 & 1 & \frac{1}{2} & \frac{1}{2} & 0 & 3 \\ 0 & 0 & \frac{5}{2} & -\frac{1}{2} & 1 & 6 \end{bmatrix}$$

Remarks: We have assumed that the origin is a feasible extreme point. If it is not, then an extreme point must be found before the simplex method, as presented, can be used. We have also assumed that the linear program is not "degenerate" in the sense that no more than two constraints intersect at the same point. These and other topics are studied in more advanced optimization courses.

Example 4.10. Building and Selling Wooden Soldiers.
Problem: Maximize the profits from making and selling wooden soldiers. (See pg. 163.)

Assumptions: We sell all the soldiers that we make each week. We also assume that we can make a fractional number of soldiers. Unfinished soldiers are completed in the following week.

Define Variables:

$x_1 =$ number of Confederate soldiers made per week

$x_2 =$ number of Union soldiers made per week

Objective Function: Maximize $Z = 28x_1 + 30x_2$.

Resource Constraints:

$$\begin{aligned} 2x_1 + 3x_2 &\leq 100 && \textit{(Units of lumber available)} \\ 4x_1 + 3.5x_2 &\leq 120 && \textit{(Hours of carpentry available)} \\ 2x_1 + 3x_2 &\leq 90 && \textit{(Hours of finishing available)} \\ x_1, x_2 &\geq 0 && \textit{(Nonnegativity)} \end{aligned}$$

The initial tableau with slack variables s_1 (lumber), s_2 (carpentry), and s_3 (finishing) is:

Basic Variable	Z	x_1	x_2	s_1	s_2	s_3	*RHS*
Z	1	−28	−30	0	0	0	0
s_1	0	2	3	1	0	0	100
s_2	0	4	3.5	0	1	0	120
s_3	0	2	3	0	0	1	90

We observe from the tableau:

- Initial basic variables: $\{Z, s_1, s_2, s_3\}$
- Non-basic variables $\{x_1, x_2\}$
- Extreme Point $(x_1, x_2) = (0, 0)$
- Value of objective function: $Z = 0$

Optimality test. Not optimal as the Z-row contains negative values. The entering variable is x_2 (corresponding to -30 in the Z-row).

Feasibility test. Compute the ratios for the RHS divided by the coefficients in the x_2 column to determine the minimum positive ratio.

Basic Variable	Z	x_1	x_2	s_1	s_2	s_3	*RHS*	Min Positive Ratio Test
Z	1	−28	−30	0	0	0	0	–
s_1	0	2	3	1	0	0	100	$100/3 = 33.33$
s_2	0	4	3.5	0	1	0	120	$120/3.5 = 34.29$
s_3	0	2	3	0	0	1	90	$90/3 = 30$

Choose s_3 to leave corresponding to the minimum positive ratio $90/3 = 30$.

Pivot: Pivot on the s_3 row, x_2 column.

Basic Variable	Z	x_1	x_2	s_1	s_2	s_3	*RHS*
Z	1	−8	0	0	0	10	900
s_1	0	0	0	1	0	−1	10
s_2	0	1.67	0	0	1	−1.17	15
x_2	0	0.67	1	0	0	0.33	30

Observe:

- Basic variables $\{Z, s_1, s_2, x_2\}$
- Non-basic variables $\{x_1, s_3\}$
- Extreme Point $(x_1, x_2) = (0, 30)$
- Value of objective function: $Z = 900$

Optimality test. The entering variable is x_1 corresponding to -8 in the Z-row.

Feasibility Test Compute the ratios for the RHS divided by the coefficients in the column labeled x_1 to determine the minimum positive ratio.

Basic Variable	Z	x_1	x_2	s_1	s_2	s_3	*RHS*	Min Positive Ratio Test
Z	1	−8	0	0	0	10	900	–
s_1	0	0	0	1	0	−1	10	*undefined*
s_2	0	1.67	0	0	1	−1.17	15	9
x_2	0	0.67	1	0	0	0.33	30	45

Choose s_2 corresponding to the minimum positive ratio 9 as the leaving variable.

Pivot. Pivot on the s_2 row, x_1 column.

Basic Variable	Z	x_1	x_2	s_1	s_2	s_3	*RHS*
Z	1.0	0.0	0.0	0.0	4.8	4.4	972.0
s_1	0.0	0.0	0.0	1.0	0.0	−1.0	10.0
x_1	0.0	1.0	0.0	0.0	0.6	−0.7	9.0
x_2	0.0	0.0	1.0	0.0	−0.4	0.8	24.0

Observe:

- Basic variables $\{Z, s_1, x_1, x_2\}$
- Non-basic variables $\{s_2, s_3\}$
- Extreme Point $(x_1, x_2) = (9, 24)$
- Value of objective function: $Z = 972$

Optimality test. Because there are no negative coefficients in the Z-row we are optimal. The optimal solution is $Z = 972$ at $(x_1, x_2) = (9, 24)$.

Thus we make 9 Confederate soldiers and 24 Union soldiers to yield the maximum profit of $972.

Information about the resources is also readily available in the final tableau. Since $s_1 = 10$ and $s_2 = s_3 = 0$, we have 10 units of lumber left over; i.e., we are only truly constrained by hours of carpentry and hours of finishing.

Revised Simplex for the Wooden Soldiers.

We finish this section with a Maple worksheet showing the solution to the Wooden Soldiers problem using the revised simplex method.

```
> with(LinearAlgebra);
> MRT := (A, k) → zip((x, y) → (x/y), A[[2.. − 1], −1], A[[2.. − 1], k]) :
> c := ⟨28 | 30⟩ :
> A := Matrix([[2, 3], [4, 3.5], [2, 3]]) :
> b := ⟨100, 120, 90⟩ :
> n, m := 2, 3 :
> Labels := Vector[row]([Z, (x_i$ i = 1..n), (s_i$ i = 1..m), RHS]) :
```

$$\begin{bmatrix} Z & x_1 & x_2 & s_1 & s_2 & s_3 & RHS \end{bmatrix}$$

```
> UV := UnitVector(1, m + 1) :
  ZV := ZeroVector[row](m) :
  I2 := IdentityMatrix(m) :
> LPMatrix := <UV, <-c, A>, <ZV, I2>, <0, b>> :
  <Labels, LPMatrix>;
```

$$\begin{bmatrix} Z & x_1 & x_1 & s_1 & s_2 & s_3 & RHS \\ 1 & -28 & -30 & 0 & 0 & 0 & 0 \\ 0 & 2 & 3 & 1 & 0 & 0 & 100 \\ 0 & 4 & 3.5 & 0 & 1 & 0 & 120 \\ 0 & 2 & 3 & 0 & 0 & 1 & 90 \end{bmatrix}$$

```
> MRT(LPMatrix, 3);
```

$$\begin{bmatrix} 33.33 \\ 34.29 \\ 30. \end{bmatrix}$$

```
> B1 := LPMatrix[.., [1, 4, 5, 3]];
```

$$\begin{bmatrix} 1 & 0 & 0 & -30 \\ 0 & 1 & 0 & 3 \\ 0 & 0 & 1 & 3.5 \\ 0 & 0 & 0 & 3 \end{bmatrix}$$

```
> LPMatrix1 := fnormal ~ (B1^(-1) . LPMatrix, 3);
```

$$LPMatrix1 := \begin{bmatrix} 1.0 & -8.0 & 0.0 & 0.0 & 0.0 & 10.0 & 900.0 \\ 0.0 & 0.0 & 0.0 & 1.0 & 0.0 & -1.0 & 10.0 \\ 0.0 & 1.67 & 0.0 & 0.0 & 1.0 & -1.17 & 15.0 \\ 0.0 & 0.667 & 1.0 & 0.0 & 0.0 & 0.333 & 30.0 \end{bmatrix}$$

```
> MRT(LPMatrix1, 2);
```

$$\begin{bmatrix} \text{Float}(\infty) \\ 8.982 \\ 44.98 \end{bmatrix}$$

```
> B2 := LPMatrix[.., [1, 4, 2, 3]];
```

$$B2 := \begin{bmatrix} 1 & 0 & -28 & -30 \\ 0 & 1 & 2 & 3 \\ 0 & 0 & 4 & 3.5 \\ 0 & 0 & 2 & 3 \end{bmatrix}$$

```
> LPMatrix2 := fnormal ~ (B2^(-1).LPMatrix, 2);
```

$$LPMatrix2 := \begin{bmatrix} 1.0 & 0.0 & 0.0 & 0.0 & 4.8 & 4.4 & 970.0 \\ 0.0 & 0.0 & 0.0 & 1.0 & 0.0 & -1.0 & 10.0 \\ 0.0 & 1.0 & 0.0 & 0.0 & 0.0 & -0.0 & 9.0 \\ 0.0 & 0.0 & 1.0 & 0.0 & -0.0 & 0.0 & 24.0 \end{bmatrix}$$

Exercises

Use the simplex method or the revised simplex method in each exercise. Assume nonnegativity, $x, y \geq 0$, in each problem.

1. Maximize $Z = 2x + 3y$
   subject to:

$$\begin{aligned} 2x + 3y &\leq 6 \\ 3x - y &\leq 15 \\ -x + y &\leq 4 \\ 2x + 5y &\leq 27 \end{aligned}$$

2. Maximize $Z = 6x + 4y$
   subject to:

$$\begin{aligned} -x + y &\leq 12 \\ x + y &\leq 24 \\ 2x + 5y &\leq 80 \end{aligned}$$

3. Maximize $Z = 6x + 5y$
   subject to:

$$\begin{aligned} x + y &\leq 6 \\ 2x + y &\leq 9 \end{aligned}$$

4. Optimize* $Z = x - y$
   subject to:

$$\begin{aligned} x + y &\geq 6 \\ 2x + y &\leq 9 \end{aligned}$$

   * Both maximize and minimize.

5. Maximize $Z = 5x + 3y$
   subject to:

$$1.2x + 0.6y \leq 24$$
$$2.0x + 1.5y \leq 80$$

Projects

Redo each project from Section 4.2 (pg. 160) using Maple.

4.4 Linear Programming with Maple's Commands

Maple provides a suite of powerful, robust routines for solving optimization problems, including linear programs (LPs). Using Maple's flexible mathematical programming language to conduct thorough sensitivity studies on solutions to optimization problems is quite simple. We will present several of Maple's built-in features for solving linear programming problems.

Method 1 uses the commands *maximize* and *minimize* from the *simplex* package. The common syntax is

$$maximize(objective\ function, constraints, NONNEGATIVE)$$

See *help*(*simplex*) and *help*(*simplex, maximize*).

Example 4.11. Using *simplex:-maximize*.

Maximize $Z = x + y$
subject to:

$$x + 3y \leq 9$$
$$2x + y \leq 8$$
$$x, y \geq 0$$

Method 1. We load the *simplex* package. Enter the objective function, calling it *Objective*. Then we enter the set of constraints, calling the set *Constraints*. Note we did not have to enter the non-negativity constraints—those are handled by the option *NONNEGATIVITY* (using all capitals). Next, we use *maximize*. The optimal results of the decision variables that satisfy the constraints are returned by Maple as a set. Substitute the results into the objective function to obtain the optimal value for the problem.

$> with(simplex):$

$> Objective := x + y:$

$> Constraints := \{x + 3y \le 9, 2x + y \le 8\}:$

$> Soln := maximize(Objective, Constraints, NONNEGATIVE);$

$$Soln := \{x = 3, y = 2\}$$

$> Z = subs(Soln, Objective);$

$$Z = 5$$

The solution is easily read as x=*3*, y=*2*, and Z = *5.*

Method 2 uses *LPSolve* from Maple's *Optimization* package requiring the same inputs as Method I. The basic syntax for the *LPSolve* command is quite similar to *maximize.*

LPSolve(objective function, constraints, variable range restrictions, options)

However, non-negativity conditions are entered as *assume=nonnegative* (no capitals this time). The default operation for *LPSolve* is to minimize the objective. To maximize an LP, use the option *maximize=true* or the shorthand *'maximize'*. See *help(LPSolve)*.

Example 4.12. Using *Optimization:-LPSolve.*
Solve the LP from Example 4.11 using *LPSolve.*

Method 2. We use the objective function and constraints defined in Method 1. First load the *Optimization* package.

$> with(Optimization):$

$> Soln2 := LPSolve(Objective, Constraints, assume = nonnegative, 'maximize');$

$$Soln2 := [5., [x = 3., y = 2.00000000000000]]$$

The output provides the solution as $Z = 5$ at $x = 3$, $y = 2$.

Method 3 uses the *Optimization* package's *LPSolve* with the LP stated in matrix form. The profit or cost coefficients from the objective function are entered in the vector c. The coefficient matrix of the constraints is entered as matrix A, and the right-hand side values are entered as a vector b. We then use the LPSolve command's matrix form.

Example 4.13. Using *Optimization:-LPSolve* with Matrices.
Solve the LP from Example 4.11 using matrices in *LPSolve.*

Method 3. First load the *Optimization* package.

$> with(Optimization):$

Now define the matrices and vectors for our LP.

$> c := \langle 1 \,|\, 1 \rangle$

$> A := Matrix([[1,3],[2,1]]);$

$$\begin{bmatrix} 1 & 3 \\ 2 & 1 \end{bmatrix}$$

$> b := \langle 9, 8 \rangle :$

(Note: When using 'angle' notation to define vectors or matrices, pay close attention to commas for row separators and vertical bars for column separators as in c and b above.) We're ready to use *LPSolve*.

$> Soln := LPSolve(c, [A, b], assume = nonnegative, maximize);$

$$Soln := \left[5., \begin{bmatrix} 3. \\ 2.00000000000000 \end{bmatrix} \right]$$

The answer is read from *LPSolve*'s output as: The optimal value of the objective function is $Z = 5.0$ when $x = 3.0$ and $y = 2.0$.

On the CD, find two additional tools: a Tableau generator and a Maplet generator for a maximization LP. These can be used when the tableaus need to be seen and discussed.

In order to obtain the shadow prices after using Maple's *maximize* or *LPSolve* commands rather than the tableau form shown earlier, we recommend using *duality*. The *dual* of a linear program[2] exchanges the decision variables with the surplus variables and the objective function's coefficients with the constraints right-hand-side constants to form a new LP. Minimizing the new LP is equivalent to maximizing the original, or *primal*, LP, and vice versa. For a complete discussion see Winston and Goldberg [WG2004] or Hillier and Lieberman [HL1990].

Recall our Memory Chip problem (pg. 151). How to find the shadow prices after using Maple's simplex commands to solve it as originally presented follows.

$> with(simplex) :$

$> Constraints := \{2x_1 + 4x_2 \leq 1500, 4x_1 + 3x_2 \leq 1600\} :$

$> Soln := maximize(Objective, Constraints, NONNEGATIVE);$

$$Soln := \{x_1 = 190, x_2 = 280\}$$

$> Z = eval(Objective, Soln);$

$$Z = 64900$$

Now, to obtain the shadow prices, we must first generate the dual LP.

[2]Invented by Lemke in 1954, and extended by Dantzig, von Neumann, and Tucker.

```
> dualObjective, dualConstraints := dual(Objective, Constraints, s);
  dualObjective, dualConstraints := 1500s1 + 1600s2, {130 ≤ 4s1 + 3s2, 150 ≤
  2s1 + 4s2}
```

Since we maximized the primal LP, we minimize the dual LP.

```
> dualSoln := minimize(dualObjective, dualConstraints, NONNEGATIVE);
                    dualSoln := {s1 = 7, s2 = 34}
> Z = eval(dualObjective, dualSoln);
                    Z = 64900
```

We read the solution as $Z = 64900$ (as before). The shadow prices from the dual's solution are $s_1 = 7$ (Constraint 1) and $s_2 = 34$ (Constraint 2), as we saw in the tableau form.

Exercises

Redo the exercises from Section 4.3, pg. 180, using Maple's intrinsic commands.

Projects

Redo each project from Section 4.2, pg. 160, using Maple's intrinsic commands.

4.5 Sensitivity Analysis with Maple

We have looked at using Maple to solve linear programming problems and to generate tableaus for further analysis on the problems. In this section, we explore the use of Maple to perform *sensitivity analysis.* Sensitivity analysis investigates how sensitive the solution is to changes, usually small, in the values of the model's parameters. In real-world problems, parameter values used in the model are at best estimates. Therefore, sensitivity analysis is essential for good decision making. We will present methods to perform sensitivity analysis on parameters of (1) coefficients of non-basic variables, (2) coefficients of basic variables, and (3) changes in the resources' limits (the RHS values). We will examine these one at a time.

To reduce the analysis to formula form, we use the following matrix notation for our analysis:

c_{BV} : Matrix of the original profit/cost coefficients of the basic variables.

B^{-1} : Inverse of the matrix of the basic variables in the order in which they enter the basis.

a_j : Original RHS column of the resources coefficient of the jth variable.

c_j : Original column of the cost coefficient of the jth variable.

b_j : The resource limit of the jth constraint.

Let's start with the linear programming problem given by

Maximize $Z = 60d + 30t + 20c$
subject to:

$$\begin{aligned} 8d + 6t + c &\leq 48 \\ 4d + 2t + 1.5c &\leq 20 \\ 2d + 1.5t + 0.5c &\leq 8 \\ d, t, c &\geq 0 \end{aligned}$$

Solve this LP to obtain the following the solution and also to create the tableaux.

```
> with(Optimization);

> Objective := 60 · d + 30 · t + 20 · c;
  Constraints := {2 · d + 1.5 · t + 0.5 · c ≤ 8, 4 · d + 2 · t + 1.5 · c ≤
  20, 8 · d + 6 · t + c ≤ 48};
  Labels := Vector[row]([Z, d, t, c, s1, s2, s3, RHS]);
```

$$Objective := 60d + 30t + 20c$$

$$Constraints := \{2d + 1.5t + 0.5c \leq 8, 4d + 2t + 1.5c \leq 20, 8d + 6t + c \leq 48\};$$

$$Labels := \begin{bmatrix} Z & d & t & c & s_1 & s_2 & s_3 & RHS \end{bmatrix}$$

```
> Soln := LPSolve(Objective, Constraints, maximize,
    assume = nonnegative, output = solutionmodule);
```

$$Soln := \textbf{module}(\ )\ldots\textbf{end module}$$

```
> Soln:-Results( );
```

["objectivevalue" = 280.0, "solutionpoint" = [c = 8.00000000000000, d = 2.0, t = 0.0], "iterations" = 2]

```
> InitialTableau := Matrix([[1, -60, -30, -20, 0, 0, 0, 0], [0, 8, 6, 1, 1, 0, 0, 48],
    [0, 4, 2, 1.5, 0, 1, 0, 20], [0, 2, 1.5, .5, 0, 0, 1, 8]]);
  ⟨Labels, InitialTableau⟩;
```

$$\begin{bmatrix} Z & d & t & c & s_1 & s_2 & s_3 & RHS \\ 1 & -60 & -30 & -20 & 0 & 0 & 0 & 0 \\ 0 & 8 & 6 & 1 & 1 & 0 & 0 & 48 \\ 0 & 4 & 2 & 1.5 & 0 & 1 & 0 & 20 \\ 0 & 2 & 1.5 & 0.5 & 0 & 0 & 1 & 8 \end{bmatrix}$$

```
> FinalTableau := Matrix([[1., 0., 5.0, 0., 0., 10., 10., 280.0],
    [0., 0., -2.0, 0., 1., 2., -8., 24.0],
    [0., 0., -2.0, 1.000, 0., 2., -4., 8.0],
    [0., 1., 1.25, 0., 0., -.5, 1.5, 2.0]]) :
  ⟨Labels, FinalTableau⟩
```

$$\begin{bmatrix} Z & d & t & c & s_1 & s_2 & s_3 & RHS \\ 1.0 & 0.0 & 5.0 & 0.0 & 0.0 & 10.0 & 10.0 & 280.0 \\ 0.0 & 0.0 & -2.0 & 0.0 & 1.0 & 2.0 & -8.0 & 24.0 \\ 0.0 & 0.0 & -2.0 & 1.0 & 0.0 & 2.0 & -4.0 & 8.0 \\ 0.0 & 1.0 & 1.25 & 0.0 & 0.0 & -0.5 & 1.5 & 2.0 \end{bmatrix}$$

Both the *LPSolve* output and the final tableau show the solution is $Z = 280$ when $d = 2$, $t = 0$, $c = 8$. The slack variables at the optimal point are $s_1 = 24$, and $s_2 = s_3 = 0$. The importance of the initial and final tableaux is that they provide information about the variables needed in the sensitivity analysis.

Change in a Coefficient of a Non-Basic Variable.

Reduce the analysis to formula form by defining

$$\overline{c}_j = c_{BV} B^{-1} a_j - c_j.$$

For the current solution basis to remain the optimal basis, all non-basic variables coefficients must remain greater than or equal to zero in the final tableau (for a maximization problem). Thus, $\overline{c}_j \geq 0$.

```
> cBV := ⟨0 | 20 | 60⟩ :
```

```
> B := InitialTableau[[2.. -1], [5, 4, 2]];
```

$$B := \begin{bmatrix} 1 & 1 & 8 \\ 0 & 1.5 & 4 \\ 0 & 0.5 & 2 \end{bmatrix}$$

```
> a2 := InitialTableau[[2.. -1], 3];
```

$$a_2 := \begin{bmatrix} 6 \\ 2 \\ 1.5 \end{bmatrix}$$

```
> c2 := 30 + δ;
```

```
> solve(cBV . B^(-1) . a2 - c2 ≥ 0, {δ});
```

$$\{\delta \leq 5.\}$$

This result means that if the objective function coefficient of t that is currently 30 is less than $30+5 = 35$, then t will never become a basic variable. To verify this, change t's coefficient in the objective function to 34 and then to 36, and resolve the LP for each new value.

Change in a Coefficient of a Basic Variable.

First, we'll change the coefficient for d, and then the coefficient for c.

Change in d.

```
> cBV := ⟨0 | 20 | 60 + δ⟩ :
  cc := ⟨60 | 30 | 20⟩ :
```

```
> B := InitialTableau[[2.. -1], [5, 4, 2]];
  a2 := InitialTableau[[2.. -1], 3];    # t is in column 3
```

$$B := \begin{bmatrix} 1 & 1 & 8 \\ 0 & 1.5 & 4 \\ 0 & 0.5 & 2 \end{bmatrix}$$

$$a_2 := \begin{bmatrix} 6 \\ 2 \\ 1.5 \end{bmatrix}$$

$> CBi := fnormal\left(c_{BV}.B^{-1}\right);$
$cbi := fnormal\left(\{CBi_1 \geq 0, CBi_2 \geq 0, CBi_3 \geq 0\}\right);$

$$CBi := \begin{bmatrix} 0. & 10.0 + 2.000000000\,\delta & 10.0 - 4.000000000\,\delta \end{bmatrix}$$

$$cbi := \{0. \leq 0., 0. \leq 10.0 - 4.000000000\,\delta, 0. \leq 10.0 + 2.000000000\,\delta\}$$

$> Cbar := \{c_{BV} \,.\, B^{-1} \,.\, a_2 - cc_2 \geq 0\};$

$$Cbar := \{0 \leq 5. + 1.25000000000000\,\delta\}$$

$> solve(cbi$ **union** $Cbar, \{\delta\});$

$$\{-4.000000000 \leq \delta, \delta \leq 20.00000000\}$$

Interpreting these results, we find that if the value of d stays between $56 = 60 - 4$ and $80 = 60 + 20$, then d will remain a basic variable.

Change in c.

Continue the computations.

$> c_{BV} := \langle 0 \,|\, 20 + \delta \,|\, 60 \rangle :$

$> CBi := c_{BV}.B^{-1};$
$cbi := fnormal(\{CBi_1 \geq 0, CBi_2 \geq 0, CBi_3 \geq 0\});$

$$CBi := \begin{bmatrix} 0.0 & 10.0 + 2.00000000000000\,\delta & 10.0 - 4.00000000000000\,\delta \end{bmatrix}$$

$$cbi := \{0 \leq 0.0, 0 \leq 10.0 - 4.000000000\,\delta, 0 \leq 10.0 + 2.000000000\,\delta\}$$

$> Cbar := \{c_{BV} \,.\, B^{-1} \,.\, a_2 - cc_2 \geq 0\};$

$$Cbar := \{0 \leq 5. - 2.00000000000000\delta\}$$

$> solve(cbi$ **union** $Cbar, \{\delta\});$

$$\{-5.000000000 \leq \delta, \delta \leq 2.500000000\}$$

Interpreting these results, we find that if the value of c stays between $15 = 20 - 5$ and $22.5 = 20 + 2.5$, then c will remain a basic variable.

Change in a RHS value.

We will test with a change in b_1, which was originally 48.

$> b\delta := \langle 48 + \delta, 20, 8 \rangle :$

$> rhs1 := B^{-1} \,.\, b\delta;$

$$rhs1 := \begin{bmatrix} 24.0 + 1.0\,\delta \\ 8.0 \\ 2.0 \end{bmatrix}$$

$> solve(rhs1_1 \geq 0, \{\delta\});$

$$\{-24. \leq \delta\}$$

This result is interpreted as follows: if the resource of Constraint 1, currently at 48 units, is at least 24, then both d and c remain basic variables.

Analysis of the other RHS coefficients is left as an exercise.

Exercises

Perform sensitivity analysis using Maple for Exercises 1 to 5 from Section 4.3 (pg. 180).

4.6 Integer and Mixed Integer Problems with Maple

In many real-life situations, fractional answers are not possible. Farmers do not sell half of a live steer at one time, waiting to sell the other half. Locating a new warehouse isn't realistically spread over several countries, scheduling airline crews to cover flights, and so forth. In these instances, only integer values are allowed, hence the name *integer programming.* At times some decision variables may be required to be integers, while others aren't—this type of problem is called a *mixed integer program.*

4.6.1 Examples for Integer, Binary Integer, and Mixed-Integer problems.

Example 4.14. Revisiting the Memory Chip Problem (pg. 151).
Let's assume that we must modify the first constraint in this problem for available assembly time for Memory Chip B from 4 hours to 3 hours. Further, we assume we want to have integer results; we can only produce whole numbers of memory chips.

First, we re-solve the LP as a linear program in order to compare the solutions.

Maximize profit $Z = 150x_1 + 130x_2$
subject to:

$$2x_1 + 3x_2 \leq 1500 \quad \text{(assembly time)}$$
$$4x_1 + 3x_2 \leq 1600 \quad \text{(installation time)}$$
$$x_1 \geq 0, x_2 \geq 0$$

```
> with(Optimization) :
```

> $Objective := 150 \cdot x_1 + 130 \cdot x_2;$

$$Objective := 150x_1 + 130x_2;$$

> $Constraints := \{2 \cdot x_1 + 3 \cdot x_2 \leq 1500, 4 \cdot x_1 + 3 \cdot x_2 \leq 1600\};$

$$Constraints := \{2x_1 + 3x_2 \leq 1500, 4x_1 + 3x_2 \leq 1600\}$$

> $LPSolve(Objective, Constraints, maximize, assume = nonnegative);$

$$[68166.6666666667, [x_1 = 49.9999999999998, x_2 = 466.666666666667]]$$

The output from *LPSolve* shows that our LP's solution is $Z = 68166.66$ when $x_1 = 49.9999$ and $x_2 = 466.6666$. Our solution is not an integer solution, so we add *integer* to the *assume* options, as shown below. We also have to add a *depthlimit* because the default solution search will not terminate.

> $LPSolve(Objective, Constraints, maximize,$
> $assume = \{nonnegative, integer\}, depthlimit = 10);$

$$[68120, [x_1 = 52, x_2 = 464]]$$

Our integer solution $(52, 464)$ is certainly not rounded from the non-integer solution $(49.99, 466.67)$. We now find $Z = 68120$ when $x_1 = 52$ and $x_2 = 464$, a little bit less due to the integer restriction, but not badly so.

Example 4.15. Emergency Services Ambulance Location.
The Emergency Service Coordinator (ESC) for a county is interested in locating the county's three ambulances to maximize the number of residents that can be reached within 8 minutes in emergency situations.[3] The county is divided into 6 zones; the average times required to travel from one region to the next under semi-perfect conditions are summarized in Table 4.10.

TABLE 4.10: Average Travel Times from Zone i to Zone j in Semi-Perfect Conditions

From: \ To:	Zone 1	Zone 2	Zone 3	Zone 4	Zone 5	Zone 6
Zone 1	1	8	12	14	10	16
Zone 2	8	1	6	18	16	16
Zone 3	12	18	1.5	12	6	4
Zone 4	16	14	4	1	16	12
Zone 5	18	16	10	4	2	2
Zone 6	16	18	4	12	2	2

The population (in thousands) in each zone is given in Table 4.11.

[3] Adapted from W. P. Fox, *Mathematical Modeling for Business Analytics* [Fox2017].

TABLE 4.11: Population in Each Zone (in thousands)

	Zone 1	Zone 2	Zone 3	Zone 4	Zone 5	Zone 6	Total
Population (1000s)	50	80	30	55	35	20	270

Understanding the Decision and Problem: We want better coverage and to improve the ability to take care of patients requiring an ambulance to go to a hospital. The objective is to determine the location for placement of the ambulances to maximize coverage within the predetermined allotted time.

Assumptions: We initially assume that time travel between zones is negligible. We further assume that the times in the data are averages under nearly ideal circumstances.

Solution: We assume that due to the nature of the problem, a facility location problem, we should decide to employ integer programming to solve the problem.
Decision Variables:

$$x_i = \begin{cases} 1 & \text{if ambulance is located on Zone } i \\ 0 & \text{if not} \end{cases}$$

$$y_i = \begin{cases} 1 & \text{if Zone } i \text{ is covered by an ambulance} \\ 0 & \text{if not} \end{cases}$$

We will let the parameter m be the number of ambulances available, even though the problem stated 3, to get a comprehensive analysis.

$$m = \text{ number of ambulances}$$

The way we have defined the decision variables as 1 or 0 is called *binary integer* values. Our goal is to maximize the covered population. It's good practice to scale large values in computations, so we will use 1000's as the unit of population as was given in Table 4.11.

The Complete LP Formulation:

Maximize $Z = 50y_1 + 80y_2 + 30y_3 + 55y_4 + 35y_5 + 20y_6$
subject to:

$$\begin{aligned} x_1 + x_2 &\geq y_1 \\ x_1 + x_2 + x_3 &\geq y_2 \\ x_3 + x_5 + x_6 &\geq y_3 \\ x_3 + x_4 + x_6 &\geq y_4 \\ x_4 + x_5 + x_6 &\geq y_5 \\ x_3 + x_5 + x_6 &\geq y_6 \end{aligned}$$

and

$$m = x_1 + x_2 + x_3 + x_4 + x_5 + x_6$$

Solve the problem using Maple.

> $with(Optimization):$

> $Objective := 50 \cdot y_1 + 80 \cdot y_2 + 30 \cdot y_3 + 55 \cdot y_4 + 35 \cdot y_5 + 20 \cdot y_6 :$

> $Constraints := \{x_1 + x_2 + x_3 + x_4 + x_5 + x_6 = m,$
> $x_1 + x_2 - y_1 \geq 0, x_1 + x_2 + x_3 - y_2 \geq 0, x_3 + x_5 + x_6 - y_3 \geq 0,$
> $x_3 + x_4 + x_6 - y_4 \geq 0, x_4 + x_5 + x_6 - y_5 \geq 0, x_3 + x_5 + x_6 - y_6 \geq 0\} :$

We'll display the constraints as a vector to make the output much easier to check for errors.

> $\langle op(Constraints) \rangle;$

$$\begin{bmatrix} x_1 + x_2 + x_3 + x_4 + x_5 + x_6 = m \\ 0 \leq x_1 + x_2 - y_1 \\ 0 \leq x_1 + x_2 + x_3 - y_2 \\ 0 \leq x_3 + x_5 + x_6 - y_3 \\ 0 \leq x_3 + x_4 + x_6 - y_4 \\ 0 \leq x_4 + x_5 + x_6 - y_5 \\ 0 \leq x_3 + x_5 + x_6 - y_6 \end{bmatrix}$$

Now it's time to use *LPSolve* with each of the values for the number of ambulances $m = 1$, 2, and 3.

```
> m := 3 :
  LPSolve(Objective, Constraints, maximize,
    assume = {binary, nonnegative});
```
$$[270, [x_1 = 0, x_2 = 1, x_3 = 0, x_4 = 1, x_5 = 1, x_6 = 0, y_1 = 1, y_2 = 1, y_3 = 1, y_4 = 1, y_5 = 1, y_6 = 1]]$$

```
> m := 2 :
  LPSolve(Objective, Constraints, maximize,
    assume = {binary, nonnegative});
```
$$[270, [x_1 = 1, x_2 = 0, x_3 = 0, x_4 = 0, x_5 = 0, x_6 = 1, y_1 = 1, y_2 = 1, y_3 = 1, y_4 = 1, y_5 = 1, y_6 = 1]]$$

```
> m := 1 :
  LPSolve(Objective, Constraints, maximize,
    assume = {binary, nonnegative});
```
$$[185, [x_1 = 0, x_2 = 0, x_3 = 1, x_4 = 0, x_5 = 0, x_6 = 0, y_1 = 0, y_2 = 1, y_3 = 1, y_4 = 1, y_5 = 0, y_6 = 1]]$$

Solution and Analysis: We find we can cover all 270,000 potential patients with three ambulances posted in Zones 2, 4, and 5. We can cover all 270,000 potential patients with only two ambulances posted in Zones 1 and 6. If we only had one ambulance, we can cover at most 185,000 potential patients with the ambulance located in Zone 3, leaving 85,000 people in Zones 1 and 5 not covered. We have provided the ESC with several options that meet demand. They might use the option that is the least costly, or covered the most people at least cost.

Example 4.16. Optimal Path to Transport Hazardous Material.
The Federal Emergency and Management Agency (FEMA) is requesting a two-part analysis. FEMA is concerned about the transportation of nuclear waste from the Savannah River Nuclear Plant to the appropriate disposal site. After the route is found, FEMA requests analysis as to the location and composition of clean-up sites. Figure 4.5 shows an abstract map of the routes with accident probablities.

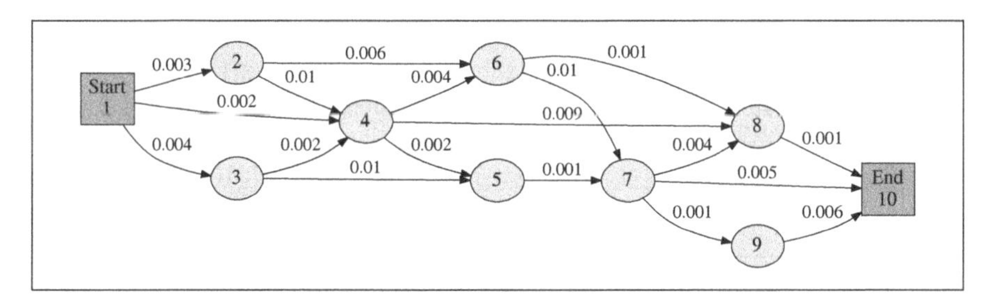

FIGURE 4.5: Available Transport Routes (Abstracted) with Accident Probabilities

In this example, we only discuss the optimal path portion of the model using generic data in Figure 4.5.

Consider a model whose requirement is to find the route from Node 1 to Node 10 that minimizes the probability of a vehicle accident. A primary concern is the I-95 and I-20 corridor where both interstate highways meet and converge in Florence, SC.

To simplify the problem using technology, we transform the model to maximize the probability of not having an accident after using *LPSolve.* If $p_{i,j}$ is the probability of an accident on the route from Node i to Node j, then $(1 - p_{i,j})$ is the probability of not having an accident. We also assume that the probabilities of an accident on the different routes are independent of each other. Let the decision variable $x_{i,j}$ be 1 if the route $i \to j$ is taken, and 0 otherwise.

The Complete LP Formulation:

$$\text{Minimize } Z = \sum_{\text{all routes } i \to j} p_{i,j} x_{i,j}$$

subject to:

$$\begin{aligned}
-x_{1,2} - x_{1,3} - x_{1,4} &= -1 && \textit{(Node 1, Start: only flow out)}\\
x_{1,2} - x_{2,4} - x_{2,6} &= 0 && \textit{(Node 2: flow in − flow out)}\\
x_{1,3} - x_{3,4} - x_{3,5} &= 0 && \textit{(Node 3: flow in − flow out)}\\
x_{1,4} + x_{2,4} + x_{3,4} - x_{4,5} - x_{4,6} - x_{4,8} &= 0\\
x_{3,5} + x_{4,5} - x_{5,7} &= 0\\
x_{2,6} + x_{4,6} - x_{6,7} - x_{6,8} &= 0\\
x_{5,7} + x_{6,7} - x_{7,8} - x_{7,10} &= 0\\
x_{4,8} + x_{6,8} + x_{7,8} - x_{8,10} &= 0\\
x_{7,9} - x_{9,10} &= 0\\
x_{7,10} + x_{8,10} + x_{9,10} &= 1 && \textit{(Node 10. End: only flow in)}
\end{aligned}$$

Using LPSolve: To make it easier to enter the terms, we'll use bracket notation for the subscripts.

> *Objective* := $p[1,2] \cdot x[1,2] + p[1,3] \cdot x[1,3] + p[1,4] \cdot x[1,4] + p[2,4] \cdot x[2,4] + p[2,6] \cdot x[2,6] + p[3,4] \cdot x[3,4] + p[3,5] \cdot x[3,5] + p[4,5] \cdot x[4,5] + p[4,6] \cdot x[4,6] + p[4,8] \cdot x[4,8] + p[5,7] \cdot x[5,7] + p[6,7] \cdot x[6,7] + p[6,8] \cdot x[6,8] + p[7,8] \cdot x[7,8] + p[7,9] \cdot x[7,9] + p[7,10] \cdot x[7,10] + p[8,10] \cdot x[8,10] + p[9,10] \cdot x[9,10]$:

[3] Adapted from W. P. Fox, *Mathematical Modeling for Business Analytics* [Fox2017].

```
> Constraints := {
  - x[1,2] - x[1,3] - x[1,4] = -1,
    x[1,2] - x[2,4] - x[2,6] = 0,
    x[1,3] - x[3,4] - x[3,5] = 0,
    x[1,4] + x[2,4] + x[3,4] - x[4,5] - x[4,6] - x[4,8] = 0,
    x[3,5] + x[4,5] - x[5,7] = 0,
    x[2,6] + x[4,6] - x[6,7] - x[6,8] = 0,
    x[5,7] + x[6,7] - x[7,8] - x[7,10] = 0,
    x[4,8] + x[6,8] + x[7,8] - x[8,10] = 0,
    x[7,9] - x[9,10] = 0
    x[7,10] + x[8,10] + x[9,10] = 1} :
> Soln := LPSolve(Objective, Constraints,
     assume = {binary, nonnegative});
```

$Soln := [0.008, [x_{1,2} = 0, x_{1,3} = 0, x_{1,4} = 1, x_{2,4} = 0, x_{2,6} = 0, x_{3,4} = 0, x_{3,5} = 0, x_{4,5} = 0, x_{4,6} = 1, x_{4,8} = 0, x_{5,7} = 0, x_{6,7} = 0, x_{6,8} = 1, x_{7,8} = 0, x_{7,9} = 0, x_{7,10} = 0, x_{8,10} = 1, x_{9,10} = 0]]$

We *select* the x's equal to 1 in *Soln* to make it easier to see the route.

```
> Route := select(X → (rhs(X) = 1), Soln[2])
```

$Route := [x[1,4] = 1, x[4,6] = 1, x[6,8] = 1, x[8,10] = 1]$

Our safest path is from $1 \to 4 \to 6 \to 8 \to 10$. The minimal accident probability is 0.008 giving the probability of no accident as $1 - 0.008 = 0.992$.

Example 4.17. Memory Chip Problem with Mixed Integer Solution. Let's assume that we must modify Constraint 1 in this problem for the assembly time of Memory Chip B from 4 hours to 3 hours. Further, we assume we only want to have x_1 as an integer, but will not require x_2 to be an integer. Recall the LP:

Maximize profit $Z = 150x_1 + 130x_2$
subject to:

$$
\begin{aligned}
2x_1 + 3x_2 &\leq 1500 \quad \text{(assembly time)}\\
4x_1 + 3x_2 &\leq 1600 \quad \text{(installation time)}\\
x_1 \geq 0, x_2 &\geq 0
\end{aligned}
$$

```
> with(Optimization) :
> Objective := 150 · x1 + 130 · x2;
> Constraints := {2 · x1 + 3 · x2 ≤ 1500, 4 · x1 + 3 · x2 ≤ 1600};
> LPSolve(Objective, Constraints, maximize, assume = nonnegative,
  integervariables = [x1]);
```

$[68166.6666666667, [x[1] = 50, x[2] = 466.666666666667]]$

We read the solution as $Z = \$68{,}166.66$ when $x_1 = 50$ and $x_2 = 466.66$.

Compare this result to the previous unrestricted solution $Z = \$68{,}166.66$ when $x_1 = 49.9999$ and $x_2 = 466.6666$, and the integer only solution $Z = \$68{,}120$ when $x_1 = 52$ and $x_2 = 464$.

Case Studies

Case 4.1. Modeling of Ranking Units Using Data Envelopment Analysis (DEA).

Data envelopment analysis (DEA), also called *frontier analysis*, was invented by Charnes, Cooper and Rhodes in 1978 using ideas on relative efficiency from combustion engineering. It is a relative performance measurement technique which, as we shall see, can be used for evaluating the *relative efficiency* of *decision-making units* (DMU's) in organizations. Here a DMU is a distinct unit within an organization that has flexibility with respect to some of the decisions it makes, but not necessarily complete freedom with respect to these decisions.

Examples of organizations to which DEA has been applied are: banks, police stations, hospitals, tax offices, prisons, defense bases (Army, Navy, Air Force), schools and university departments. Note here that one advantage of DEA is that it can be applied to non-profit making organizations.

Example 4.18. Manufacturing.

Consider the following manufacturing process (modified from Trick, 2014) where we have three DMUs each of which has 2 inputs and 3 outputs as shown in Table 4.12.

TABLE 4.12: DEA Input Output Data Table

	Input #1	Input #2	Output #1	Output #2	Output #3
DMU 1	5	14	9	4	16
DMU 2	8	15	5	7	10
DMU 3	7	12	4	9	13

No units are given and the scales are similar, so we decide not to normalize the data. Define the following decision variables:

$$t_i = \text{ value of a single unit of output of DMU } i$$
$$w_i = \text{ cost or weight for one unit of inputs of DMU } i$$
$$\textit{efficiency}_i = DMU_i = \frac{(\text{total value of DMU } i\text{'s outputs})}{(\text{total cost of DMU } i\text{'s inputs})}$$

all for $i = 1, 2, 3$.

We make the following modeling assumptions:

(a) No DMU will have an efficiency greater than 100%.
(b) If any efficiency is less than 1, then it is inefficient.
(c) We will scale the costs so that the costs of the inputs equals 1 for each linear program. For example, we will use $5w_1 + 14w_2 = 1$ in our program for DMU_1.
(d) All values and weights must be strictly positive, so, if needed, we use a small constant such as 0.0001 with $x > 0.0001$ in lieu of $x \geq 0$.

To calculate the efficiency of DMU_1, we define the linear program as

Maximize $DMU_1 = 9t_1 + 4t_2 + 16t_3$
subject to:

$$\begin{aligned}
-9t_1 - 4t_2 - 16t_3 + 5w_1 + 14w_2 &\geq 0\\
-5t_1 - 7t_2 - 10t_3 + 8w_1 + 15w_2 &\geq 0\\
-4t_1 - 9t_2 - 13t_3 + 7w_1 + 12w_2 &\geq 0\\
5w_1 + 14w_2 &= 1
\end{aligned}$$

To calculate the efficiency of DMU_2, we define the linear program as

Maximize $DMU_2 = 5t_1 + 7t_2 + 10t_3$
subject to:

$$\begin{aligned}
-9t_1 - 4t_2 - 16t_3 + 5w_1 + 14w_2 &\geq 0\\
-5t_1 - 7t_2 - 10t_3 + 8w_1 + 15w_2 &\geq 0\\
-4t_1 - 9t_2 - 13t_3 + 7w_1 + 12w_2 &\geq 0\\
8w_1 + 15w_2 &= 1
\end{aligned}$$

To calculate the efficiency of DMU_3, we define the linear program as

Maximize $DMU_1 = 4t_1 + 9t_2 + 13t_3$
subject to:

$$\begin{aligned}
-9t_1 - 4t_2 - 16t_3 + 5w_1 + 14w_2 &\geq 0\\
-5t_1 - 7t_2 - 10t_3 + 8w_1 + 15w_2 &\geq 0\\
-4t_1 - 9t_2 - 13t_3 + 7w_1 + 12w_2 &\geq 0\\
7w_1 + 12w_2 &= 1
\end{aligned}$$

Note: the objective functions and the cost scaling inequalities are the only changes between the three LPs.

Use *LPSolve* to obtain the efficiency ratings.

```
> with(Optimization):
> NumDMU, NumInputs, NumOutputs := 3, 2, 3 :
> W := Vector(NumInputs, symbol = w);
  T := Vector(NumOutputs, symbol = t);
```

$$W := \begin{bmatrix} w_1 \\ w_2 \end{bmatrix}$$

$$T := \begin{bmatrix} t_1 \\ t_2 \\ t_3 \end{bmatrix}$$

```
> Inputs := Matrix([[5, 14], [8, 15], [7, 12]]);
  Outputs := Matrix([[9, 4, 16], [5, 7, 10], [4, 9, 13]]);
```

$$Inputs := \begin{bmatrix} 5 & 14 \\ 8 & 15 \\ 7 & 12 \end{bmatrix}$$

$$Outputs := \begin{bmatrix} 9 & 4 & 16 \\ 5 & 7 & 10 \\ 4 & 9 & 13 \end{bmatrix}$$

```
> Constraints := seq((Inputs . W − Outputs . T)_i ≥ 0, i = 1..3);
```

$$Constraints := 0 \leq 5w_1 + 14w_2 - 9t_1 - 4t_1 - 16t_3, 0 \leq 8w_1 + 15w_2 - 5t_1 - 7t_2 - 10t_3, 0 \leq 7w_1 + 12w_2 - 4t_1 - 9t_2 - 13t_3$$

```
> ConsDMU := Inputs.W;
```

$$ConsDMU := \begin{bmatrix} 5\,w_1 + 14\,w_2 \\ 8\,w_1 + 15\,w_2 \\ 7\,w_1 + 12\,w_2 \end{bmatrix}$$

```
> ObjDMU := Outputs.T;
```

$$ObjDMU := \begin{bmatrix} 9\,t_1 + 4\,t_2 + 16\,t_3 \\ 5\,t_1 + 7\,t_2 + 10\,t_3 \\ 4\,t_1 + 9\,t_2 + 13\,t_3 \end{bmatrix}$$

> $LPSolve(ObjDMU_1, \{Constraints, ConsDMU_1 = 1\}, maximize,$
> $assume = nonnegative);$

$[1.00000000000000, [t_1 = 0., t_2 = 0., t_1 = 0.0625000000000000, w_1 = 0.,$
$w_2 = 0.0714285714285714]]$

> $LPSolve(ObjDMU_2, \{Constraints, ConsDMU_2 = 1\}, maximize,$
> $assume = nonnegative);$

$[0.773333333333333, [t_1 = 0.0800000000000000, t_2 = 0.0533333333333333,$
$t_3 = 0., w_1 = 0., w_2 = 0.0666666666666667]]$

> $LPSolve(ObjDMU_3, \{Constraints, ConsDMU_3 = 1\}, maximize,$
> $assume = nonnegative);$

$[1., [t_1 = 0., t_2 = 0.00905797101449278, t_3 = 0.0706521739130435, w_1 = 0.,$
$w_2 = 0.0833333333333333]]$

The linear programming solutions show the relative efficiencies as $DMU_1 = DMU_3 = 1$, and $DMU_2 = 77.33\%$.

Interpretation: DMU_2 is operating at 77.33% of the efficiency of DMU_1 and DMU_3. Management could concentrate some improvements or best practices from DMU_1 or DMU_3 to help DMU_2 improve.

The dual analysis is conducted after the efficiency rankings of all DMUs have been determined. The Maple commands for solving the dual LP for DMU_2 follow.

> $with(simplex):$

Maple's *dual* cannot handle equality constraints, so we replace $8w_1 + 15w_2 = 1$ with an equivalent pair of inequalities $8w_1 + 15w_2 \leq 1$, $8w_1 + 15w_2 \geq 1$.

> $theDual := dual(DMUObj_2, \{Constraints, DMUCons_2 \geq 1,$
> $DMUCons_2 \leq 1\}, s);$

> $DualSoln := LPSolve(theDual, assume = nonnegative);$

$DualSoln := [0.773333333264493, [s1 = 0., s2 = 0.773333333264493, s3 = 0.,$
$s4 = 0.661538461538462, s5 = 0.261538461538461]]$

An examination of the dual prices (shadow prices) for the linear program of DMU_2 yields $s_5 = 0.261538$, $s_3 = 0$, and $s_4 = 0.661538$. Then, the average output vector for DMU_2 can be written as

$$0.261538 \begin{bmatrix} 9 \\ 4 \\ 16 \end{bmatrix} + 0.0 \begin{bmatrix} 5 \\ 7 \\ 10 \end{bmatrix} + 0.661538 \begin{bmatrix} 4 \\ 9 \\ 13 \end{bmatrix} = \begin{bmatrix} 5.0 \\ 7.0 \\ 12.785 \end{bmatrix},$$

and the average input vector can be written as

$$0.261538 \begin{bmatrix} 5 \\ 14 \end{bmatrix} + 0.0 \begin{bmatrix} 8 \\ 15 \end{bmatrix} + 0.661538 \begin{bmatrix} 7 \\ 12 \end{bmatrix} = \begin{bmatrix} 5.938 \\ 11.60 \end{bmatrix}.$$

Our data for DMU_2's outputs are $[5, 7, 10]$ units. Thus, we may clearly see the inefficiency is in Output 3 where 12.785 units are required on average. We find that they are short 2.785 units $(12.785 - 10 = 2.785)$. This helps focus on treating the inefficiency found for Output 3.

Sensitivity Analysis: Sensitivity analysis in a linear program is sometimes referred to as "what if" analysis. Let's assume that without management engaging some additional training for DMU_2 that DMU_2's Output 3 dips from 10 to 9 units, while the Input 2 increases from 15 to 16 hours. We find that these changes in the *technology coefficients* are easily handled in resolving the LPs. Since only DMU_2 is affected, we might only modify and solve the LP concerning DMU_2. We find with these changes that DMU_2's efficiency is down by 3.1%, now only 74.23% as effective as DMU_1 and DMU_3.

This analysis invokes issues of highlighting and disseminating examples of best practices. Equally important are the issues relating to identification of poor practices.

In DEA, the concept of the reference set can be used to identify the best performing branches with which to compare poorly performing branches. If you use this procedure, use it wisely.

Case 4.2. Supply Chain Operations.[4]

In this case study, we present a linear program for a supply chain design. We consider producing a new mixture of gasoline, and desire to minimize the total cost of manufacturing and distributing the new formula. There is a supply chain involved with a product that must be modeled. The product is made up of components listed in Table 4.13 that are produced separately. Demand information is given in Table 4.14.

TABLE 4.13: Production Components

Crude Oil Type	A (%)	B (%)	C (%)	Cost per Barrel	Barrels Available (1000's)
X10	35	25	35	$26	15,000
X20	50	30	15	$32	32,000
X30	60	20	15	$55	24,000

Let i = crude type 1, 2, 3 (X10, X20, X30, respectively)
Let j = gasoline type 1, 2, 3 (Premium, Super, Regular. respectively)

[4] Adapted from William P. Fox and Fausto P. Garcia (March 6th 2013). *Modeling and Linear Programming in Engineering Management* [FoxGarcia2013].

TABLE 4.14: Mixture and Demand Information

Gasoline	Compound A (%)	Compound B (%)	Compound C (%)	Expected Demand (1000's)
Premium	≥ 55	≤ 23		14,000
Super		≥ 25	≤ 35	22,000
Regular	≥ 40		≤ 25	25,000

We define the following decision variables:

G_{ij} = amount of crude type i used to produce gasoline type j, for i, $j = 1, 2, 3$.

For example,

$$G_{11} = \text{ amount of crude X10 used to produce Premium gasoline}$$
$$G_{23} = \text{ amount of crude X20 used to produce Regular gasoline}$$

and so forth. We are ready for the
Complete LP Formulation:
Minimize $Cost = \$86(G_{11} + G_{21} + G_{31}) + \$92(G_{12} + G_{22} + G_{32}) + \$95(G_{13} + G_{23} + G_{33})$
subject to:
Gasoline Demand:

$$G_{11} + G_{21} + G_{31} \geq 14000 \quad \text{(Premium)}$$
$$G_{12} + G_{22} + G_{32} \geq 22000 \quad \text{(Super)}$$
$$G_{13} + G_{23} + G_{33} \geq 25000 \quad \text{(Regular)}$$

Availability of Components:

$$G_{11} + G_{12} + G_{13} \leq 15000 \quad \text{(X10)}$$
$$G_{21} + G_{22} + G_{23} \leq 32000 \quad \text{(X20)}$$
$$G_{31} + G_{32} + G_{33} \leq 24000 \quad \text{(X30)}$$

Mixture Format:

$$(0.35G_{11} + 0.50G_{21} + 0.60G_{31}) \geq 0.55 \cdot (G_{11} + G_{21} + G_{31}) \text{ (X10 in Premium)}$$
$$(0.25G_{11} + 0.30G_{21} + 0.20G_{31}) \leq 0.23 \cdot (G_{11} + G_{21} + G_{31}) \text{ (X20 in Premium)}$$
$$(0.25G_{12} + 0.30G_{22} + 0.20G_{32}) \geq 0.25 \cdot (G_{12} + G_{22} + G_{32}) \text{ (X20 in Super)}$$
$$(0.35G_{12} + 0.15G_{22} + 0.15G_{32}) \leq 0.35 \cdot (G_{12} + G_{22} + G_{32}) \text{ (X30 in Super)}$$
$$(0.35G_{13} + 0.50G_{23} + 0.60G_{33}) \geq 0.40 \cdot (G_{13} + G_{23} + G_{33}) \text{ (X10 in Regular)}$$
$$(0.35G_{13} + 0.15G_{23} + 0.15G_{33}) \leq 0.25 \cdot (G_{13} + G_{23} + G_{33}) \text{ (X30 in Regular)}$$

Maple's *LPSolve* returns

```
> LPSolve(Cost, Constraints, assume = [nonnegative, integer]);
  [5603000, [G11 = 2800, G12 = 0, G13 = 12200, G21 = 0, G22 = 19200,
  G23 = 12800, G31 = 11200, G32 = 2800, G33 = 0]]
```

which gives the result minimum $Cost = \$5{,}603{,}000$ at

$$\begin{aligned} G_{11} &= 2{,}800 \\ G_{12} &= 0 \\ G_{13} &= 12{,}200 \\ G_{21} &= 0 \\ G_{22} &= 19{,}200 \\ G_{23} &= 12{,}800 \\ G_{31} &= 11{,}200 \\ G_{32} &= 2{,}800 \\ G_{33} &= 0 \end{aligned}$$

Case 4.3. Raleigh Recruiting Office (adapted from McGrath, 2007). Although this is a simple model, it was adopted by the US Army Recruiting Command for operations. The model determines the optimal mix of prospecting strategies that a recruiter should use in a given week. The two prospecting strategies initially modeled and analyzed are phone and email prospecting. The data came from the Raleigh Recruiting Company United States Army Recruiting Command in 2006. On average, each phone lead yields 0.041 enlistments and each email lead yields 0.142 enlistments. The forty recruiters assigned to the Raleigh Recruiting Office prospected a combined 19,200 minutes of work per week via phone and email. The company's weekly budget is $60,000. Table 4.15 shows the relevant recruiting data.

TABLE 4.15: Recruiting Data

	Phone per lead	Email per lead
Prospecting Time	60 min	1 min
Budget	$10	$37

The decision variables are

$$\begin{aligned} x_1 &= \text{Number of phone leads} \\ x_2 &= \text{Number of email leads} \end{aligned}$$

Then the LP is

Maximize $Z = 0.041x_1 + 0.142x_2$
subject to:

$$\begin{aligned} 60x_1 + x_2 &\le 19200 \quad \text{(Prospecting minutes available)} \\ 10x_1 + 37x_2 &\le 60000 \quad \text{(Budget dollars available)} \\ x_1, x_2 &\ge 0 \end{aligned}$$

Solving with Maple gives

```
> with(Optimization) :
  with(simplex) :
> Objective := 0.041x1 + 0.142x2 :
  Constraints := {60x1 + x2 ≤ 19200, 10x1 + 37x2 ≤ 60000} :
> LPSolve(Objective, Constraints, maximize, assume = nonnegative);
  [231.041809954751, [x1 = 294.298642533937, x2 = 1542.08144796380]]
```

We find an optimal point, $x_1 = 294.3$, $x_2 = 1542.1$, achieving 231.04 recruitments.

```
> LPSolve(dual(Objective, Constraints, s), assume = nonnegative);
        [231.041809954752, [s1 = 0.00383665158371041,
  s2 = 0.0000438914027149628]]
```

Clearly, sensitivity analysis is important. First, we see we maintain a mixed solution over a fairly large range of values for the coefficients of x_1 and x_2. Further, the shadow prices provide additional information. A one unit increase in prospecting minutes available yields an increase of approximately 0.000044 in recruits, while an increase in budget of $1 yields an additional 0.0038 recruits. At first look, it appears that increasing budget resources is best.
Let's assume that it cost only $0.01 for each additional prospecting minute. Thus we could get $100 \cdot 0.000044 = 0.0044$ increase in recruits for the same unit cost increase. In this case, we would be better off obtaining the additional prospecting minutes.
Although the problem is quite simple, linear programming gives us a good deal of insight into allocating resources and where increases may give the most benefit.

Exercises

1. The Rating Departments at college has provided the data in Table 4.16.

TABLE 4.16: College Ratings Data

DMU	Inputs	Outputs		
Dep't	No Faculty	Student Cr Hr	Num of Students	Total Degrees (MS & PhD)
Unit 1	25	18,341	9,086	63
Unit 2	15	8,190	4,049	23
Unit 3	10	2,857	1,255	31
Unit 4	33	22,277	6,102	31
Unit 5	12	6,830	2,910	19

Formulate and solve the DEA model, and rank order the five departments by relative efficiency.

2. In the Raleigh Recruiting Case Study, assume the data has been updated as shown in Table 4.17. Re-solve the LP and redo the sensitivity analysis.

TABLE 4.17: New Recruiting Data

	Phone per lead	Email per lead
Prospecting Time	45 min	1.5 min
Budget	$15	$42

3. Re-solve the Supply Chain Case Study with the following new data from Table 4.18. Table 4.19 repeats demand information from the case study.

TABLE 4.18: New Production Components

Crude Oil Type	A (%)	B (%)	C (%)	Cost per Barrel	Barrels Available (1000's)
X10	45	35	45	$26.50	18,000
X20	60	40	25	$32.85	35,000
X30	70	30	25	$55.97	26,000

TABLE 4.19: Mixture and Demand Information (Table 4.14 repeated)

Gasoline	Compound A (%)	Compound B (%)	Compound C (%)	Expected Demand (1000's)
Premium	≥ 55	≤ 23		14,000
Super		≥ 25	≤ 35	22,000
Regular	≥ 40		≤ 25	25,000

4. Consider ranking companies within a Task Force. As a simplification, we will consider only 6 companies.

TABLE 4.20: Companies Ratings Data

Companies	Inputs	Outputs		
	Size of Unit	Output 1	Output 2	Output 3
Unit 1	120	18,341	9,086	63
Unit 2	110	8,190	4,049	23
Unit 3	100	2,857	1,255	31
Unit 4	135	22,277	6,102	31
Unit 5	120	6,830	2,910	19
Unit 6	95	5,050	1,835	12

Further Reading

[Albright2014] Brian Albright, *Mathematical Modeling with Excel*, Jones & Bartlett Publishers, 2011.

[Apaiah2006] R. K. Apaiah, *Linear programming for supply chain design: A case on novel protein foods*, PhD thesis, Wageningen University, 2006.

[BRS2007] N. Balakrishnan, B. Render, and Stair R., *Managerial Decision Making*, 2nd ed., Prentice Hall, 2007.

[Bazaraa2011] Mokhtar S. Bazaraa, John J. Jarvis, and Hanif D. Sherali, *Linear Programming and Network Flows*, John Wiley & Sons, 2011.

[EK1988] Joseph G. Ecker, Michael Kupferschmid, et al., *Introduction to Operations Research*, John Wiley & Sons, New York, 1988.

[Fishback2009] Paul E. Fishback, *Linear and Nonlinear Programming with Maple: An Interactive, Applications-Based Approach*, Chapman and Hall/CRC, 2009.

[Fox2017] William P. Fox, *Mathematical Modeling for Business Analytics*, Chapman and Hall/CRC, 2017.

[FoxGarcia2013] William P. Fox and Fausto P. Garcia, *Modeling and Linear Programming, Engineering Management* (Fausto Pedro García Márquez and Benjamin Lev, eds.), vol. *Engineering Management*, InTech, 2013.

[GFH2013] Frank Giordano, William P. Fox, and Steven Horton, *A First Course in Mathematical Modeling*, 5th ed., Nelson Education, 2014.

[HL1990] Frederick S. Hillier and Gerald J. Lieberman, *Introduction to Mathematical Programming*, McGraw-Hill, 1990.

[McG2007] George F. McGrath III, *Email Marketing for US Army and Special Operations Forces (SOF) Recruiting*, Tech. Report, Naval Postgraduate School, Monterey, CA, 2007.

[Trick2014] M. A. Trick, *Data Envelopment Analysis*, 2014. Chapter 12 from: http://mat.gsia.cmu.edu/classes/QUANT/NOTES/chap12.pdf

[W1994] Wayne L. Winston. *Operations Research: Applications and Algorithms,* 3rd ed., Duxbury Press, 1994.

[WG2004] Wayne L. Winston and Jeffrey B. Goldberg, *Operations Research: Applications and Algorithms*, Vol. 3, Thomson Brooks/Cole Belmont, 2004.

[WVG2003] Wayne L. Winston, Munirpallam Venkataramanan, and Jeffrey B. Goldberg, *Introduction to Mathematical Programming*, Vol. 1, Thomson/Brooks/Cole Duxbury; Pacific Grove, CA, 2003.

5

Model Fitting and Linear Regression

Objectives

(1) Understand the concept of model fitting and curve fitting.

(2) Understand least squares regression, its model, and interpretation.

(3) Understand residuals and the information residuals and their plots give.

(4) Understand correlation and how it is used in the interpretation of regression.

5.1 Introduction

Consider a spring mass system such as the one in Figure 5.1.

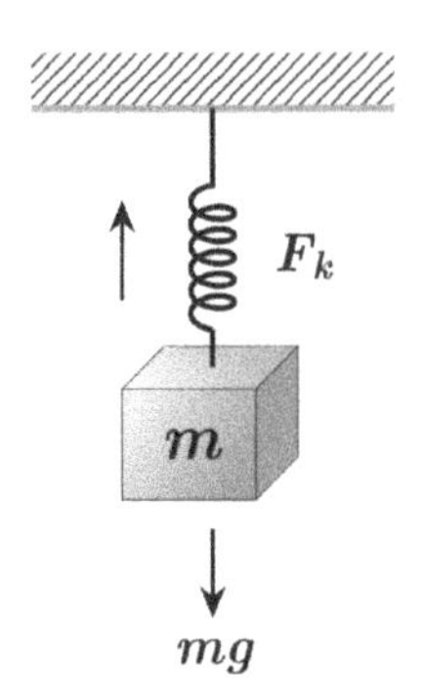

FIGURE 5.1: Spring-Mass System

We conducted an experiment to measure the stretch of the spring as a function of mass placed on the spring. We recorded only how far the spring stretched from its original position; the data collected is displayed in Table 5.1. The plotted data, seen in Figure 5.2, looks reasonably like a straight line through the origin. The image suggests we might want to fit a line using "least squares."

TABLE 5.1: Spring-Mass System Force (mass) v. Stretch (distance in cm)

Mass (g)	0	50	100	150	200	250
Distance (cm)	0	0.375	0.75	1.125	1.50	1.875

Mass (g)	300	350	400	450	500	550
Distance (cm)	2.25	2.625	3.00	3.375	3.75	4.125

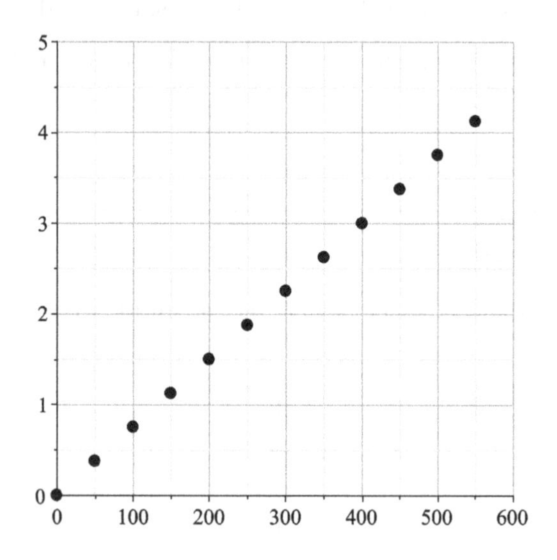

FIGURE 5.2: Scatter Plot of Spring-Mass Data

Correlation

Correlation is a measure of the linear relationship between variables. The correlation coefficient, also called *Pearson's* r for data-based variables, between X and Y, denoted as ρ_{XY}, is

$$\rho_{XY} = \frac{\text{cov}(X, Y)}{\sigma_X \sigma_Y} = \frac{E[XY] - E[X]E[Y]}{\sigma_X \sigma_Y} \tag{5.1}$$

The values of correlation range from -1 to $+1$. A value of $+1$ corresponds to a perfect line with a positive slope, and a value of -1 corresponds to a perfect line with a negative slope. A value of 0, or close to zero, indicates that there is no linear relationship between the variables.

For samples $X = \{x_1, x_2, \ldots, x_n\}$ and $Y = \{y_1, y_2, \ldots, y_n\}$, we often use $\rho_{XY} = r$ and calculate

$$r = \frac{\sum\left((x_i - \bar{x}) \cdot (y_i - \bar{y})\right)}{\sqrt{\sum(x_i - \bar{x})^2} \cdot \sqrt{\sum(y_i - \bar{y})^2}} \tag{5.2}$$

where $\bar{x}$ and $\bar{y}$ are the means, respectively.

We present two rules of thumb for correlation from the literature. First, from Devore [Devore2011], for math, science, and engineering data, we have the following:

$0.8 < \lvert\rho\rvert \leq 1.0$	Strong linear relationship
$0.5 < \lvert\rho\rvert \leq 0.8$	Moderate linear relationship
$0 \leq \lvert\rho\rvert \leq 0.5$	Weak linear relationship

According to Johnson [Johnson2012] for non-math, non-science, and non-engineering data, we find a more liberal interpretation of ρ:

$0.5 < \lvert\rho\rvert \leq 1.0$	Strong linear relationship
$0.3 < \lvert\rho\rvert \leq 0.5$	Moderate linear relationship
$0.1 < \lvert\rho\rvert \leq 0.3$	Weak linear relationship
$0 \leq \lvert\rho\rvert \leq 0.1$	No linear relationship

Further, in our modeling efforts we emphasize the interpretation of $\rho \approx 0$. This can be interpreted as either no linear relationship or may indicate the existence of a nonlinear relationship. Most students and many researchers fail to pick up on the importance of the nonlinear relationship aspect of the interpretation.

This chapter focuses on the analytical methods to arrive at a model for a given data set using a prescribed criterion. For example, from the family $y = kx^2$, the parameter k can be determined analytically by using a curve-fitting criterion, such as least-squares, Chebyshev's criterion, or minimizing the sum of the absolute error, and then solving the resulting optimization problem. Although we briefly present these other criterion, we concentrate on least squares. We then present the Maple commands that solve the least-squares optimization problem and provide an analysis of the adequacy of the resulting model.

5.2 The Different Curve Fitting Criterion

We will briefly describe three curve fitting criterion: least squares, Chebyshev's criterion, and minimizing the sum of the absolute error. We start with least squares.

Criterion 1: Least Squares.

The method of least-squares curve fitting, also known as *ordinary least squares* and *linear regression*, is simply the solution to a model that minimizes the sum of the squares of the deviations between the observations and predictions. In Maple, the *Fit* command from the *Statistics* package fits a model curve to a set of data points using least-squares methods. Least squares will find the parameters of the function $f(x)$ that will minimize the sum of squared differences between the real data and the proposed model. The least-squares criterion is

$$\text{Minimize } S = \sum_{j=1}^{m} (y_j - f(x_j))^2 . \tag{5.3}$$

For example, to fit a proposed proportionality model $y = kx^2$ to a set of data, the least-squares criterion requires the minimization

$$\text{Minimize } S = \sum_{j=1}^{m} (y_j - k\,x_j^2))^2 . \tag{5.4}$$

Minimizing (5.4) is achieved using basic calculus: take the first derivative with respect to k, set it equal to zero, and solve for the unknown parameter k as follows.

$$\frac{dS}{dk} = -2\sum_{j=1}^{n} x_j^2\,(y_j - k\,x_j^2) = 0$$

Solving for k yields

$$k = \left(\sum_{j=1}^{n} x_j^2 y_j\right) \Big/ \left(\sum_{j=1}^{n} x_j^4\right) \tag{5.5}$$

Given the data set in Table 5.2, we will find the least-squares fit to the model $y = kx^2$.

Solve for k using (5.5); the model $y = kx^2$ becomes $y = 3.1869x^2$.

Let's redo fitting the quadratic model using Maple.

Maple's *Fit* command will fit models that are linear in their parameters. For instance, *Fit* will fit the type $Y = b + b_1X_1 + b_2X_2 + \cdots + b_kX_k$. to

TABLE 5.2: Least-Squares Data Points

X	0.5	1.0	1.5	2.0	2.5
Y	0.7	3.4	7.2	12.4	20.1

the data set given in k specified columns. To fit a complete quadratic model $y = a + bx + cx^2$ (which is quadratic in x, but *linear in the parameters* a, b, and c) to a data set X_v of x values and Y_v of y values, use

$> Fit(a + b \cdot x + c \cdot x^2, X_v, Y_v, x);$

The syntax is $Fit(\langle model \rangle, \langle x_1\ data \rangle, \langle x_2\ data \rangle, \ldots, \langle y\ data \rangle, \langle model\text{'s}\ variable \rangle)$. To fit our quadratic proportionality example $y = kx^2$, just use

$> Fit(k \cdot x^2, X_v, Y_v, x);$

Since *Fit* is part of the *Statistics* package, remember to first load the package.

$> with(Statistics):$

$> X := [0.5, 1.0, 1.5, 2.0, 2.5]:$
$Y := [0.7, 3.4, 7.2, 12.4, 20.1]:$

$> QuadFit := Fit(k \cdot x^2, X, Y, x);$

$$QuadFit := 3.18692543411645\, x^2$$

We see Maple's answer matches ours (to our 4 decimals).

Let's overlay the data points with the model's graph to check for a good visual fit.

$> with(plots):$

$> DataPlot := plot([seq([X_i, Y_i], i = 1..5)], style = point, symbol = solidcircle,$
$symbolsize = 14):$

$> FitPlot := plot(QuadFit, x = 0..2.6):$

> $display(DataPlot, FitPlot);$

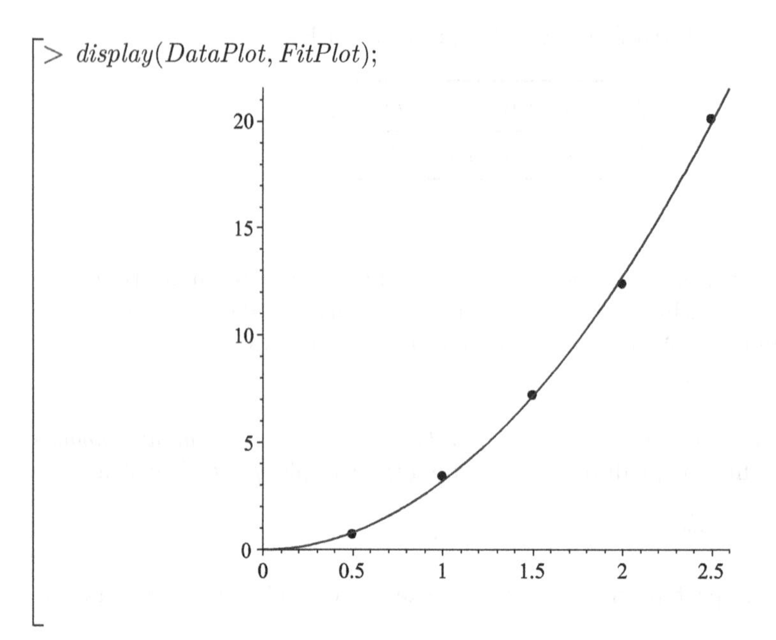

The *Fit* command will provide more information. For example, consider

> $xpts := [1..5]:$
> $ypts := [3, 5, 8, 9, 12]:$

> $Fit(b_0 + b_1 \cdot x, xpts, ypts, x, summarize = true);$

```
Summary:
--------
Model: .80000000+2.2000000*x
--------
Coefficients:
    Estimate  Std. Error  t-value  P(>|t|)
b[0] 0.8000    0.5416      1.4771   0.2362
b[1] 2.2000    0.1633     13.4722   0.0009
--------
R-squared: 0.9837, Adjusted R-squared: 0.9783
```

$$0.799999999999999 + 2.20000000000000\,x$$

Now we use the option *summarize=embed* to get even more information.

> $Fit(b_0 + b_1 \cdot x, xpts, ypts, x, summarize = embed);$

$$0.799999999999999 + 2.20000000000000\,x$$

Summary

Model: $0.80000000 + 2.2000000x$

Coefficients	Estimate	Standard Error	t-value	P(> |t|)
b[0]	0.800000	0.541603	1.47710	0.236154
b[1]	2.20000	0.163299	13.4722	**0.000884317**

R-squared: 0.983740
Adjusted R-squared: 0.978320

Residuals

Residual Sum of Squares	Residual Mean Square	Residual Standard Error	Degrees of Freedom
0.800000	0.266667	0.516398	3

Five Point Summary

Minimum	First Quartile	Median	Third Quartile	Maximum
−0.600000	−0.333333	$2.30391\,10^{-15}$	0.333333	0.600000

We need to see the information contained in a standard *ANOVA Table* in order to check the adequacy of the model. (See, e.g., Devore [Devore2011] for background.) There is a procedure, *SLRReport*, in our *PSM* Maple package for the text, that produces tables required for diagnostic checking of the model. To load the routine, enter:

$> with(PSM, SLRReport);$

$$[SLRReport]$$

A call to *SLRReport* returns the linear regression equation and shows a window, seen in Figure 5.3, with the information we want. Let's apply *SLRReport* to our data set.

$> \langle X, Y\rangle = Matrix([xpts, ypts]);$

$$\begin{bmatrix} X \\ Y \end{bmatrix} = \begin{bmatrix} 1 & 2 & 3 & 4 & 5 \\ 3 & 5 & 8 & 9 & 12 \end{bmatrix}$$

$> SLRReport(xpts, ypts);$

$$y = 2.20000000000000 \cdot x + 0.799999999999996$$

A summary of the main calculations follows. First, find the linear regression equation coefficients.

$$b_1 = S_{xy}/S_{xx} = \frac{\left(\sum x_i y_i\right) - (n\,\bar{x}\,\bar{y})}{\left(\sum x_i^2\right) - (n\,\bar{x}^2)}$$

$$b_0 = \bar{y} - b_1 \cdot \bar{x}$$

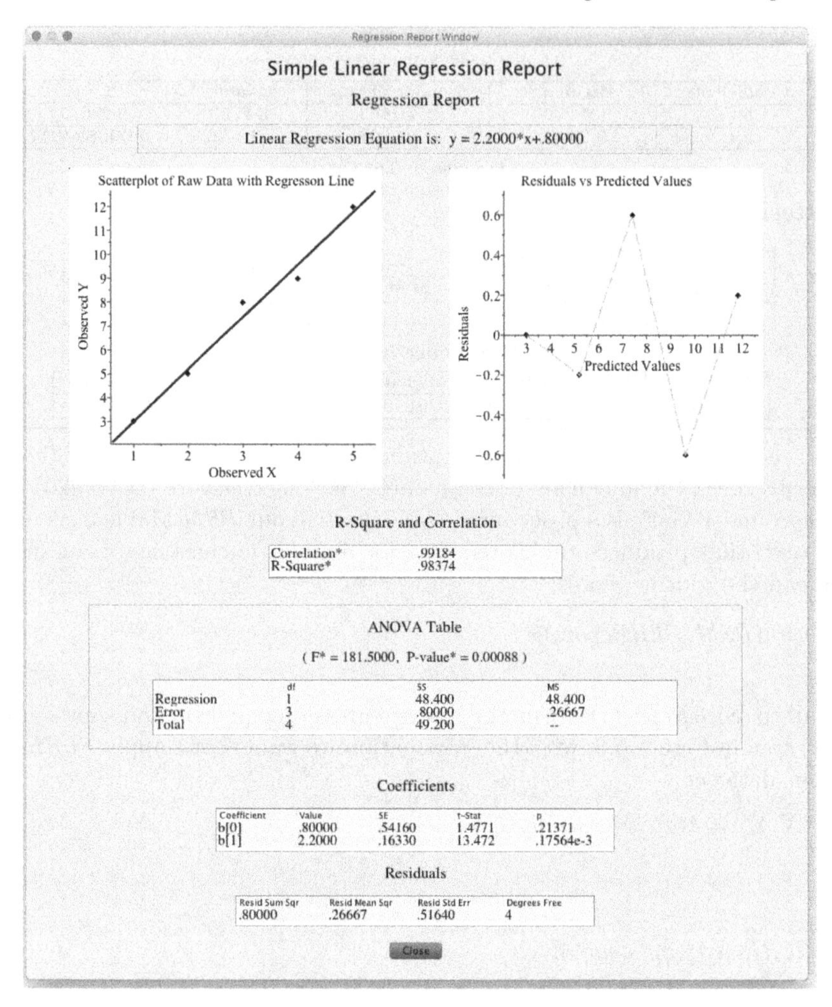

FIGURE 5.3: The *SLRReport* Window

Now, compute the diagnostic values.

$$\begin{aligned}
\text{SST} &= \sum (y_i - \bar{y})^2 \\
\text{SSE} &= \sum (y_i - (b_0 + b_1\, x_i))^2 \\
\text{SSR} &= \sum ((b_0 + b_1\, x_i) - \bar{y})^2 \\
r^2 &= 1 - \frac{\text{SSE}}{\text{SST}} = \frac{\text{SSR}}{\text{SST}}
\end{aligned}$$

then

$$\begin{aligned} \text{MSE} &= \text{SSE}/(n-2) \\ \text{MSR} &= \text{SSR}/1 \\ F_{\text{stat}} &= \text{MSR}/\text{MSE} \end{aligned}$$

Using these formulas gives $r^2 = 0.98374$ for our data. Graph the residuals versus the predicted values.

```
> f := x -> 0.8 + 2.2x :
> Pred := map(f, xpts) :
  Resid := zip((x, y) -> (x - y), ypts, Pred) :
> pointplot([seq([Pred_i, Resid_i], i = 1..5)], symbol = solidcircle,
    symbolsize = 14, view = [2..12, -0.8..0.8],
    title = "Residuals vs Predicted Values",
    labels = ["Predicted Values", "Residuals"],
    labeldirections = [horizontal, vertical]);
```

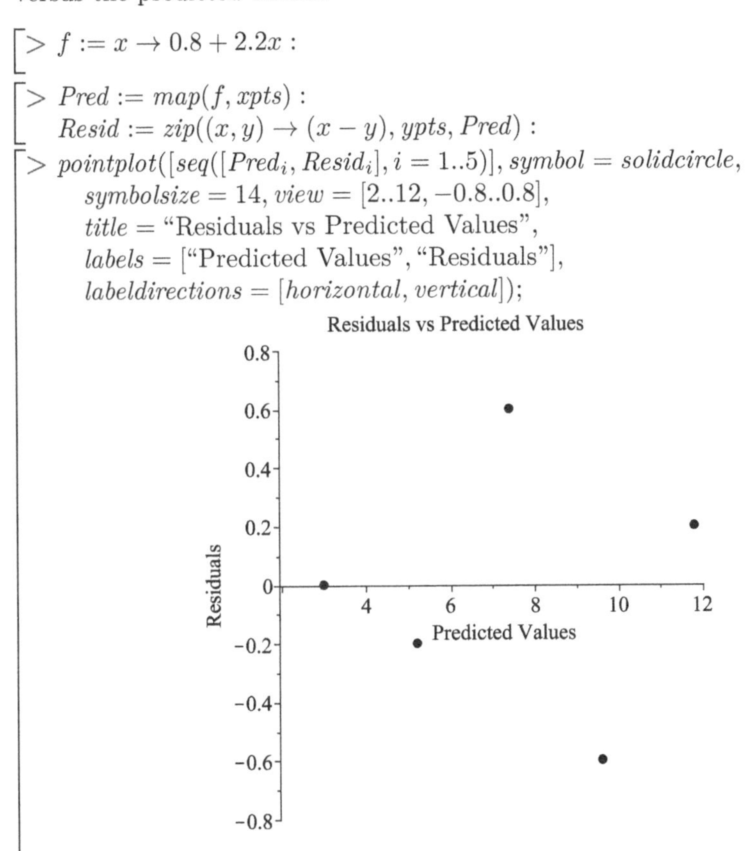

The plot appears to be random points. If this graph showed patterns, we would need to carefully investigate our model looking for more relationships in the data. We conclude our model is adequate and it passes the "commonsense" test for its intended use.

Another way to obtain the values for the estimators is to use a Hessian matrix (see Chapter 3, Volume II). We might also use linear algebra, (which we did in *SLRReport*). Create a matrix X consisting of a column of all 1s and a column of the original x values. The product $(X^{\mathrm{T}}X)^{-1}$ produces the variance-covariance matrix. The SEs we want are the square root of the values along the main diagonal of this matrix. To change from a linear to a quadratic or cubic fitting function only requires that the $(X^{\mathrm{T}}X)^{-1}$ changes as we add new columns to X. We also point out the degrees of freedom *df* also change accordingly with the number of parameters estimated. The function *SimpleRegression* which does linear, quadratic, or cubic regression and prints a diagnostic report, is included in the *PSM* package included with this text; load the function via *with*(*PSM*). The function's syntax is:

SimpleRegression(⟨vector or list of x data⟩, ⟨vector or list of y data⟩);

An optional third argument specifying the fitting degree can be added: *deg*=1, 2, or 3. The default is linear regression.

Example 5.1. A Least-Squares Fit of Snook Fish Data.
Consider snook fishing where we want to construct a simple model to predict the weight of a fish that we catch as function of its length. Assume we have developed a proportionality model, $W = kL^3$. We have a (modified) data set from the Florida Fish and Wildlife Conservation Commission[1] giving snook length and weight data.

TABLE 5.3: Snook Length and Weight Data

Length	20	21	22	23	24	25	26	27	28	29	30
Weight	2.8	3.2	4.4	4.2	4.7	5.7	6.5	7.5	10	11	10.4

This example illustrates the *Fit* command analytically fitting the model $W = kL^3$ to our data set using the least-squares criterion.

```
> interface(rtablesize = 100) :    #To view arrays with any dimension > 10

> FishData := Matrix([[$20..30],
    [2.8, 3.2, 4.4, 4.2, 4.7, 5.7, 6.5, 7.5, 10, 11, 10.4]]) :
FishData := [ 20   21   22   23   24   25   26   27   28   29   30   ]
            [ 2.8  3.2  4.4  4.2  4.7  5.7  6.5  7.5  10   11   10.4 ]

> Length, Weight := FishData[1, [1.. - 1]], FishData[2, [1.. - 1]]) :

> RegressionFit := Fit(k · t^3, Length, Weight, t);
                RegressionFit := 0.000398536180975772 t^3
```

Put together a graph of the data overlaid by the fit.

```
> FishScatterPlot := pointplot(FishData, symbol = solidcircle,
    symbolsize = 14) :
  FitPlot := plot(RegressionFit, t = 19..32) :
```

[1] See http://myfwc.com/research/saltwater/fish/snook/length-weight/.

> *display*(*FitPlot*, *FishScatterPlot*,
title = "Model and Scatterplot of Snook Data");

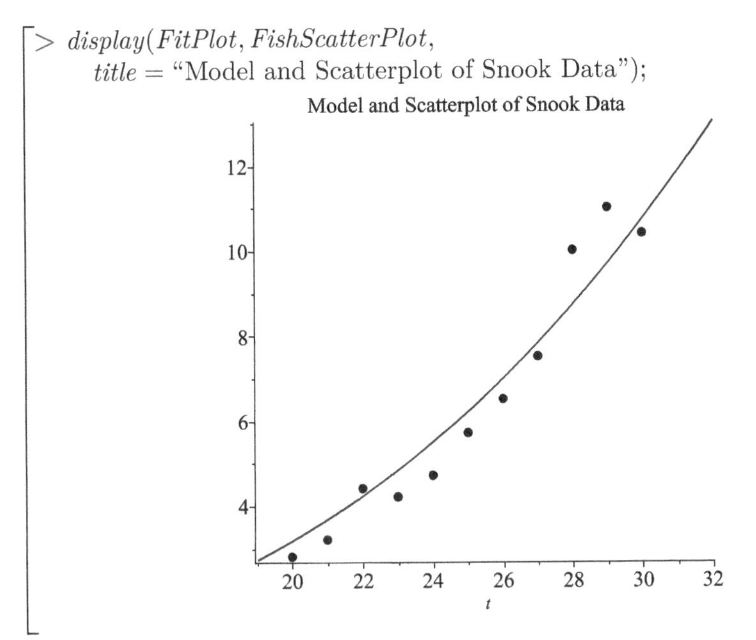

The least-squares estimate of the proportionality constant in this model is $k = 0.0003985$. Thus, the model is $W = 0.0003985\,L^3$. The graph of the least-squares fit with the original data above shows that the model does capture the trend of the data.

Criterion 2: Minimize the Sum of the Absolute Deviations.

Our next criterion for 'best' is to have the minimum sum of the absolute value of the errors, as opposed to the sum of the squares of the errors. That is, choose the parameter values in f that

$$\text{Minimize } SAE = \sum_{i=1}^{n} |y_i - f(x_i)|\,.$$

This approach is sometimes called a "minimax" criterion.

Suppose we have the following data and want to fit the model $W = k\,L^3$ using the minimax criterion.

TABLE 5.4: Minimax Example Data

Length L	12.5	12.625	12.625	14.125	14.5	14.5	17.27	17.75
Weight W	17	16	17	23	26	27	43	49

We formulate the model as:

$$\text{Minimize } S = \left|17 - k \cdot 12.5^3\right| + \left|16 - k \cdot 12.625^3\right| + \cdots + \left|49 - k \cdot 17.75^3\right|$$

The standard calculus techniques that we used for least-squares regression won't work here since the absolute value function is nondifferentiable at 0. We'll use Maple's *minimize* function to find the value of k.

```
> L := [12.5, 12.625, 12.625, 14.125, 14.5, 14.5, 17.27, 17.75] :
  W := [17, 16, 17, 23, 26, 27, 43, 49] :
  Matrix([L, W]);
```

$$\begin{bmatrix} 12.5 & 12.625 & 12.625 & 14.125 & 14.5 & 14.5 & 17.27 & 17.75 \\ 17 & 16 & 17 & 23 & 26 & 27 & 43 & 49 \end{bmatrix}$$

In this case, it will be easier to build the function S using an *unapply* statement that turns an expression into a function.

```
> S := unapply(sum(|W_i - k · L_i^3|, i = 1..8), k);
```

$$S := k \mapsto |1953.125\,k - 17| + |2012.306641\,k - 16| + |2012.306641\,k - 17| + |2818.158203\,k - 23| + |3048.625\,k - 26| + |3048.625\,k - 27| + |5150.827583\,k - 43| + |5592.359375\,k - 49|$$

Now use *minimize* with the *location* option so that Maple tells us both the minimum value and where it occurs.

```
> minimize(S(k), location);
```

$$5.935456832, \{[\{k = 0.008528434950\}, 5.935456832]\}$$

We see that $k = 0.008528$ gives the minimum value of 5.935 for the sum of the absolute values of the errors. Our model is then

$$W = 0.008528\,L^3.$$

Compare this model to the one you find using least-squares minimization.

We could also have found the value of k that minimizes S by a numerical search technique called the *Golden Section Search* which is presented in Chapter 3 of Volume II.

Criterion 3. Chebyshev's Criterion or Minimize the Maximum Error.

Our third criterion for 'best' is to have the minimum of the largest absolute value of the errors. That is, choose the parameter values in f that

$$\text{Minimize } R = \max_{i=1..n} |y_i - f(x_i)|\,.$$

Let R_i be the ith error term $R_i = y_i - f(x_i)$. Then we can formulate the problem as the linear program

Minimize $R = \max |y_j - f(x_j)|$
subject to:
$R - R_i \geq 0$
$R + R_i \geq 0$
for $i = 1, 2, \ldots, n$

The LP formulation with our length-weight data from above is

Minimize R
subject to:

$$
\begin{aligned}
R - (17 - k \cdot 12.5^3) &\geq 0 \\
R + (17 - k \cdot 12.5^3) &\geq 0 \\
&\vdots \\
R - (49 - k \cdot 17.75^3) &\geq 0 \\
R + (49 - k \cdot 17.75^3) &\geq 0 \\
R &\geq 0 \\
k &\geq 0
\end{aligned}
$$

This is a linear programming problem that finds the value of k that minimizes the largest error R. Using the techniques we studied for linear programming models in Chapter 4 gives the value of k to be 0.008547. The Maple work follows.

$> with(Optimization):$

$> Objective := R:$

$> Constraints := \mathbf{union}(seq(\{R-(W_i-k \cdot L_i^3) \geq 0, R+(W_i-k \cdot L_i^3) \geq 0\}, i = 1..8), R \geq 0, k \geq 0);$

$$
\begin{aligned}
Constraints := \{&0 \leq R, 0 \leq k, 0 \leq R - 5592.359375\,k + 49, \\
&0 \leq R - 5150.827583\,k + 43, 0 \leq R - 3048.625\,k + 26, \\
&0 \leq R - 3048.625\,k + 27, 0 \leq R - 2818.158203\,k + 23, \\
&0 \leq R - 2012.306641\,k + 16, 0 \leq R - 2012.306641\,k + 17, \\
&0 \leq R - 1953.125\,k + 17, 0 \leq R + 1953.125\,k - 17, \\
&0 \leq R + 2012.306641\,k - 17, 0 \leq R + 2012.306641\,k - 16, \\
&0 \leq R + 2818.158203\,k - 23, 0 \leq R + 3048.625\,k - 27, \\
&0 \leq R + 3048.625\,k - 26, 0 \leq R + 5150.827583\,k - 43, \\
&0 \leq R + 5592.359375\,k - 49\}
\end{aligned}
$$

$> LPSolve(Objective, Constraints, assume = nonnegative);$

$[1.19995741953699, [R = 1.19995741953699, k = 0.854738391709009e - 2]]$

The model by this method is

$$W = 0.008547\,L^3$$

Exercises

Fit the following data with the model specified using least squares. Compare the different fits using diagnostic values.

1.

x	1	2	3	4	5
y	1	1	2	2	4

(a) $y = ax + b$
(b) $y = ax^2$

2. Stretch of a spring data:

$x(\times 10^{-3})$	5	10	20	30	40	50	60	70	80	90	100
$y(\times 10^{5})$	0	19	57	94	134	173	216	256	297	343	390

(a) $y = ax$
(b) $y = ax + b$
(c) $y = ax^2$

3. Ponderosa pine data:

x	17	19	20	22	23	25	28	31	32	33	36	37	39	42
y	19	25	32	51	57	71	113	140	153	187	192	205	250	260

(a) $y = ax + b$
(b) $y = ax^2$
(c) $y = ax^3 + bx^2 + c$

4. Kepler's planetary data:

Body	*Period (s)*	*Distance from sun (m)*
Mercury	7.60×10^6	5.79×10^{10}
Venus	1.94×10^7	1.08×10^{11}
Earth	3.16×10^7	1.50×10^{11}
Mars	5.94×10^7	2.28×10^{11}
Jupiter	3.74×10^8	7.79×10^{11}
Saturn	9.35×10^8	1.43×10^{12}
Uranus	2.64×10^9	2.87×10^{12}
Neptune	5.22×10^9	4.50×10^{12}

(a) Fit the model $y = ax^{3/2}$. (Use appropriate scaling on the data.)

Projects

Project 5.1. Least-Squares Program
Write a Maple program using the least-squares criterion to find the general proportionality model: y is proportional to x^n.

Further Reading

[BF2005] Richard L. Burden and J. Douglas. Faires, *Numerical Analysis*, 8th ed., Thomson Books, 2005.

[CK2012] E. Ward Cheney and David R. Kincaid, *Numerical Mathematics and Computing*, Cengage Learning, 2012.

[Devore2011] Jay L. Devore, *Probability and Statistics for Engineering and the Sciences*, Cengage Learning, 2011.

[Johnson2012] I. Johnson, *I'll Give You a Definite Maybe, An Introductory Handbook on Probability, Statistics, and Excel* (2000). https://malvma.viu.ca/~johnstoi/maybe/maybe4.htm

[MRBSB2019] E. Marland, G. Rhoads, M. Bossé, J. Sanqui, and W. Bauldry, "Q^2: A Measure of Linearity," (submitted, Feb., 2019).

[NKNW1996] John Neter, Michael H. Kutner, Christopher J. Nachtsheim, and William Wasserman, *Applied Linear Statistical Models*, Vol. 4, Irwin Chicago, 1996.

5.3 Plotting the Residuals for a Least-Squares Fit

In the previous section, we learned how to obtain a regression fit of a model, most often done with the least-squares criterion, and we plotted the model's predictions on the same graph as the observed data points in order to get a visual indication of how well the model matches the trend of the data. We also plotted residuals versus the predicted values looking for randomness in the plot. A powerful technique for quickly determining whether the model breaks down or is adequate is to graph the actual deviations, the residuals, between the observed and predicted values as a function of the independent variable of the model. We plot the residuals on the y-axis and the model's independent variable on the x-axis. The deviations should be randomly distributed and contained in a reasonably small band that is commensurate with the accuracy required by the model. Any excessively large residual warrants further investigation of the data point in question to discover the cause of the large deviation. A pattern or trend in the residuals indicates that a predictable effect is unexplained and remains to be modeled. The nature of a pattern gives clues on how to refine the model when a refinement is needed. We illustrate several possible patterns for residuals in Figures 5.4 (a) through (e). Our intent is to provide the modeler with knowledge concerning the adequacy of the model chosen. We will leave further investigations into correcting the patterns to courses in statistical regression.

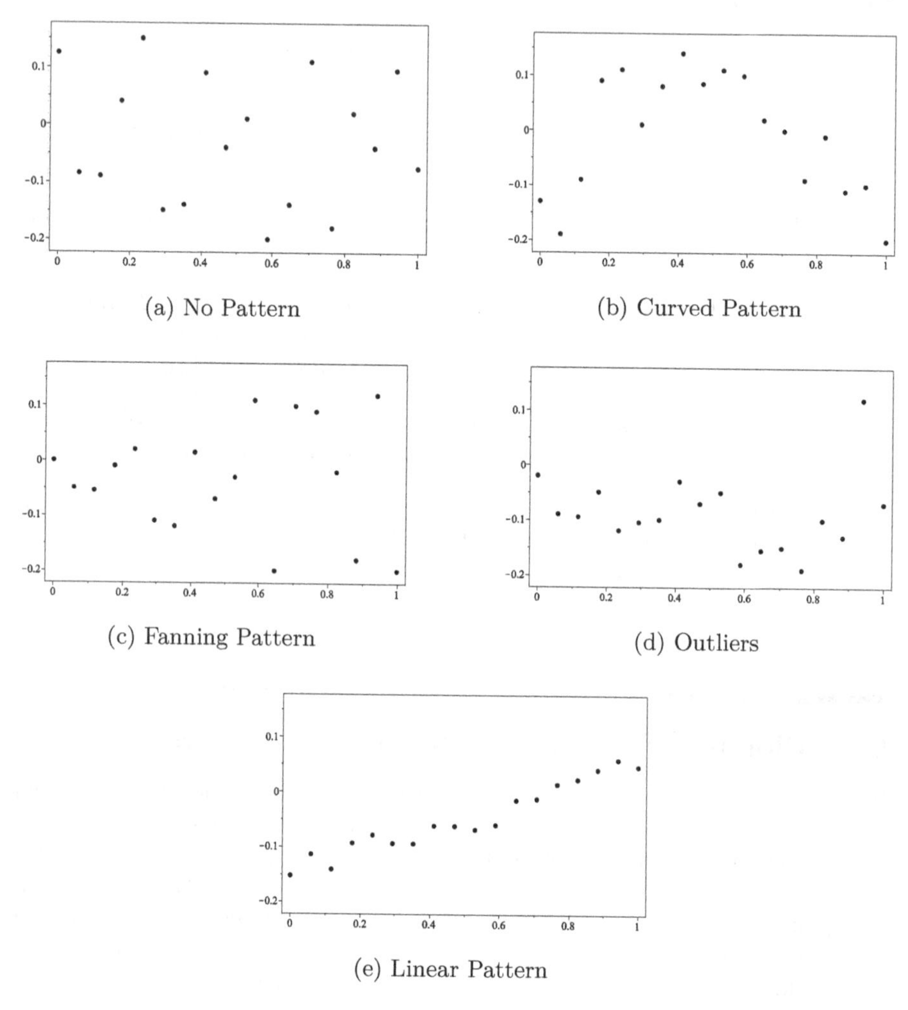

(a) No Pattern

(b) Curved Pattern

(c) Fanning Pattern

(d) Outliers

(e) Linear Pattern

FIGURE 5.4: Patterns in the Residual Plot

We'll now use Maple to compute and graph the residuals. After fitting a specified model, the difference between the observed and predicted values can be easily calculated.

Example 5.2. Snook Revisited.
We return to our snook weight v. length example from the previous section. We found (pg. 217) the relationship between the weight of a snook and its length could be reasonably modeled by the following expression:

$$W = 0.000398536\, L^3.$$

The residuals are the differences between the predicted and observed values. Mathematically, for the model $y = f(x)$, residuals, also called errors or deviations, are found by $(y_i - f(x_i))$, the actual value minus the predicted value. In our snook example, the residuals are

$$residual_i = W_i - 0.000398536\, L_i^3.$$

We will use Maple to collect and plot this information. The observed values for the weight of the snook fish are in the lists L and W, while the predicted values must be calculated using the formula above. The residuals are calculated by subtracting each observed value from each predicted value; we can do this in one step using matrix arithmetic.

```
> L := [12.5, 12.625, 12.625, 14.125, 14.5, 14.5, 17.27, 17.75] :
  W := [17, 16, 17, 23, 26, 27, 43, 49] :
  Matrix([L, W]);
```

$$\begin{bmatrix} 12.5 & 12.625 & 12.625 & 14.125 & 14.5 & 14.5 & 17.27 & 17.75 \\ 17 & 16 & 17 & 23 & 26 & 27 & 43 & 49 \end{bmatrix}$$

```
> f := x -> 0.000398536 x^3 :
```

```
> Residuals := W - map(f, L);
```

Residuals := [−0.388288000, −0.490841896, 0.156388672, −0.648987512, −0.809361664, −0.527125000, −0.504668736, −0.344384088, 1.251337728, 1.280105496, −0.36047200]

To analyze the results, the residuals are plotted, and the modeler must interpret the resulting graph. We combine the data into a matrix with L in the first column and the residuals in the second column.

```
> RvL := ⟨⟨L⟩ | ⟨Residuals⟩⟩ :
```

> $pointplot(RvL, symbol = solidcircle, symbolsize = 14);$

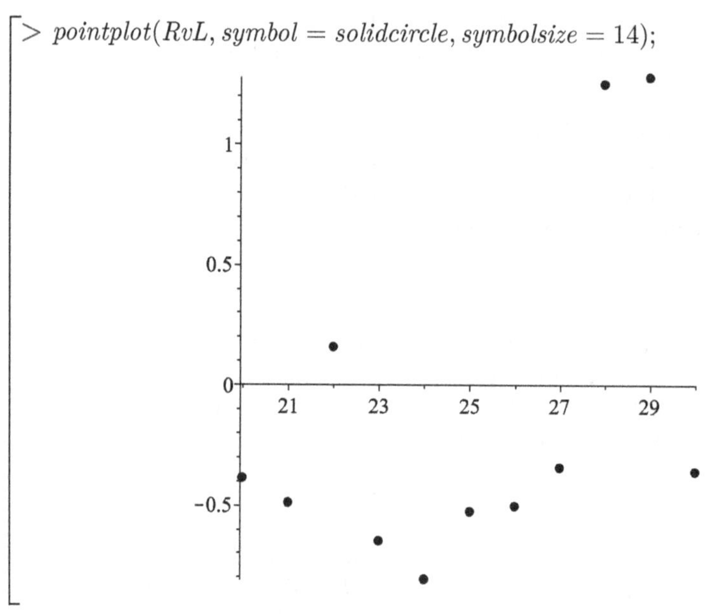

Note that the residuals appear to have a slight upward curvature—they are not randomly distributed and contained in a relatively small band about zero. There are comparatively no outliers, or unusually large residuals. There appears to be a curved pattern in the residuals. Based on these aspects of the graph of the residuals, the model may not be adequate.

Example 5.3. Fitting Population Data.
Given certain existing conditions, we decide to fit a cubic model to the data of Table 5.5 where T represents years and P represents the size of the population.

TABLE 5.5: Population Data

T	7	14	21	28	35	42
P	125	275	800	1200	1700	1650

> $xdata := [7, 14, 21, 28, 35, 42]:$
$ydata := [125, 275, 800, 1200, 1700, 1650]:$

> $pointplot([xdata, ydata], symbol = solidcircle, symbolsize = 14,$
> $title =$ "Scatterplot of Population Data");

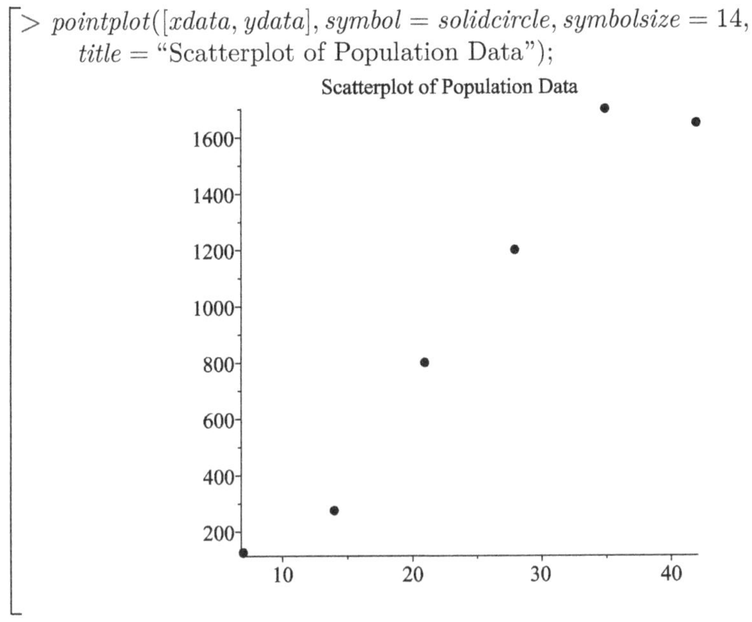

As there appears to be a change in concavity in the data, the cubic equation fit for our least-squares model appears appropriate.

> $with(Statistics):$

> $Fit(ax^3 + bx^2 + cx + d, xdata, ydata, x):$
> $CubicFit := unapply(fnormal(\%), x);$

$$CubicFit := x \mapsto -0.1066299536\, x^3 + 7.436426952\, x^2 - 95.78136810\, x + 466.6666667$$

> $predict := map(CubicFit, xdata);$

$$predict := [124.0079365, 290.6746037, 747.2222227, 1274.206349, 1652.182538, 1661.706345]$$

> $residuals := ydata - predict;$

$$residuals := [0.9920635, -15.6746037, 52.7777773, -74.206349, 47.817462, -11.706345]$$

> *pointplot*([*predict*, *residuals*], *symbol* = *soliddiamond*,
symbolsize = 14, *title* = "Residual Plot");

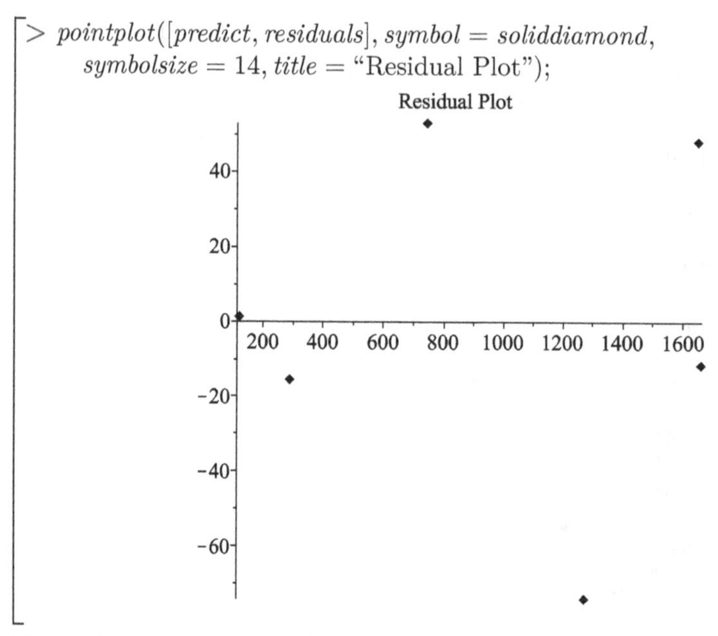

There does not appear to be any pattern to the residual plot, so we conclude the model is adequate.

Exercises

Obtain and interpret the residual plots for the models previously found by using least-squares regression in the exercises (pg. 220) for Section 5.2.

1.

x	1	2	3	4	5
y	1	1	2	2	4

(a) $y = ax^2$

2. Stretch of a spring data:

$x(\times 10^{-3})$	5	10	20	30	40	50	60	70	80	90	100
$y(\times 10^{5})$	0	19	57	94	134	173	216	256	297	343	390

(a) $y = ax$

(b) $y = ax^2$

3. Ponderosa pine data:

x	17	19	20	22	23	25	28	31	32	33	36	37	39	42
y	19	25	32	51	57	71	113	140	153	187	192	205	250	260

   (a) $y = ax^2$

   (b) $y = ax^3$

4. Use the model $y = ax^{3/2}$ for Kepler's planetary data.

Body	*Period (s)*	*Distance from sun (m)*
Mercury	7.60×10^6	5.79×10^{10}
Venus	1.94×10^7	1.08×10^{11}
Earth	3.16×10^7	1.50×10^{11}
Mars	5.94×10^7	2.28×10^{11}
Jupiter	3.74×10^8	7.79×10^{11}
Saturn	9.35×10^8	1.43×10^{12}
Uranus	2.64×10^9	2.87×10^{12}
Neptune	5.22×10^9	4.50×10^{12}

5.4 Case Studies

5.4.1 Competitive Bidding

Consider two companies bidding competitively for a future job. For each, we have data that we can examine and, in this case, build a regression model and compare results. Does one company have a longer setup time per call and a larger time per data byte?

TABLE 5.6: Companies' Bidding Data

(a) Company 1

Data Bytes	64	64	64	64	234	590	846
Time	26.4	26.4	26.4	26.2	33.8	41.6	50

Data Bytes	1060	1082	1088	1088	1088	1088
Time	48.4	49	42	41.8	41.8	42

(b) Company 2

Data Bytes	92	92	92	92	348	604	860
Time	32.8	34.2	32.4	34.4	41.4	51.2	76

Data Bytes	1074	1074	1088	1088	1088	1088
Time	80.4	79.8	58.6	57.6	59.8	57.4

We build the two regression models.

$$\text{Company 1:} \quad \mathit{Time} = 26.898 + 0.017 \cdot (\mathit{Data\ Bytes})$$
$$\text{Company 2:} \quad \mathit{Time} = 31.068 + 0.034 \cdot (\mathit{Data\ Bytes})$$

The diagnostics for the two separate models look adequate. Thus, we build a confidence interval for the parameters for each. To compute the confidence interval we will use the formulas:

$$b_1 \pm \mathrm{SE}_{b_1} t\left(1 - \frac{\alpha}{2}, n-2\right)$$
$$b_0 \pm \mathrm{SE}_{b_0} t\left(1 - \frac{\alpha}{2}, n-2\right)$$

TABLE 5.7: Parameters' Confidence Intervals

Coefficient	Company 1	Confidence Interval	Company 2	Confidence Interval
Intercept	26.898	[23.297, 30.499]	31.068	[22.608, 39.528]
Slope	0.017	[0.013, 0.0292]	0.034	[0.023, 0.044]

Table 5.7 clearly shows that the intervals for the intercepts overlap while those for the slopes do not. We conclude that the setup times are not significantly different between Company 1 and Company 2, while the per-data-byte times

are significantly different. We might prefer the company with the smaller interval and select Company 1.

5.4.2 Index Prediction

We have data over 5 weeks of an index (acceptable values are between 0 and 100). We want to create a model to predict the values after 10 and 20 weeks. We assume that conditions and service do not change from the first 5 weeks through the next 5 to 15 weeks. Our collected data is $x = [0, 1, 2, 3, 4, 5]$ (week) and $y = [15, 11.5, 8.5, 7, 6, 3]$ (index value). We'll use the *SLRReport* function from our *PSM* package.

```
> with(PSM) :
```

```
> xpts := [0, 1, 2, 3, 4, 5] :
  ypts := [15, 11.5, 8.5, 7, 6, 3] :
> pointplot([xpts, ypts], symbol = solidcircle, symbolsize = 14,
    title = "Scatterplot of Raw Data");
```


The data appears linear so we continue the linear regression process.

```
> SLRReport(xdata, ydata);
```

$$y = 14.0714285714286 - 2.22857142857143\,x$$

The *SLRReport*'s report window is shown in Figure 5.5.

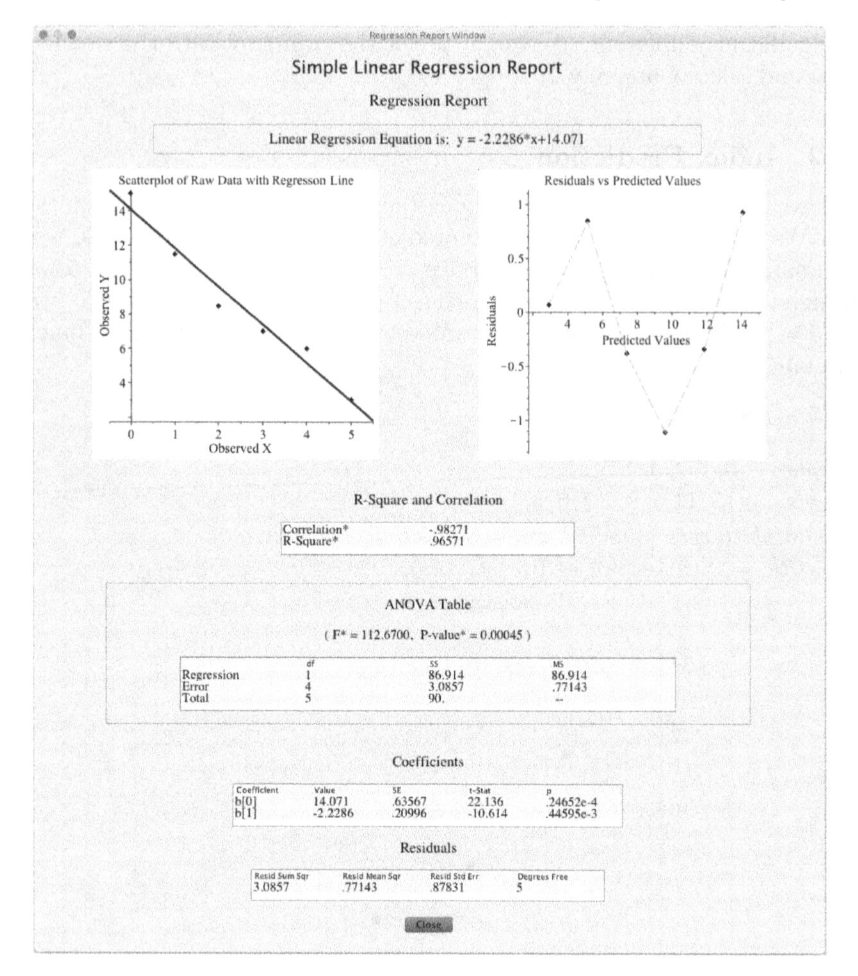

FIGURE 5.5: SLRReport for Index Data

The correlation is -0.9827, which is a strong negative linear relationship. All the p-values of the model and the coefficient are significant (less than 0.05). The model appears to be looking good, so far.

We see plots of the data and model as well as the residuals in the report window, Figure 5.5. The residual plot has a curve trend, so the model may not be adequate. However, we might use it to predict the index after 10 and 20 weeks.

$$y(10) = --8.214 \quad \text{and} \quad y(20) = -30.5.$$

Both values are less than zero, so the model predictions do not pass the "commonsense test."

As a corrective action to the curved residual pattern, we might try fitting a quadratic polynomial, $y = b_0 + b_1 x + b_2 x^2$. We leave this to the exercises.

Exercises

1. Solve the regression problem in Case Study 5.4.2, Index Prediction, using a quadratic fitting function $y = b_0 + b_1 x + b_2 x^2$.

   (a) Predict the index at $x = 10$ and 20 weeks.

   (b) Do the results pass the commonsense test?

2. Solve the regression problem in Case Study 5.4.2, Index Prediction, using a cubic fitting function $y = b_0 + b_1 x + b_2 x^2 + b_3 x^3$.

   (a) Predict the index at $x = 10$ and 20 weeks.

   (b) Do the results pass the commonsense test?

References and Further Reading

[Devore2011] Jay L. Devore, *Probability and Statistics for Engineering and the Sciences*, Cengage Learning, 2011.

[GFH2014] Frank Giordano, William P. Fox, and Steven Horton, *A First Course in Mathematical Modeling*, 5th ed., Nelson Education, 2014.

[Johnson2012] I. Johnson, *I'll Give You a Definite Maybe, An Introductory Handbook on Probability, Statistics, and Excel* (2000). https://malvma.viu.ca/~johnstoi/maybe/maybe4.htm

[NKNW1996] John Neter, Michael H. Kutner, Christopher J. Nachtsheim, and William Wasserman, *Applied Linear Statistical Models*, Vol. 4, Irwin Chicago, 1996.

Appendix

To view the Maple code of the *SimpleRegression* function/program, use a *print* statement.

```
> with(PSM) :
> print(SimpleRegression);
```

6

Statistical and Probabilistic Problem Solving with Maple

Objectives

(1) Understand statistical concepts and applications.

(2) Understand probabilistic concepts and applications.

(3) Solve discrete and continuous probability problems.

(4) Solve discrete and continuous reliability problems.

(5) Solve confidence intervals and hypothesis testing problems.

6.1 Introduction

Why Do Airlines Overbook Flights?

Airlines routinely overbook flights to compensate for no-shows: people who reschedule or opt not to fly. An empty seat on a plane means a loss of revenue to an airline. Being removed from an overbooked flight is often referred to as "bumping."

What Is Bumping?

Bumping occurs when there are more passengers with confirmed reservations who show up for a flight than there are seats on the plane. This is called "overbooking" the plane. Airlines overbook because they know from experience that not all passengers who have made reservations show up for the flight. To have any chance of filling the plane, airline computers estimate the number of passengers likely to be no-shows and accept reservations accordingly. (Airlines also ignore the overbook limit when a customer is buying a full-fare ticket, because the cost of compensating volunteers with a 'bump ticket' is usually less than the additional income derived from a full-fare ticket.)

In an overbooking situation, agents will make a gate area announcement asking for volunteers with flexible schedules to give up their seats. Typically, the initial amount offered is based on two factors: the length of the flight, and how long the volunteer must wait in order to be rescheduled on a later flight. Usually, the gate agent's first offer is for $250 (this varies a good deal).

If not enough passengers volunteer to take up the first offer, agents will usually only increase the offer just once or twice more. For example, Delta Airlines says that it tries to limit increases to just two rounds only in order to get flights out on time. If agents are unable to find enough volunteers, they will begin to involuntarily bump a few unlucky passengers, based on a variety of factors, such as the time the passenger arrived for the flight, the amount they paid for their ticket, and their frequent flyer status

Overbooking is a standard practice and is perfectly legal. Many airlines regularly overbook busy routes by as much as 200 percent. By law, all bumped passengers are entitled to some form of compensation, usually in the form of a free ticket.

Thus, the airlines have to balance the *risk* of a no-show with the compensation they have to pay to bumped passengers. They overbook according to a number of variables: whether it's a holiday season, how the airline market is doing in general, and perhaps most importantly, a specific flight's history of no-shows.

According to several published articles, an average of 50,000 passengers are bumped by the nation's ten largest airlines every year. If you do lose your seat, you should try to get the 'most bang from your bump.' We will solve this problem later in the chapter.

6.2 Basic Statistics: Univariate Data

6.2.1 Data

Statistics is the *science of reasoning from data* — a natural place to begin your study is by examining what is meant by the term "data." The most fundamental principle in statistics is that of *variability.* If the world were perfectly predictable and showed no variability, there would be no need to study statistics. You will need to discover the statistical notion of a **variable**, and then first learn how to classify variables.

Any characteristic of a person or thing that can be expressed as a number is called a *variable.* A *value* of that variable is the actual number that describes that person or thing. Think of the variables that might be used to describe you: height, weight, income, grade, job, and gender.

Data can be *quantitative* or *categorical (qualitative).*

Quantitative means that the data are numerical where the number has relative meaning. Examples could be a list of heights of students in your class, weights of players on a football team, or batting averages of the starting line-up for the 1998 New York Yankees. See Table 6.1

TABLE 6.1: Quantitative Data

(a) Heights in Class

Height	5′10″	6′2″	5′5″	5′2″	6′	5′9″

(b) Weights on Team

Weight	135	155	215	192	173	170	165	142

(c) New York Yankees Starting Lineup Batting Average 1998

Batting Average	.276	.320	.345	.354	.269	.275	.300	.254	.309

These data elements provide numerical information. We can determine from the data which height is the tallest or smallest, or which batting average is the greatest or the smallest. We can also compare and contrast these values "mathematically."

Quantitative data can be either discrete (counting data) or continuous. These types become important as we analyze them and use them in models later in the course. Quantitative data allows us to "do meaningful mathematics." $(+, -, \times, \div)$

Categorical or **qualitative** data can describe objects, such as recording the people with a particular hair color as: blonde = 1 or brunette = 0. If we had four colors of hair: blonde, brunette, black and red; we could use as codes: brunette = 0, blonde = 1, black = 2, and red = 3. We certainly cannot have an average hair color from these numbers — it would not make sense. Another example is categories by gender: male = 0 and female = 1. In general, it may not make sense to do any arithmetic using categorical variables.

After distinguishing between quantitative and categorical data, we need to move on to a fundamental principle of data analysis: "Begin by looking at a visual display of the data set."

6.2.2 Statistical Measures

Measures of Central Tendency or Location

Describing the Data

In addition to plots and tables, numerical descriptors are often used to summarize data. Three numerical descriptors, the *mean*, the *median*, and the *mode* offer different ways to describe and compare data sets. These are generally known as the *measures of location* or *measures of central tendency.*

The Mean

The mean is the arithmetic average, with which you are probably very familiar. For example, your academic average in a course is generally the arithmetic average of your graded work. The mean of a data set is found by summing all the data and dividing this sum by the number of data elements.

The following data represent ten scores earned by a student in a college algebra course: 55, 75, 92, 83, 99, 62, 77, 89, 91, 72.

To compute the student's average sum the 10 scores

$$55 + 75 + 92 + 83 + 99 + 62 + 77 + 89 + 91 + 72 = 795$$

and then divide by the number of data elements, 10, giving $795/10 = 79.5$.

To describe this process in general, we represent each data element by a letter with a numerical subscript. Thus, for a class of n tests, the scores can be represented by $a_1, a_2, \ldots, a_n$. The mean of these n values is found by adding all the values, and then dividing the sum by n, the number of values. The Greek letter Σ (called 'sigma') is used to represent the sum of all the terms in a certain group or the addition operator. Thus, we may see this written as $\sum_{i=1}^{n} a_i = a_1 + a_2 + \cdots + a_n$. The mean, often written as $\bar{a}$ for $\{a_i\}$, is then

$$\text{mean} = \bar{a} = \frac{1}{n} \cdot \sum_{i=1}^{n} a_i$$

Think of the mean as the average. Notice that the mean does not have to equal any specific value of the original data set. The mean value of 79.5 was not a score ever earned by our student.

Batting average is the total number of hits divided by the total number of official at bats. Is batting average a mean? Explain.

There are other measures of central tendency, the geometric mean and harmonic mean, suitable for different situations. Central tendency will be studied further in a basic statistics course.

The Median

The median locates the true middle of a numerically ordered list. The hint here is to make sure that the data is in numerical order listed from smallest to largest along the x-number line. There are two cases for finding the median (middle value) of an ordered list depending on n, the number of data elements:

- *Odd Number of Data.*
  The median (middle) is the exact data element that is the middle value. For example, in 5 ordered math grades earned by a student: 55, 63, 76, 84, 88. The middle value is 76 since there are exactly two scores on each

side (lower and higher) of 76. Thus the median is 76. Notice that with an odd number of data values that the median is one of the data elements.

- *Even Number of Data.*
  There is no true middle value within the data itself. In this case, we need to find the mean of the two middle numbers in the ordered list. This value, most likely not a value of the data set, is the median. Let's illustrate with two examples.

  1. Six math scores for Student A are 56, 62, 75, 77, 82, 85. The middle two scores are 75 and 77, because there are exactly two scores below 75 and exactly 2 scores above 77. We average 75 and 77 to compute the median: $(75+77)/2 = 152/2 = 76$. Note that 76 is *not* one of the original data values.
  2. Eight scores for Student B are 72, 80, 81, 84, 84, 87, 88, 89. The middle two scores are 84 (fourth score) and 84 (fifth score), because there are exactly 3 scores lower than 84 and 3 scores higher than 84. The average of these two scores is 84. Note that this median is one of our data elements.

It is also very possible for the mean to be equal to the median.

The Mode

The data value that occurs the most often is called the *mode*. The mode is one of the numbers in our original data. The mode does not have to be unique. It is possible for there to be more than one mode in a data set. As a matter of fact, if every data element is different from the other data elements, then every element is a mode.

Consider the following test scores for a mathematics class: 75, 80, 80, 80, 80, 85, 85, 90, 90, 100. The number of occurrences for each value is

TABLE 6.2: Data Frequencies

Value	75	80	85	90	100
Number of Occurrences	1	4	2	2	1

Since 80 occurred 4 times and that is the largest value among the number of occurrences, then 80 is the mode of this data set.

Measures of Dispersion

Variance and Standard Deviation

Measures of variation or *measures of the spread* of the data include the *variance* and *standard deviation.* These two values measure the spread in the data, how far the data are from the mean. The *sample*[1] *variance* has notation S^2, and the *sample deviation* has notation S. For n data elements x_i with mean $\bar{x}$,

$$S^2 = \frac{\sum_{i=1}^{n}(x_i - \bar{x})^2}{n-1} = \frac{\sum_{i=1}^{n}\left(x_i^2 - \bar{x}^2\right)}{n-1} \qquad \text{and} \qquad S = \sqrt{\frac{\sum_{i=1}^{n}(x_i - \bar{x})^2}{n-1}}$$

Example 6.1. Statistical Measures.
Consider the following 10 data elements: 50, 54, 59, 63, 65, 68, 69, 72, 90, 90.

The mean is

$$\bar{x} = \frac{1}{10} \cdot \sum_{i=1}^{10} x_i = 68$$

The variance is found by subtracting the mean, 68, from each point, squaring the result, adding, and then dividing the total by $9 = n - 1$. Sample deviation is the square root of the sample variance.

$$\begin{aligned} S^2 &= \frac{1}{9} \cdot \Big[(50-68)^2 + (54-68)^2 + (59-68)^2 + (63-68)^2 + (65-68)^2 \\ &\qquad + (68-68)^2 + (69-68)^2 + (72-68)^2 + (90-68)^2 + (90-68)^2\Big] \\ &= 180 \end{aligned}$$

$$S = \sqrt{S^2} = 6\sqrt{5} \approx 13.42$$

Example 6.2. Metabolic Rate.
A person's metabolic rate is the rate at which the body consumes energy. Seven men took part in a study of dieting giving the rates in Table 6.3. The units are calories in a 24-hour period.

TABLE 6.3: Metabolic Rates

Rate	1792	1666	1362	1614	1460	1867	1439

The researchers reported both $\bar{x}$ and S for these men.

[1] "Sample" indicates we have a subset of data taken from a full population.

The mean is

$$\bar{x} = \frac{1}{7} \cdot (1792 + 1666 + 1362 + 1614 + 1460 + 1867 + 1439) = \frac{11200}{7} = 1600$$

To see clearly the nature of the variance, start with a table of the deviations of the observations from the mean.

TABLE 6.4: Metabolic Rates Deviation Table

Observations x_i	Deviations $x_i - \bar{x}$	Squared Deviations $(x_i - \bar{x})^2$
1792	$1792 - 1600 = 192$	36,864
1666	$1666 - 1600 = 66$	4,356
1362	$1362 - 1600 = -238$	56,644
1614	$1614 - 1600 = 14$	196
1460	$1460 - 1600 = -140$	19,600
1867	$1867 - 1600 = 267$	71,289
1439	$1439 - 1600 = -161$	25,921
	Sum $= 0$	Sum $= 214{,}870$

The variance $S^2 = 214{,}870/6 = 35{,}811.67$; the standard deviation $S = \sqrt{35{,}811.67} = 189.24$.

Some properties of the standard deviation are:

- S measures how widely spread the data is about the mean.
- $S = 0$ only when there is no spread.
- S is strongly influenced by extreme outliers (because deviations are squared).

Measures of Symmetry and Skewness

The *coefficient of skewness*, S_k, is determined from the mean $\bar{x}$, the median $\tilde{x}$, and the standard deviation S by the formula

$$S_k = \frac{3 \cdot (\bar{x} - \tilde{x})}{S}.$$

We use the following rules of thumb for skewness and symmetry.

- If $S_k > 0$, the data are positively skewed (skewed right).
- If $S_k \approx 0$, the data are symmetric.

- If $S_k < 0$, the data are negatively skewed (skewed left).

We use the bell-shaped curve for easy-to-see symmetry. Figure 6.1 illustrates skewness.

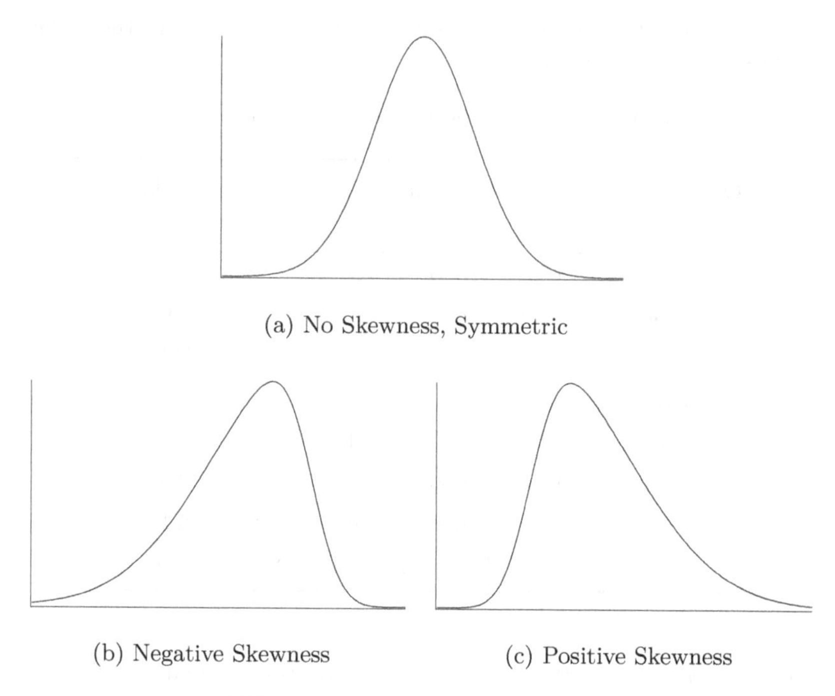

(a) No Skewness, Symmetric

(b) Negative Skewness

(c) Positive Skewness

FIGURE 6.1: Skewness Examples

For negative skewness, Figure 6.1(b), we see the distribution's tail is larger to the left (the negative direction) giving more area to the left. For positive skewness, Figure 6.1 (c), we see the distribution's tail is larger to the right (the positive direction) giving more area to the right.

Range

Range is a measure of the distance between the maximum and minimum values of the data. Often, this is provided a single number; however, more information is given when the range is specified as an interval. In the data of Table 6.5, the maximum value is 1867 and the minimum value is 1362. Taking the difference, $1867 - 1362 = 505$ gives 505 for the range. But what does 505 represent?

TABLE 6.5: Showing Range

1792	1666	1362	1614	1260	1867	1439

Giving the range as the interval $[1362, 1867]$ gives more information.

6.2.3 Displays of Data

For categorical data, think about what you want to display. We can show size or relative size only.

Categorical Displays

Pie Chart: The circle (pie) represents the whole or 100%. The wedges or pieces of the pie represent the proportion or part of the total for that category.

Bar Chart: Bars can be horizontal or vertical; they should be uniformly spaced and of the same width. The lengths or heights of the bars represent the values of the categories we wish to compare.

Quantitative Displays

Stem and Leaf: To produce a stem and leaf plot:

Step 1. Order the data.

Step 2. Separate according to the one or more leading digits. The list stems in a vertical column.

Step 3. The leading digit is the stem, and the trailing digit is the leaf. For example, for 32, the stem is 3 and the leaf is 2. Separate the stem and leaf by a vertical line.

Step 4. Indicate the units for the stem and leaf in the display.

Histogram: To produce a histogram:

Step 1. Determine and select the classes, 5-15 classes are usually best. Find the range (lowest to highest value). Classes should be evenly spaced if possible.

Step 2. Tally the data in the classes.

Step 3. Find the numerical (relative) frequencies from the tallies.

Step 4. Find the cumulative frequencies.

Step 5. Set the class intervals as a base and the tallies (or relative frequencies) as the height of a rectangle. The rectangles are centered at the mid-point of each class interval.

Boxplot: To produce a boxplot:

Step 1. Draw a horizontal measurement scale that includes all data within the range of data.

Step 2. Construct a rectangle whose left edge is the lower quartile value and whose right edge is the upper quartile value.

Step 3. Draw a vertical line segment in the box for the median value.

Step 4. Extend line segments from the rectangle to the smallest and largest data values (these are called whiskers).

Let's construct a boxplot for the data set

$$53, 55, 66, 69, 71, 73, 74, 75, 75, 76, 77, 78, 79, 82, 83, 85, 90, 92, 95, 99.$$

The data are already listed in numerical order. The values needed now are the range, the quartiles, and the median. We'll present the range as the interval [*smallest, largest*] = [53, 99]. As there are 20 values, an even number, the median is the average of the 10th and 11th elements: $(76 + 77)/2 = 76.5$. The quartile values are the medians of the lower and upper half of the data. The lower quartile is

$$53, 55, 66, 69, 71, 73, 74, 75, 75, 76.$$

The lower quartile's median is $(71 + 73)/2 = 72$. The upper quartile is

$$77, 78, 79, 82, 83, 85, 90, 92, 95, 99.$$

The upper quartile's median is $(83 + 85)/2 = 84$.

Draw a rectangle from 72 to 84 with a vertical line at 76.5, then draw a whisker to the left to 53 and to the right to 99. Figure 6.2 shows the boxplot.

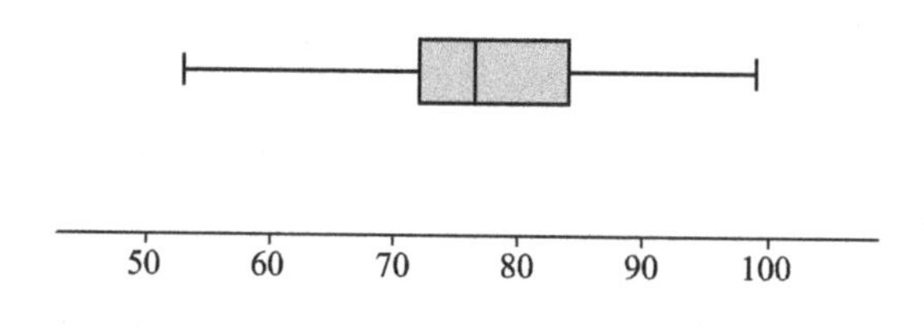

FIGURE 6.2: A Boxplot

This boxplot shows the data is skewed right.

Each plot type is shown below displaying the same data. The program *StemandLeafPlot* from Maplesoft is included in our *PSM* package; load it via a *with* command as usual.

```
> with(Statistics) :
  with(PSM, StemandLeafPlot) :
```

```
> theData := [12, 127, 28, 42, 39, 113, 42, 18, 44, 118, 44, 37, 113, 124, 37, 48,
    127, 36, 29, 31, 125, 139, 131, 115, 105, 132, 104, 123, 35, 113, 122, 42, 117,
    119, 58, 109, 23, 105, 63, 27, 44, 105, 99, 41, 128, 121, 116, 125, 32, 61, 37,
    127, 29, 113, 121, 58] :
> StemandLeafPlot(theData);
1  | 28
2  | 37899
3  | 12567779
4  | 12224448
5  | 88
6  | 13
7  |
8  |
9  | 9
10 | 45559
11 | 333356789
12 | 11234557778
13 | 129
> Histogram(theData, color = grey);
```

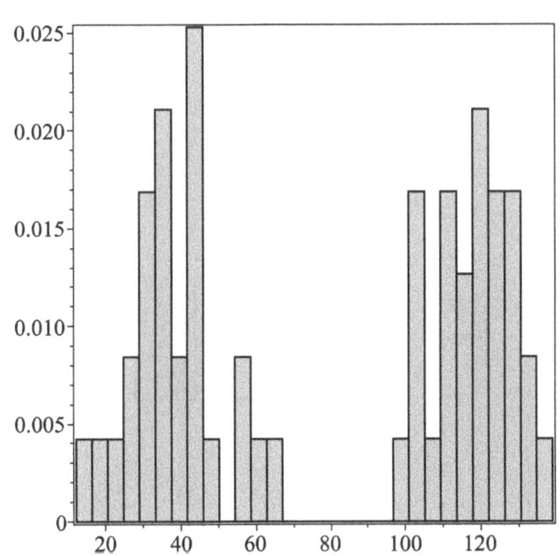

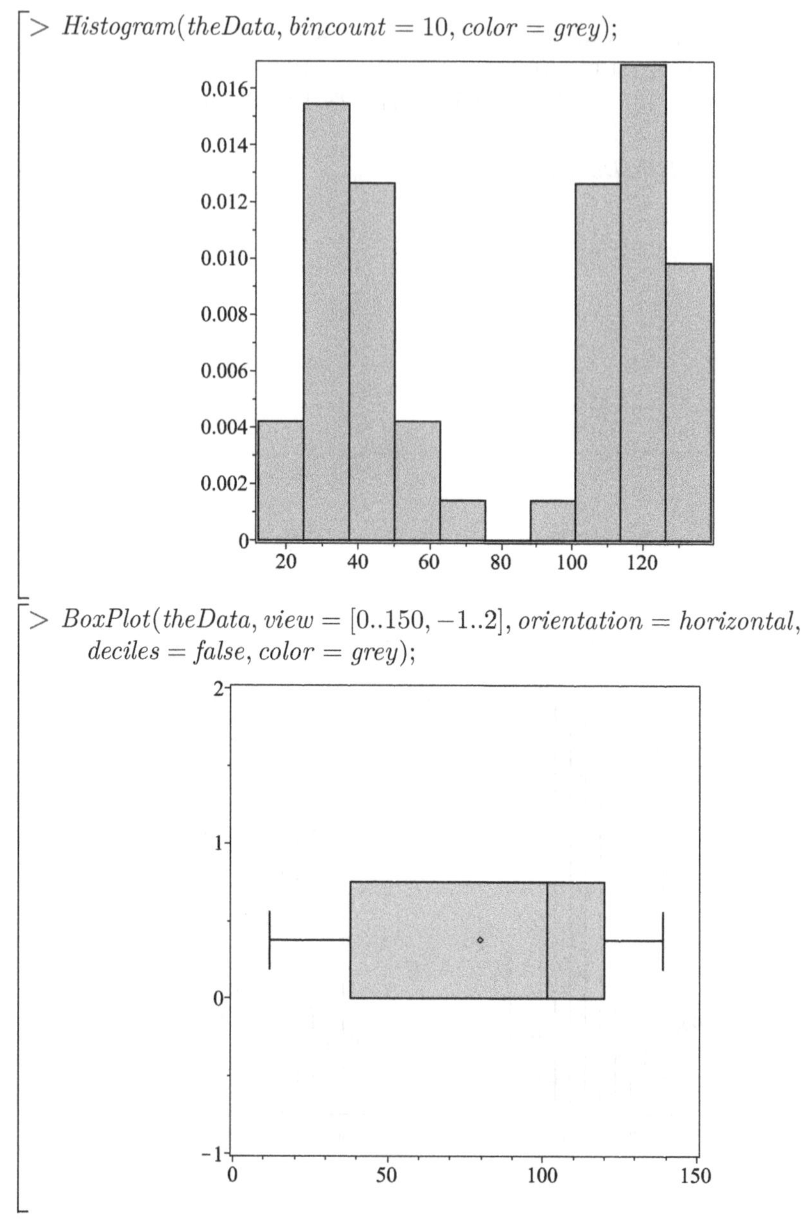

Compare the types of plots to see the advantages and strengths of each for illustrating features of the data.

Boxplots for Comparisons

Boxplots make a good way to compare data sets from multiple sources. For example, let's look at violence in regions in Afghanistan. Putting the 10 boxplots together into a single image allows us to compare many aspects such as medians, ranges, and dispersions quite easily, and to present comparisons intuitively.

Boxplots for data of casualties in Afghanistan through the years 2002–2009 is shown in the array in Figure 6.3. This figure is presented to the Senate. What information can the Senators interpret from this array of boxplots?

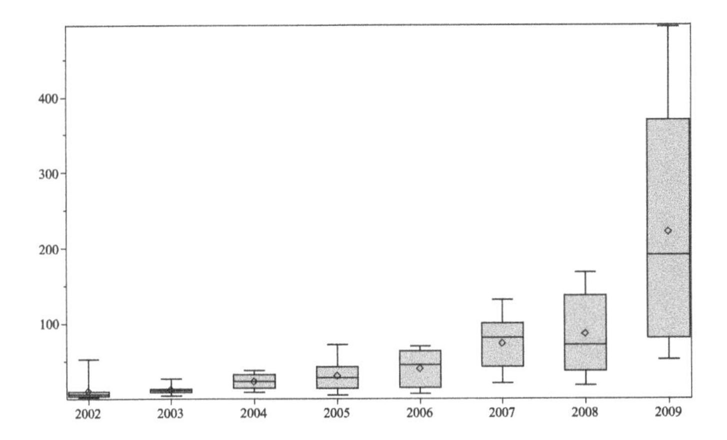

FIGURE 6.3: Afghanistan Casualties from 2002–2009

The classic text on graphically presenting data is Tufte's *The Visual Display of Quantitative Information.*[2]

6.2.4 Statistics and Maple

Maple can be used to obtain these useful statistical measures as well as produce displays of the data. The commands we will use are in the *Statistics* package. We illustrate many of the commands here, but suggest looking closely at the help page and contents of the *Statistics* package to see all that is available.

Example 6.3. Descriptive Statistics from Maple.

```
> with(Statistics) :
> weights := [100, 190, 168, 112, 125, 130, 175, 170, 290, 130, 176, 125, 106, 170,
    143.3, 145, 155, 150, 156, 189] :
```

[2] See https://www.edwardtufte.com/tufte/books_vdqi.

> *DataSummary*(*weights*);

$$\begin{bmatrix} mean = 155.264999999999958 \\ standarddeviation = 41.4554202538226875 \\ skewness = 1.54496270218672627 \\ kurtosis = 6.30858506444997058 \\ minimum = 100.0 \\ maximum = 290.0 \\ cumulativeweight = 20.0 \end{bmatrix}$$

> *FivePointSummary*(*weights*);

$$\begin{bmatrix} minimum = 100.0 \\ lowerhinge = 127.500000000000000 \\ median = 152.500000000000000 \\ upperhinge = 172.500000000000000 \\ maximum = 290.0 \end{bmatrix}$$

> *Mean*(*weights*);

$$155.265000000000$$

> *Mode*(*weights*);

$$\{125., 130., 170.\}$$

> *Variance*(*weights*);

$$1718.55186842105$$

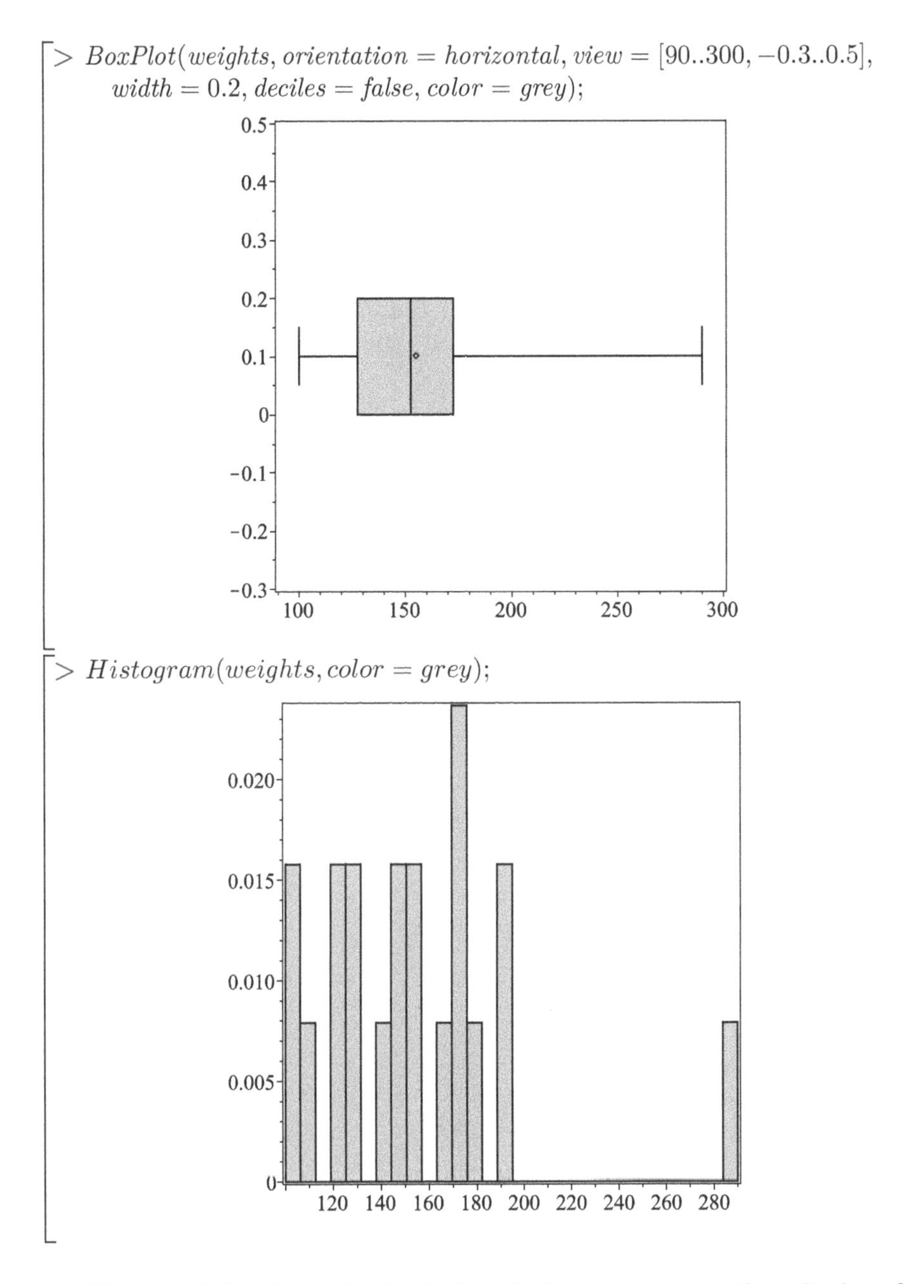

We entered the data, obtained statistical measures as well as displays for weight. We see that the data is skewed (positively) right.

Example 6.4. Homicide Data for Two Cities.
Problem: Data for homicides for two cities in Virginia are presented. We see clearly that Newport News has more killing than Hampton. The question is, is it twice as violent.[3]
Assumption: The data is accurate.

In this example, we only find the descriptive statistics and produce a few displays for comparison. Later in this chapter, we will answer the posed problem question using *hypothesis testing* techniques.

We will use Maple's implementation of *DataFrames*[4] — a data structure that makes it quite easy to build statistical reports and graphics from data sets.

Enter the data and create a *DataFrame.*

```
> Hampton := [6, 7, 13, 17, 9, 31, 12, 6, 8, 6, 9, 17, 6, 10, 10, 6, 6, 10, 6, 11, 14, 14,
    10, 14, 11, 14, 10, 6, 5, 13, 8, 9, 8, 12, 13, 13, 15, 8, 9, 11, 17, 9, 14, 22, 10, 15,
    23, 18];
  NewportNews := [14, 13, 22, 28, 33, 25, 22, 21, 12, 26, 24, 19, 24, 20, 20, 14, 15,
    16, 14, 19, 27, 18, 33, 24, 26, 27, 27, 18, 20, 17, 20, 34, 21, 29, 20, 20, 20, 30,
    17, 24, 23, 17, 21, 15, 25, 25, 31, 24];
> DataMatrix := ⟨⟨Hampton⟩ | ⟨NewportNews⟩⟩ :
> Homicides := DataFrame(DataMatrix,
    columns = ["Hampton", "NewportNews"])
```

HomicidesVA :=

	"Hampton"	"NewportNews"
1	6	14
2	7	13
3	13	22
4	17	28
5	9	33
6	31	25
7	12	22
8	6	21
...	...	...

[3] Source: *Daily Press*, March 4, 2018.
[4] For background on data frames, see, e.g., https://github.com/mobileink/data.frame/wiki/ What-is-a-Data-Frame%3F.

> *BoxPlot*(*Homicides*, *deciles* = *false*, *color* = *grey*, *width* = 0.15);

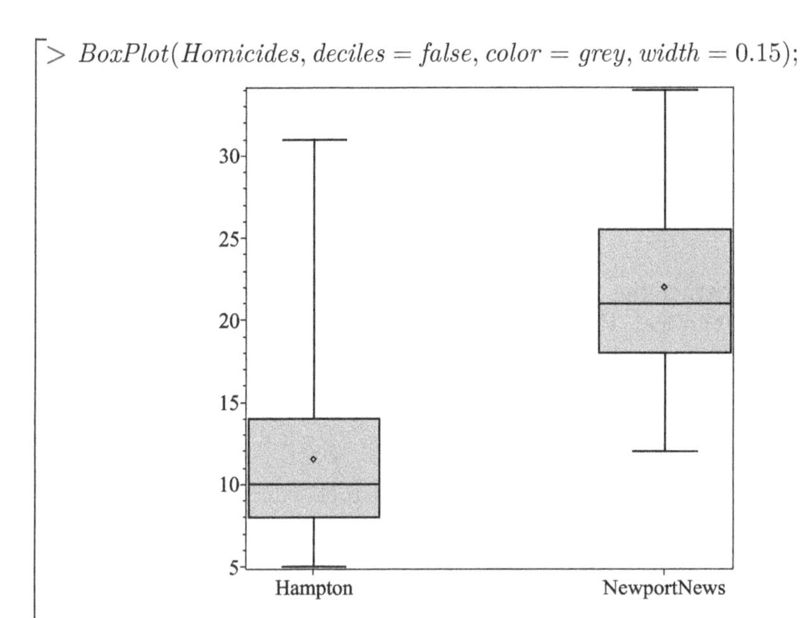

Comparing the boxplots clearly shows the homicides in Newport News are greater than the homicides in Hampton.

We obtain descriptive statistics for use later.

> *DataSummary*(*Homicides*);
FivePointSummary(*Homicides*);

	"Hampton"	"NewportNews"
mean	11.4791666666666679	21.9583333333333321
standarddeviation	5.09480161989642966	5.52348307214088852
skewness	1.47275726686510344	0.279189796393610579
kurtosis	6.01201992131035823	2.40124339167407497
minimum	5.0	12.0
maximum	31.0	34.0
cumulativeweight	48.0	48.0

	"Hampton"	"NewportNews"
minimum	5.0	12.0
lowerhinge	8.0	18.0
median	10.0	21.0
upperhinge	14.0	25.5000000
maximum	31.0	34.0

We see the mean in Hampton is 11.479 and the mean in Newport News in 21.958.

Exercises

1. Determine whether the following describe variables that would be quantitative or categorical. Provide an example of the value of such a variable and include the units, if any exist.

   (a) Flip a penny that lands as a "head" or "tail."
   (b) The color of M&M's.
   (c) The number of calories in the local fast food selections.
   (d) The life expectancy for males in the United States.
   (e) The life expectancy for females in the United States.
   (f) The number of babies born on New Year's Eve.
   (g) The dollars spent each month out of the allocated supply budget.
   (h) The number of hours that a soldier works per week.
   (i) The amount of car insurance paid per year.
   (j) Whether the bride is older, younger, or the same age as the groom.
   (k) The difference in ages of a couple at a wedding.
   (l) Average low temperature in Monterey, CA, in January.
   (m) The eye color of a student.
   (n) The gender of a soldier.
   (o) The number of intramural sports a person plays per year.
   (p) The distance a bullet travels from a specific weapon.
   (q) The number of deployments of a soldier in 3 years.
   (r) The location a bullet hits on a target.

2. Table 6.6 below presents the numbers of sports-related injuries treated in U.S. hospital emergency rooms in 2001, along with an estimate of the number of participants in that sport.

TABLE 6.6: Sports Injuries

Sport	Injuries	Participants	Sport	Injuries	Participants
Basketball	646,678	26,200,000	Fishing	84,115	47,000,000
Bicycling	600,649	54,000,000	Skateboard	56,435	8,000,000
Baseball	459,542	36,100,000	Hockey	54,601	1,800,000
Football	453,684	13,300,000	Golf	38,626	24,700,000
Soccer	150,449	10,000,000	Tennis	29,936	16,700,000
Swimming	130,362	66,200,000	Water skiing	26,663	9,000,000
Weightlifting	86,398	39,200,000	Bowling	25,417	40,400,000

(a) If we want to use the number of injuries as a measure of the hazardousness of a sport, which sport is more hazardous between bicycling and football? Between soccer and hockey?

(b) Use either a calculator or a computer to calculate the *rate* of injuries per thousand participants. *Rate* is defined as the average number of injuries out of the total participants.

(c) Rank order the rate measure for the sports.

(d) How do your answers in part (a) compare if we do the hazard analysis using the *rates* in (b). If different, why are the results different?

3. Make a boxplot of each data set below and comment about the shape of the plot.

   (a) $\{100, 105, 111, 115, 121, 129, 131, 131, 133, 135, 137, 145, 146, 150, 160\}$,

   (b) $\{0.10, 0.15, 0.22, 0.23, 0.50, 0.62, 0.62, 0.65, 0.66, 0.69, 0.72\}$.

   (c) $\{63, 65, 72, 81, 83, 85, 92, 93, 94, 105, 106, 121, 135\}$.

4. Make a pie chart for the following data.

TABLE 6.7: Marriage Data

Never Married	Married	Widowed	Divorced
43.9 million	116.7 million	13.4 million	17.6 million

What type of information is best displayed by a pie chart?

5. Make a bar chart for the data in Table 6.8: female doctorates as a percentage of graduates in that field that are females. Can you make a pie chart? What do you have to do first?

TABLE 6.8: Female Doctorates Percentages

Computer Science	15.4%
Education	60.8%
Engineering	11.1%
Life Sciences	40.7%
Physical Sciences	21.7%
Psychology	62.2%
Mathematics	10.0%

6. Display the following data: In 1995, there were 90,402 deaths from accidents in the US. Among these there were 43,363 from motor vehicles, 10,483 from falls, 9072 from poisoning, 4350 from drowning, and 4235 from fires. How many deaths were due to other unknown causes?

7. In Math I class last semester, the final averages were: 88, 63, 82, 98, 89, 72, 86, and 70. Display the data as a stem-and-leaf plot. Are they symmetric? Are they skewed?

8. In Math II class last semester, the final averages were: 66, 61, 78, 54, 75, 40, 78, 91, 84, 82, 76, and 65. Display the data as a histogram. Are they symmetric? Are they skewed?

9. Using the grades in Math I (7.) and Math II (8.), display the data as two boxplots side by side. Is each display symmetric? Is each display skewed? Can you compare the two data sets? Which class had the higher grades? Which class has grades that are more spread out? Does the symmetry and skewness of the data sets tell us anything about the grades?

6.3 Introduction to Classical Probability

Elementary probability theory is required for understanding this chapter. We will provide a quick review of some important concepts in probability.

We define the *probability* that *event* A occurs, $P(A)$, as the number of times A occurs out of the total number of possible outcomes in the *sample space*. An event is any collection of results or outcomes of an experiment. A sample space is a listing of all possible outcomes from an *experiment*. An experiment is any process that allows a researcher to obtain observations (data). In a flip of a fair coin 2 times, the sample space is the set of all possible outcomes of 2 flips. If we label a head, H, and a tail, T, then the possible outcomes (assuming the flips are recorded consecutively) are

$$\{HH, HT, TH, TT\}$$

This set constitutes the entire sample space. Let's call event A the event that exactly one head appeared in the two flips. That occurred in flip HT and flip TH, or 2 times. Since there were four possible outcomes, the probability of event A occurring is $P(A) = 2/4 = 0.5$. How does this change if we flip two coins at one time?

Let's consider a tennis match between Player A and Player B where the winner is the first to win three sets. The sample space for the winner is:

$$\{AAA, AABA, AABBA, AABBB, ABAA, ABABA, ABABB, ABBAA,$$

$$ABBAB, ABBB, BAAA, BAABA, BAABB, BABAA, BABAB,$$

$$BABB, BBAAA, BBAAB, BBAB, BBB\}$$

If A and B were equally likely to win a set, then we can compute the probability of each event in the sample space. The probability that A wins in 3 straight sets is $1/20$, in 4 sets is $3/20$, and in 5 sets $6/20$. Thus, the probability that A wins is $1/20 + 3/20 + 6/20 = 0.5$.

Now let's assume that A is a higher ranked player with odds to win a set of 3:1. We can re-compute the probabilities that A wins from the given sample space in 3, 4 or 5 sets.

$$P(A \text{ wins in 3 sets}) = (0.75^3) = 0.421875$$
$$P(A \text{ wins in 4 sets}) = 3(.75^3)(0.25) = 0.3164065$$
$$P(A \text{ wins in 5 sets}) = 6(0.75^3)(0.25^2) = 0.158203125$$

The probability that A wins the match $P(A \text{ wins the match}) = 0.896$.

Rules of Probability

There are many important rules for probability, The rules we will use include the following:

1. The *Law of Large Numbers:* If an experiment is repeated again and again, the relative frequency of an event approaches its probability. We see this in simulations.

2. *Addition Rule:* $P(A \text{ or } B) = P(A) + P(B) - P(A \text{ and } B)$

3. For any event A in the sample space S,

   (a) $P(A) > 0$,

   (b) $P(A) \leq 1$,

   (c) $P(S) = 1$,

   (d) $P(\text{not } A) = P(\neg A) = 1 - P(A)$.

4. For any events A and B in the sample space S,
   (a) if A and B are mutually exclusive, $P(A \text{ and } B) = 0$,
   (b) if A and B are independent, $P(A \text{ and } B) = P(A) \cdot P(B)$,
   (c) otherwise, $P(A \text{ and } B) = P(A) + P(B) - P(A \text{ or } B)$.

5. *Conditional Probability:* $P(A \text{ given } B \text{ has occurred already}) = P(A|B) = P(A \text{ and } B)/P(B)$.

6.3.1 Discrete Distributions in Problem Solving

We will use several probability distributions for *discrete random variables.* A random variable is a rule that assigns a number to every outcome of a sample space. A discrete random variable takes on counting numbers 0, 1, 2, 3, ..., etc. The possible values are either finite or countable. Then, a probability distribution gives the probability for each value of the random variable.

Let's return to our coin flipping example earlier. Let the random variable F be the number of heads of the two flips of the coin. The possible values of the random variable F are 0, 1, and 2. We can count the number of outcomes that fall into each category of F as shown in the *probability mass function* in Table 6.9.

TABLE 6.9: Probability Table

Random Variable	0	1	2
Occurrences	1	2	1
Corresponding Events	TT	TH, HT	HH
$P(F)$	1/4	2/4	1/4

Note that the $\sum P(F) = 1/4 + 2/4 + 1/4 = 1$. This is a rule for any probability distribution: the sum of all probabilities must be 1. Summarizing:

1. $P(\text{each event}) > 0$
2. $\sum_{\text{all events}} P(\text{event}) = 1$

Thus, the coin flip experiment is a discrete probability distribution.

All probability distributions have means, μ, and variances, σ^2. We can find the mean and the variance for a random variable X using the following formulas:

$$\mu = E[X] = \sum_{\text{all } x} x \cdot P(X = x)$$

$$\sigma^2 = E[X^2] - (E[X])^2$$

For our example, we compute the mean and variance as follows:

$$\mu = E[X] = \sum_{\text{all } x} x \cdot P(X = x) = 0 \cdot \frac{1}{4} + 1 \cdot \frac{2}{4} + 2 \cdot \frac{1}{4} = 1$$

$$\sigma^2 = E[X^2] - (E[X])^2$$
$$= \left[\sum_{\text{all } x} x^2 \cdot P(X = x)\right] - E[X]^2 = \left(0 \cdot \frac{1}{4} + 1 \cdot \frac{2}{4} + 4 \cdot \frac{1}{4}\right) - 1^2 = 0.5$$

That is, $\mu = 1$ and $\sigma^2 = 0.5$.

There are several discrete distributions that often arise in modeling: Bernoulli, Binomial, and Poisson.

Binomial Distribution

Consider an experiment made up of a repeated number of independent and identical trials having only two outcomes, such as tossing a fair coin {*Head, Tail*}, or a {*red, green*} stoplight. Experiments with only two possible outcomes are called *Bernoulli trials.* Often they are found by assigning either an S (success) or an F (failure), or a 0 or 1, to an outcome. Something either happened (1) or did not happen (0).

A *binomial experiment* is found by counting the number of successes in N trials.

Let's define a binomial experiment formally:

1. Consists of n trials where n is fixed in advance.
2. Trials are identical and can result in either a success or a failure.
3. Trials are independent.
4. The probability of a success is constant from trial to trail.

The formula (PDF) for a binomial distribution is

$$b(x; n, p) = p(X = x) = \binom{n}{x} p^x (1 - p)^{n-x}$$

for $x = 0, 1, \ldots n$. The cumulative distribution formula (CDF) of a binomial distribution is

$$B(x; n, p) = p(X \leq x) = \sum_{y=0}^{x} \binom{n}{y} p^y (1 - p)^{n-y}$$

for $x = 0, 1, \ldots n$. For a binomial distribution, we have

$$\text{Mean: } \mu = n\,p$$
$$\text{Variance: } \sigma^2 = n\,p\,(1-p)$$

For example, our coin flip experiment follows the rules for a binomial experiment. The probability that we get 1 head in 2 flips is

$$P(X=1) = \binom{2}{1} 0.5^1\,(1-0.5)^{2-1} = 0.5$$

Or if we wanted 5 heads in 10 flips of a fair coin, then we can compute

$$P(X=5) = \binom{10}{1} 0.5^5\,(1-0.5)^{10-5} = 0.246$$

Example 6.5. Light Bulb Problems.
Light bulbs are manufactured in a small local plant. In testing the light bulbs, prior to packaging and shipping, they either work, S, or fail to work, F. The company cannot test all the light bulbs, but does test a random batch of 100 light bulbs per hour. In this batch, they found 2% did not work, but all batches were shipped to distributors.

As a distributor, you are worried about past performance of the light bulbs which you sell individually off the shelf. If a customer buys 20 light bulbs, what is the probability that all work?

Problem: Predict the probability that x light bulbs out of N work.

Assumption: The light bulbs follow the binomial distribution rules stated earlier.

We will use Maple to calculate the probabilities according to the binomial PDF.

$$> \mathit{Binomial} := (x, n, p) \to \mathit{binomial}(n, x) \cdot p^x \cdot (1-p)^{n-x};$$

$$> \mathit{Binomial}(20, 20, 1-0.02);$$

$$0.6676079718$$

Poisson Distribution

A random variable is said to have a *Poisson distribution* if the probability of a certain number of events happening over a fixed time interval occurs with a specific constant rate and independently of the time elapsing since the last event. The probability distribution function (PDF) of X for a Poisson distribution is

$$p(x; \lambda) = \frac{e^{-\lambda}\lambda^x}{x!}$$

for $x = 1, 2, \ldots, n$ and with $\lambda > 0$. We consider λ as a *rate per unit time* or *per unit area*. For a Poisson distribution, we have

$$\text{Mean: } \mu = \lambda$$
$$\text{Variance: } \sigma^2 = \lambda$$

For example, let X represent the number of flaws on the surface of a randomly selected crystal glass. It has been found that on average 5 flaws are found per glass surface. Find the probability that a randomly selected glass has exactly 2 flaws.

$$> Poisson := (x, \lambda) \to evalf\left(\frac{e^{-\lambda}\lambda^x}{x!}\right):$$

$$> Poisson(2, 5);$$

$$0.08422433750$$

A *Poisson process* is a Poisson distribution that varies over time (generally over time, but may also vary over space). There exists a rate, called α for a short time period. Over a longer period of time, λ becomes αt. For example, suppose your pulse is read by an electronic machine at a rate of five times per minute. Find the probability that your pulse is read 15 times in a 4-minute interval. With $\alpha = 5$, we have

$$\lambda = \alpha t = 5 \text{ reads } \cdot 4 \text{ minutes } = 20 \text{ pulse reads in a 4 minute period,}$$

so that

$$p(X = 15) = \frac{e^{-20}20^{15}}{15!} \approx 0.0516$$

Exercises

1. If 75% of all purchases at Walmart are made with a credit card and X is the number among ten randomly selected purchases made with a credit card, then find the following:

   (a) $p(X = 5)$
   (b) $p(X \leq 5)$
   (c) μ and σ^2

2. Russell Stover produces fine chocolates, and it's known from experience that 10% of its chocolate boxes have flaws and must be classified as "seconds."

   (a) Among six randomly selected chocolate boxes, how likely is it that one is a second?

(b) Among the six randomly selected boxes, what is the probability that at least two are seconds?

(c) What is the mean and variance for "seconds"?

3. Consider the following TV ad for an exercise program: 17% of the participants lose 3 pounds, 34% lose 5 pounds, 28% lose 6 pounds, 12% lose 8 pounds, and 9% lose 10 pounds. Let X = the number of pounds lost on the program.

   (a) Give the probability mass function of X in a table.

   (b) What is the probability that the number of pounds lost is at most 6? At least 6?

   (c) What is the probability that the number of pounds lost is between 6 and 10?

   (d) What are the values of μ and σ^2?

4. A machine fails on average 0.4 times a month (30 consecutive days). Determine the probability that there are 10 failures in the next year.

6.4 Reliability in Engineering and Business

Consider your cell phone, graphing calculator, or automobile. How often do you replace them? Probably not that often because they perform well over a reasonably long period of time. If they do last for a long period of time, we say that these systems are *reliable.* The reliability of a system or its components is the probability that it will not fail over a specific time period t. Let's define $f(t)$ to be the failure rate of an item, component, or system over time t where $f(t)$ is a probability distribution. Let $F(t)$ be the cumulative distribution function corresponding to $f(t)$ as discussed in the last section. We define the reliability of an item, component, or system by

$$R(t) = 1 - F(t)$$

Man-machine systems, whether electrical, mechanical, analog, digital, etc., consist of components which interact to form systems. Consider your personal computer, stereo, PlayStation, or automobile. We want to build simple models to examine the reliability of complex systems. We consider relationships in series, parallel, and combinations of these. Although individual item failure rates can follow a wide variety of distributions, we consider only a few elementary examples using distributions we have already reviewed.

Modeling Series Systems

Let's investigate a system with n components where each of the individual components must work in order for the system to function. Figure 6.4 shows a model of this type of system.

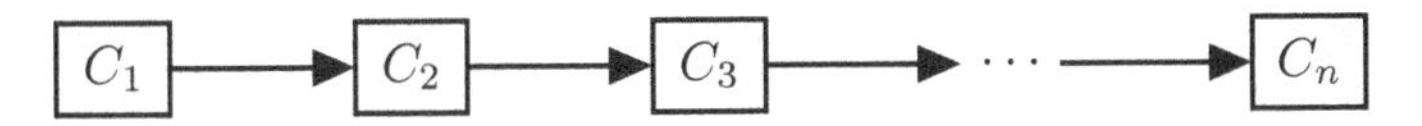

FIGURE 6.4: Series System

If we assume the components are mutually independent, the reliability of this type of system is easy to compute. We denote the reliability of component i at time t by $R_i(t)$. In other words, $R_i(t)$ is simply the probability that component i will function continuously from time 0 until time t. We are interested in the reliability of the entire system of n components. Since these components are mutually independent, the system reliability is

$$R(t) = R_1(t) \cdot R_2(t) \cdot \cdots \cdot R_n(T)$$

Example 6.6. Series System.
A *series system* is one that performs well as long as every item or component is performing well. Consider a video game with four components: the game, the controller, the CPU (such as a PlayStation), and the television as displayed in Figure 6.5.

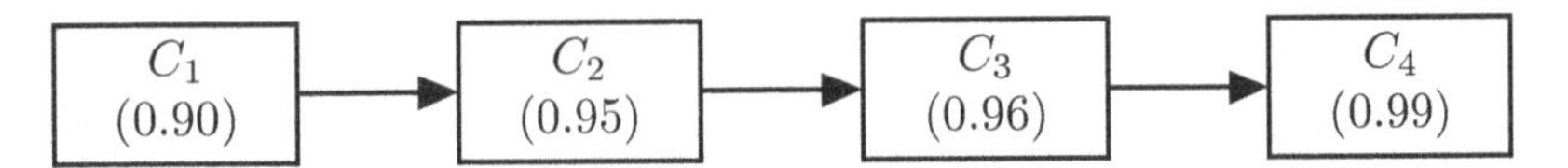

FIGURE 6.5: Video Game Series System

This is an example of a series system because failure of any one of the independent items will result in failure of the whole system. If the component reliabilities are given by $R_1(t) = 0.90$, $R_2(t) = 0.95$, $R_3(t) = 0.96$, and $R_4(t) = 0.99$ then the system reliability is defined to be their product

$$R_s(t) = R_1(t) \cdot R_2(t) \cdot R_3(t) \cdot R_4(t) = (0.90)(0.95)(0.96)(0.99) = 0.812592$$

Thus, it is assumed that the system will work correctly 82% of the time. Note that the reliability of the system is less than that of any single component because each has a reliability that is less than 1.0.

Modeling Parallel Systems (Two Components)

Now we consider a *parallel system* with two components where only one of the components must work for the system to function. A system of this type is depicted in Figure 6.6.

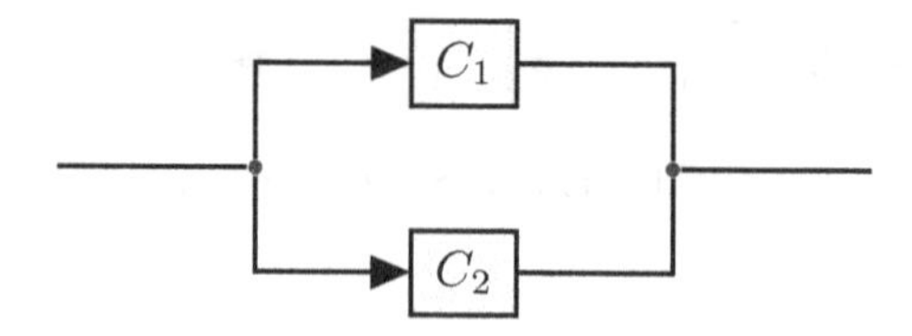

FIGURE 6.6: Parallel System

Notice that in this situation the two components are *both* put in operation at time 0; they are both subject to failure throughout the period of interest. Only when *both* components fail before time t does the system fail. Again we also assume that the components are independent. The reliability of this type of system can be found using the well-known *inclusion-exclusion principle* model[5]

$$P(A \cup B) = P(A) + P(B) - P(A \cap B)$$

In this case, A is the event that the first component functions for longer than some time t, and B is the event that the second component functions longer than the same time t. Since reliabilities *are* probabilities, we can translate the above formula into

$$R(t) = R_1(t) + R_2(t) - R_1(t) \cdot R_2(t)$$

Example 6.7. Parallel System.
A *parallel system* is one that performs as long as a single one of its components remains operational. Your home communications system of cordless telephone and conventional telephone is a system like Figure 6.6. Note that there are two separate and independent routes to transverse the network (input to output); either can be used to complete the call. If either component is working, the call can be completed. Let C_1 (wireless) and C_2 (conventional) have reliabilities of 0.95 and 0.96, respectively.

We apply the definition of the system reliability for parallel components

$$R_s(t) = R_1(t) + R_2(t) - R_1(t) \cdot R_2(t) = 0.95 + 0.96 - (0.95)(0.96) = 0.998$$

Note that in parallel relationships the system reliability is higher than any single component reliability.

[5] Recall the Addition Rule, (pg. 253).

Example 6.8. Series and Parallel Combination.
Now we consider a system that is a combination of series and parallel components. Consider the video game of Example 6.6 now with 2 controllers in parallel as shown in Figure 6.7. The system's reliability can be calculated as

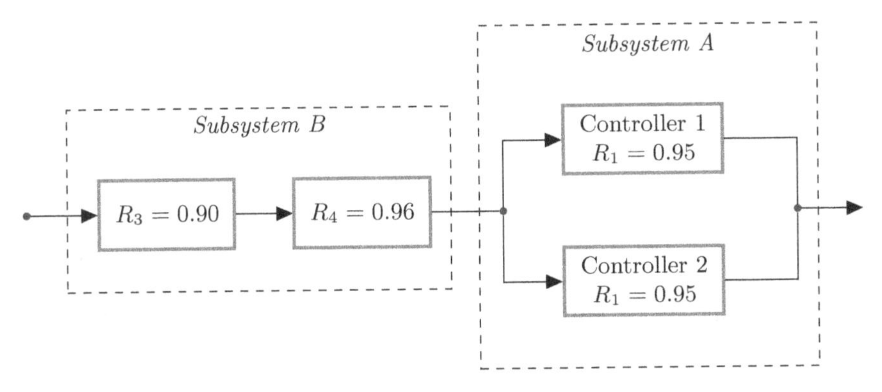

FIGURE 6.7: Series and Parallel Video Game System

two sub-systems in series. Sub-system A is the two controllers in parallel and sub-system B is the console and television in series. The subsystem technique is common in analyzing electronic circuits. First, the two subsystems:

$$R_A(t) = R_1(t) + R_2(t) - R_1(t)R_2(t) = 0.95 + 0.95 - (0.95)(0.95) = 0.9975$$
$$R_B(t) = R_3(t)R_4(t) = (0.90)(0.96) = 0.864$$

Subsystems A and B are in series, so

$$R_S(t) = R_A(t)R_B(t) = (0.9975)(0.864) = 0.86184$$

The system's reliability as a whole is 0.86184.

Modeling Active Redundant Systems

Suppose a system has n components, all of which begin operating (are active) at time $t = 0$. The system continues to function properly as long as at least k of the components do *not* fail. In other words, if $(n - k) + 1$ components fail, the system fails. This type of system is called an *active redundant system*. An active redundant system can be modeled as a parallel system of subsystems or components as shown in Figure 6.8.

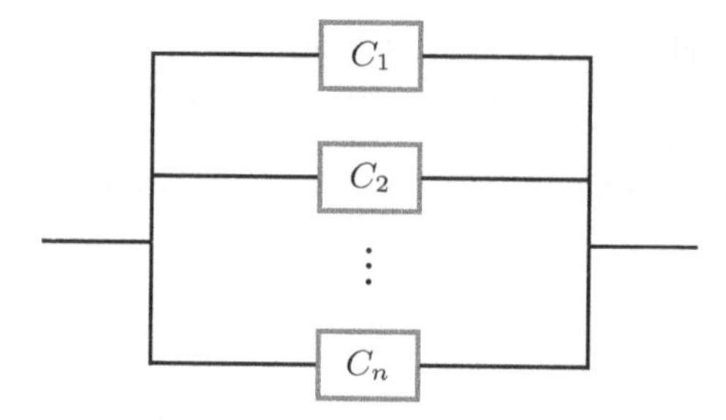

FIGURE 6.8: Active Redundant System

We assume that all n components are identical and will fail independently. If we let T_i be the time to failure of the ith component, then the T_i terms are independent and identically distributed for $i = 1, 2, 3, \ldots, n$. Thus $R_i(t)$, the reliability at time t for component i, is identical for all components.

Recall that our system operates if at least k components function properly. Define the random variables X and T as follows:

$$X = \text{number of components functioning at time } t, \text{ and}$$
$$T = \text{time to failure of the entire system.}$$

Then we have

$$R(t) = P(T > t) = P(X \geq k).$$

We now have n identical and independent components with the same probability of failure by time t. This situation corresponds to a binomial experiment, so we can solve for the system reliability using the binomial distribution with parameters n and $p = R_i(t)$.

Example 6.9. Undercover Police Stakeout.
Three undercover police officers on a stakeout have been instructed to place 15 "listening" devices to detect and listen to plans for a robbery. The experts estimate that all communications can be heard as long as at least 12 of the devices are operating. These devices are assumed to be in parallel and fail independently. Each device has a 0.75 probability of working properly for 24 hours. The reliability of the systems for 24 hours is computed as

$$R(24) = P(T > 24) = P(X \geq 12) = 1 - P(X \leq 11) = 1 - 0.5387 = 0.4613.$$

Thus, the reliability of the system for 24 hours is 0.4613.
If only 10 had to work, then $R(24) = 1 - P(X \leq 9) = 0.8517$.

Problem Solving with Standby Redundant Systems

Active redundant systems can sometimes be inefficient. These systems require only k of the n components to be operational, but all n components are initially in operation and thus subject to failure. An alternative is the use of spare components. Such systems have only k components initially in operation; exactly what we need for the whole system to be operational. When a component fails, we have a spare "standing by" which is immediately put in to operation. For this reason, we call these *Standby Redundant Systems.* Suppose our system requires k operational components and we initially have $n - k$ spares available. When a component in operation fails, a decision switch causes a spare or standby component to activate (become an operational component). The system will continue to function until there are fewer than k operational components remaining. In other words, the system works until $n - k + 1$ components have failed. We will consider only the case where one operational component is required (the special case where $k = 1$) and there are $n - 1$ standby (spare) components available. We will assume that a decision switch (DS) controls activation of the standby components instantaneously and 100% reliably. We use the model shown in Figure 6.9 to represent this situation.

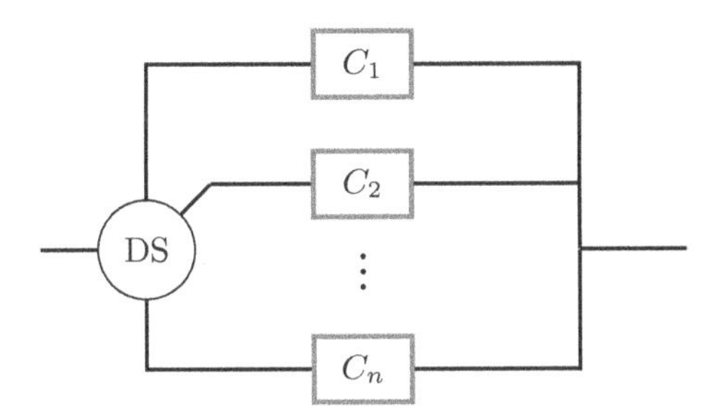

FIGURE 6.9: Standby Redundant System

If we let T_i be the time to failure of the ith component, then the T_i's are independent and identically distributed for $i = 1, 2, 3, \ldots, n$. Thus $R_i(t)$ is identical for all components. Let T = time to failure of the entire system. Since the system fails only when all n components have failed, and component is put into operation only when component i fails, it is easy to see that

$$T = T_1 + T_2 + \cdots + T_n.$$

In other words, we can compute the system failure time easily if we know the failure times of the individual components.

We assume the decision switch (DS) is 100% reliable and instantaneously switches to a standby component. Thus $R_i(t)$ is identical for all components. The reliability of the system is equal to the probability $P(X < N)$, which is the probability that fewer than N components fail during the interval $(0, t)$. This is a property of a *Poisson experiment.* In fact, we can show that the random variable X follows a **Poisson distribution** with parameter $\lambda = \alpha t$ where α is the component failure rate. The reliability for some specific time t becomes

$$R(t) = P(X < N) = \text{Poisson}(\lambda = \alpha t, N - 1).$$

Example 6.10. Game Boy Tetris.
Consider playing Tetris on Game Boy. A Game Boy runs on AA batteries. For 24 hours, the reliability of the batteries with constant use is 0.80. You hate it when the game quits due to low battery power so you carry spare battery packs. The addition of a spare should increase the systems reliability. In this case, $N = 3$ total battery packs and $\alpha = 1/30$ per hour. The reliability is

$$R(24) = P[X < 3\lambda = (1/30) \cdot 24] = P(0 \leq X \leq 2) = 0.9526.$$

The system reliability for 24 hours with the spare battery packs is 0.9526.

If we want to keep the reliability at 95% for 48 hours, how many spare battery packs are required? We use trial and error to solve this problem. Using the Poisson distribution, we find the reliability for 48 hours and then vary the number of spares until the reliability is above 95%.

TABLE 6.10: Battery Packs versus Reliability

Number of Battery Packs as Spares	System Reliability (using Poisson)
2	0.7834
3	0.9212
4	0.9763

Table 6.10 indicates that 4 spare battery packs are required for the 48-hour period to insure at least a 95% reliability of the Game Boy. The analysis is based solely on the batteries.

Exercises

1. Your college football stadium is to be the site of a televised night football game. The TV lighting experts have told you that a minimum level of lighting must be maintained throughout the game. If more than 3 of the stadium's 12 light grids go out, the field will not be illuminated enough for the TV coverage. The maintenance crew services all the light grids prior to the game; this alone yields a probability of 0.7596 of working for the required 5.5 hours. What is the reliability of the lighting system for providing minimum TV illumination throughout the game? What assumptions are made to model this problem?

2. You are the emergency preparedness (EP) coordinator for an East Coast beach resort. You fear hurricane season and you know that you must maintain power to your EP center during weather emergencies. You use generators for power and back-up power. Their failure rate has been about 1 every 7.5 hours of operation. Assuming all generators have the same time to failure and that they fail independently, find the reliability of the power system for 10 hours if 2 spares are available.

3. In Exercise 2, how many spares are needed so the system has at least a 99% reliability for the 10-hour period.

4. Consider a car stereo system with CD player, AM-FM radio turner, speakers (dual), and power supply as displayed with their reliabilities in Figure 6.10. What assumptions are required by your model? Determine the reliability of the system.

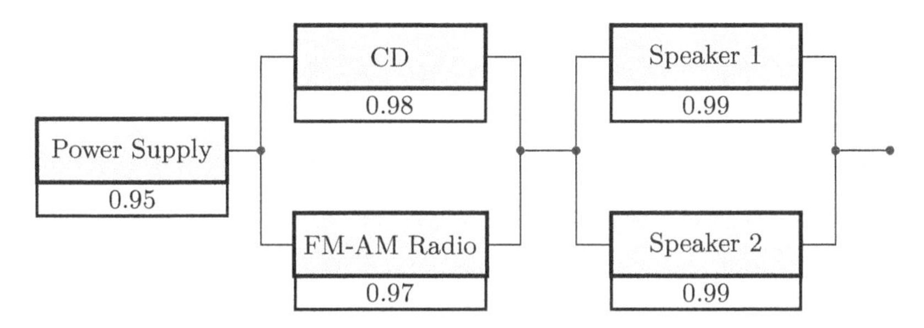

FIGURE 6.10: Car Sound System

5. Your personal computer system component block diagram with each component's reliability is shown in Figure 6.11. What assumptions are required? Determine the system's reliability.

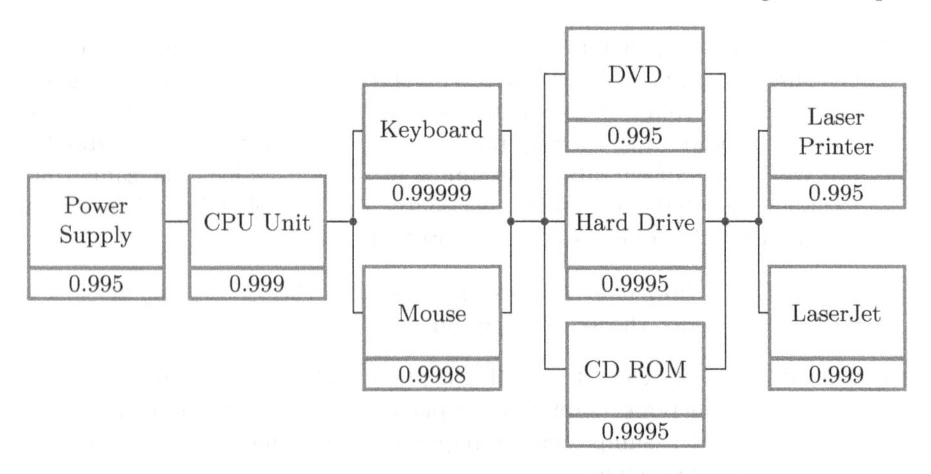

FIGURE 6.11: Computer System Block Diagram

Projects

Project 6.1. Two alternate designs are submitted for the Mars landing probe. (Figure 6.12.) The mission is to land safely on mars, collect soil and other samples, and then return to the shuttle. Determine which design you would recommend to NASA for the mission? What assumptions are required? What assumptions are reasonable?

Power, Communications, and Storage are the same for each alternative. The Landing and Rocket Modules are different for each alternative. Power is the main power supply whose reliability is 0.998. Communications is a parallel system of two radios each having a reliability of 0.995. Storage is a storage arm component whose reliability is 0.998. Each of the five major modules is in series.

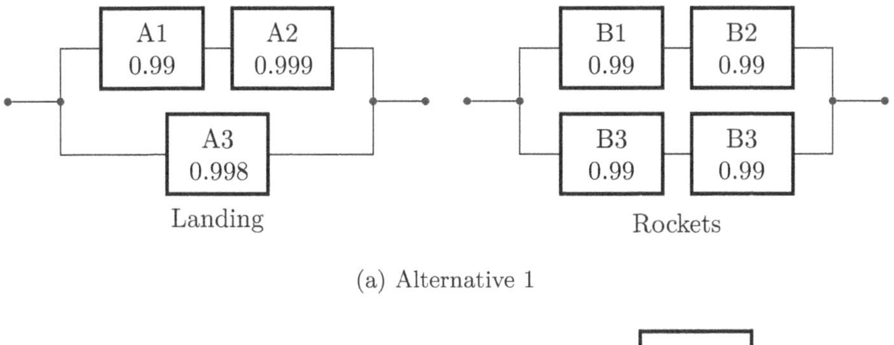

(a) Alternative 1

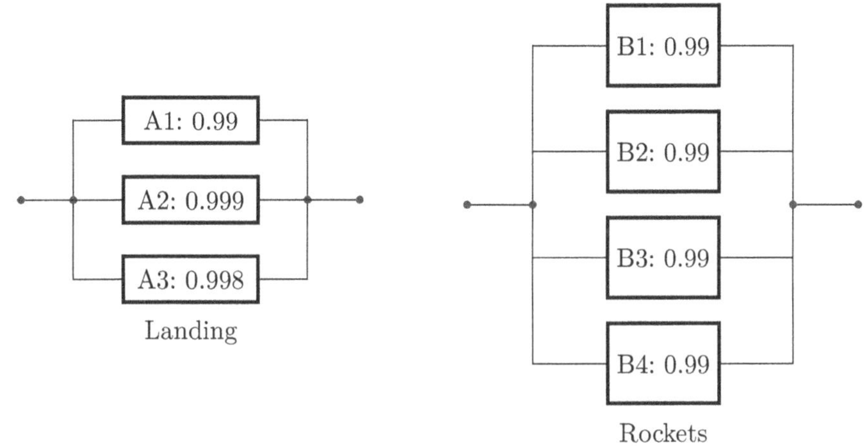

(b) Alternative 2

FIGURE 6.12: Mars Lander Alternatives

Project 6.2. You are assigned as the Transportation Coordinator for emergency evacuations. It is your job to analyze the maps below with bridges (denoted as A, B, C, D, E, F, G, H, and J in Figure 6.13) to determine the reliability of the bridge system that must be used to travel safely out of the area. Additionally you must determine the best route over which to travel. The bridges operate independently of each other; the lifetime of each bridge is distributed exponentially with reliability $= 1 - e^{at}$. You find historical records to indicate the failure rates per year (52-week period) for the bridges are

TABLE 6.11: Bridge Failure Rate

Bridge	A	B	C	D	E	F	G	H	J
Failure Rate	2	2	3	1	8	7	1	6	3

At bridge position E the Corps of Engineers has two spare bridges to use in case one fails. At bridge position J, it is determined that at least 2 of the four bridges must operate for effective transportation.

(a) Determine the reliability of the bridge system for any two-week period.

(b) Determine the best route over which to travel.

(c) Consider Points 1 and 2 as decision points for your routes. Compute the reliabilities of each route. Which has the highest reliability?

(d) If you need a three-week window, will the reliabilities increase or decrease? Why?

(e) Make a recommendation to the governor for an evacuation plan.

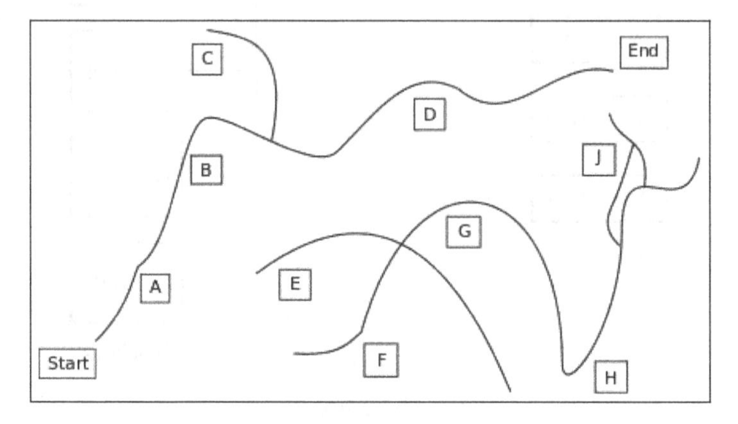

FIGURE 6.13: Map of the Bridges

6.5 Case Study: Airlines Overbooking Model

Recently on an airplane, several passengers were overheard complaining that they were "bumped off their flight" from Rome to New York City because the plane was overbooked. The passengers were told that the airlines had sold over a hundred more seats than the plane held. Due to a delay and their late arrival to the Rome airport, these passengers found that their seats had been given to other passengers. Of course, the "bumped" passengers were compensated, but they were extremely upset and threatening to sue the airlines. Is overbooking a common event? Let's build a model from the standpoint of the airlines and examine the overbooking phenomenom.

Airlines know that only a certain percentage of passengers who have made reservations on a particular flight will actually take that flight. Consequently, most airlines *overbook* — that is, they take more reservations than the capacity of the aircraft. Occasionally, more passengers will want to take a flight than the capacity of the plane leading to one or more passengers being *bumped* — voluntarily or involuntarily — and thus unable to take the flight for which they had reservations.

Airlines deal with bumped passengers in various ways in order to try to compensate them. They will be allowed to fly on a later flight, and they may be given some kind of cash settlement or vouchers for free plane tickets.

We will model the effects of the various types of overbooking strategies that an airline company might employ. In particular, we'll examine the effect that different overbooking schemes have on the revenue received by the airline company. Given the probability that a person who purchased a ticket will show up for his/her flight, we will attempt to discover an optimal overbooking strategy, i.e., the number of people by which an airline should overbook on a particular flight so that the company's revenue is maximized. As you might imagine, there are many variables to consider in analyzing these types of situations. We will simplify as many as necessary for our analysis.

Problem: Predict the number of overbooked reservations to be made on an airline in order to maximize the revenue of the airline.

Assumptions: For the purposes of this model, we are going to assume the following:

1. There is only one passenger class available, and all n passengers pay the same price for a ticket. No first-class seats are available, and there is no business class.

2. Bumped passengers are given cash settlements, all of which are of the same amount. This settlement is over and above the price of the ticket that is also refunded to the passenger.

3. The probability that a passenger will actually take the flight is the same from passenger to passenger; that is, it may be regarded as a constant. This probability is provided from historical data.

4. The decision to take the flight is independent among the passengers. That is, Passenger A's decision to take (or not to take) the flight has no effect on the decision of Passenger B.

5. The airlines have historical files which provide information about the frequency of passenger "no-shows" and about the revenues for all flights.

The assumptions that have been made are, essentially, the assumptions of the *binomial probability* function. Suppose there are n independent trials, the outcome of each of which may be regarded as a "success" or "failure." Let p be the probability of success on a single trial, and $1 - p$, the probability of failure. Then the probability of x successes in n trials is given by

$$P(X = x) = \binom{n}{x} p^x (1 - p)^{n-x}.$$

The analysis will be performed on the basis of the *expected value* of revenue. If X is a random variable which represents the possible outcomes in a random experiment, then the expected value of X is defined by $E[X] = \sum_{\text{all } x} xP(X = x)$. In the special case of a binomial random variable, the expected value of X is given by

$$E[x] = np.$$

Let's assume that the price in dollars of an airline ticket is P and the refund in dollars to bumped passengers is R, where $R > P$. Assume that n tickets are sold and $c \leq n$ is the capacity of the aircraft.

The airline's revenue is P dollars if one passenger shows up, $2P$ dollars if two take the flight, etc., up to cP dollars. If one passenger gets bumped, the revenue is reduced to $cP - R$, if two are bumped, $cP - 2R$, and so on. Thus, if M is a random variable which represents the total revenue, the following equations represent the model.

$$P(\text{at least one passenger gets bumped}) = \sum_{k=1}^{n-c} p^{c+k}(1-p)^{n-c-k}$$

and

$$E[M] = \sum_{k=1}^{c} P \cdot k \binom{n}{k} p^k (1-p)^{n-k} + \sum_{k=1}^{n-c} (cP - kR) \binom{n}{c+k} p^{c+k} (1-p)^{n-c-k}$$

To present a specific example, suppose that an airplane has room for five passengers. The airline charges all fliers \$100 per ticket, refunds \$200 to a bumped passenger, and overbooks by one, knowing the probability that a passenger shows up is 0.6. Then the expected revenue is

$$\begin{aligned} E[M] &= 100(0.036864) + 200(0.13824) + 300(0.27648) + 400(0.31104) \\ &\quad + 500(0.186624) + 300(0.046656) \\ &= \$346.00 \end{aligned}$$

Note that the \$300 in the last term of the above computation is $5(100) - 200$ as stated in the description of the model. The numbers in parentheses are the binomial probabilities for $x = 1, 2, 3, \ldots, 6$ with $n = 6$ and $p = 0.6$.

If seven passengers are booked, a similar computation yields

$$\begin{aligned} E[M] &= 100(0.0172032) + 200(0.0774144) + 300(0.193536) + 400(0.290304) \\ &\quad + 500(0.2612736) + 300(0.1306368) + 100(0.0279936) \\ &= \$364.01. \end{aligned}$$

Note that the expected revenue if there are no airline "overbookings" is $(.6)(5)(500) = \$300$. Thus, it is clear that the airlines make money by overbooking their flights. Under these conditions, the best of the three strategies is, of course, the second; that is, overbook by two passengers. On the other hand, if $p = 0.8$ these same calculations for overbooking by 0, 1, and 2 respectively, yield \$401.35, \$400.00, and \$324.07; that is, accepting zero or one overbookings are approximately equivalent, but accepting two overbookings is a bad strategy. If $p = 0.9$, the expected revenues are, respectively, \$450.00, \$380.57, and \$231.42.

Indeed, the results are intuitive. An airline should overbook if the "show up" probability is relatively low, and should not (or not much) if the probability is high.

In Table 6.12 following, expected revenues have been compiled for various values of p, using $P = 100$ and $R = 200$, with c (the airplane capacity) equal to 25. In each column, the optimal figure is boxed. Note that for $p = 0.4$, the optimal strategy is to overbook a 25-passenger plane by 33 people! This number drops to 14 for $p = 0.6$, to 5 for $p = 0.8$, and to 2 for $p = 0.9$. Note the airline has a net loss beginning at 22 overbooked seats when $p = 0.8$ and a net loss beginning at 18 when $p = 0.9$.

TABLE 6.12: Revenue versus Number Overbooked for Various p

Number	Expected Revenue ($)			
Overbooked	$p = 0.4$	$p = 0.6$	$p = 0.8$	$p = 0.9$
0	1000.00	1500.00	2000.00	2250.00
1	1040.00	1560.00	2079.09	2320.62
2	1080.00	1619.99	2153.65	2342.78
3	1120.00	1679.96	2216.43	2301.94
4	1160.00	1739.83	2257.98	2204.41
5	1200.00	1799.44	2269.83	2067.19
6	1240.00	1858.42	2247.23	1906.95
8	1320.00	1971.69	2102.60	1558.35
10	1399.98	2070.52	1860.71	1199.84
12	1479.92	2141.16	1568.82	839.99
13	1519.85	2161.25	1414.36	660.00
14	1559.71	2169.11	1257.24	480.00
16	1639.06	2144.44	939.40	120.00
18	1717.43	2065.01	619.88	−240.00
20	1793.83	1936.11	299.98	−600.00
22	1866.70	1767.64	−20.00	−960.00
24	1933.93	1570.66	−340.00	−1320.00
26	1992.89	1354.95	−660.00	−1680.00
28	2040.64	1127.87	−980.00	−2040.00
30	2074.26	894.39	−1300.00	−2400.00
32	2091.16	657.51	−1620.00	−2760.00
33	2092.69	538.36	−1780.00	−2940.00
34	2089.38	418.94	−1940.00	−3120.00

Return to our Rome to New York flight. Let's assume that for overseas flights the probability of showing up is 0.75. Further, assume the cost per ticket is $1000 and a refund for a bumped passenger costs the airlines $2000. Also assume the plane holds a total of 200 passengers. The model, calculated with Maple, shows that 62 additional tickets (overbooked) should be sold to maximize the airline profits.

```
> Overbook := (n,p) → sum(P·k·binomial(n,k)·p^k·(1−p)^(n−k), k = 1..c)
    + sum((P·c − R·k)·binomial(n, c+k)·p^(c+k)·(1 − p)^(n−c−k), k =
  1..(n − c));
```

$$Overbook := (n,p) \mapsto \left(\sum_{k=1}^{c} P\,k \binom{n}{k} p^k\,(1-p)^{n-k}\right) + \left(\sum_{k=1}^{c} (P\,c - R\,k)\binom{n}{c+k} p^{c+k}\,(1-p)^{n-c-k}\right)$$

```
> c, P, R := 200, 1000, 2000 :
> Overbook(260, 0.75);
  Overbook(261, 0.75);
  Overbook(262, 0.75);
  Overbook(263, 0.75);
  Overbook(264, 0.75);
```

$$192154.2097$$
$$192314.8454$$
$$192392.1834$$
$$192381.8571$$
$$192280.3881$$

The optimum point for the airlines is overbooking by 62 seats. The moral of the story is to check in to your flight online as soon as possible and arrive at the airport early because the odds are very high that your flight will be overbooked.

Exercises

1. Rework the calculation for $E[M]$ given in this section using R equals:

   (a) \$150

   (b) \$250

   Recall that $c = 5$, $n = 6$ or 7, and $p = 0.6$, 0.8, or 0.9. For $n = 6$ and $p = 0.6$, use the numbers in the previous section with \$350 ($350 = 500 - 150$) and \$250 ($250 = 500 - 250$) instead of \$300 for the last term in each sum.

   For $n = 7$, add an additional term of either 200 ($200 = 500 - 2 \cdot 150$) or 0 ($0 = 500 - 2 \cdot 250$). For $p = 0.8$ and $p = 0.9$, do the same kind of calculations using a table of binomial probabilities.

2. Find the optimal overbooking for $p = 0.6$ and $p = 0.8$ with $c = 5$, $P = 100$, and $R = 200$. (Hint: use the binomial tables or Maple.)

3. Redo Exercise 2. with $R = \$250$.
4. Redo Exercise 2. with $R = \$150$.

Projects

Project 6.1. Redo the overbooking analysis if the distribution were Poisson, not binomial.

Project 6.2. Perform a sensitivity analysis by varying the mean and variance of each distribution and measure the impacts on the results.

6.6 Continuous Probability Models

Suppose a lumber company is looking at a forest of ponderosa pines that appears ripe for cutting. We can build a mathematical model that could predict for the company approximately how many board-feet of lumber can be cut from the trees.

Consider another situation where a company claims that only 0.5% of the items they produce are defective and not useable. We can build a mathematical model to support or refute the claim using *hypothesis testing.*

Some random variables do not have a discrete range of values. In Section 6.3 we saw examples of discrete random variables and discrete distributions. What if we were looking at time as a random event? Time has a continuous range of values, and thus, as a continuous random variable, will have a continuous probability distribution. We define a *continuous random variable* as any random variable measured on a continuous scale. Other examples include altitude of a plane, the percent of alcohol in a person's blood, net weight of a package of frozen chicken wings, or the time to failure of an electric light bulb. We cannot list the sample space by listing all the elements because the sample space is infinite. We need to be able to define the random variable's distribution as well as its domain and range.

The probability density function (PDF) of the continuous random variable x is defined to be

$$P(a \leq x \leq b) = \int_a^b f(x)\,dx.$$

For any continuous random variable, we define the cumulative distribution function (CDF) as $F(b) = P(X \leq b)$.

To be a valid probability density function (PDF)

1. $f(x)$ must be greater than or equal to zero for all x in its domain, and
2. the integral $F(\infty) = \int_{-\infty}^{\infty} f(x)\,dx = 1 =$ area under the entire graph of f.

The expected value or average of a random variable X with PDF given above is defined to be

$$E[X] = \int_{-\infty}^{\infty} x \cdot f(x)\,dx$$

In the upcoming segments, we will see problem-solving applications using continuous distributions such as the *exponential distribution* and the *normal distribution.*

Reliability Revisited with a Continuous Distribution

You are an undercover police investigator on a stake-out. Your mission is to manage the stake-out with three police officers for at least 24 hours. All necessary meals, equipment, and supplies for the 24-hour period must be carried with your team. The stake-out is ineffective unless your team can communicate with headquarters in a timely manner. Therefore, radio communications must be reliable. The radio has several components which affect its reliability, an essential one being the battery. Batteries have a useful life, which is not deterministic — we do not know exactly how long a battery will last when we install it. The battery's lifetime is a random variable, which may depend on previous use, manufacturing defects, weather, etc. The battery that is installed in the radio prior to leaving for the stake-out could last only a few minutes or for the entire 24 hours. Since communications are so important to the mission, we are interested in modeling and analyzing the reliability of the battery.

Definition. Continuous Reliability Function.
Let T be the time to failure of a component of a system, and $f(t)$ be the probability distribution function of T. Then the components' reliability at time t is

$$R(t) = P(T > t) = 1 - F(t)$$

where $F(t)$ is the cumulative distribution function of $f(t)$. $R(t)$ is called the *reliability function.*

A measure of this reliability is the probability that a given battery will last more than 24 hours. If we know the probability distribution for the battery life, we can use our knowledge of probability theory to determine the reliability. If the battery reliability is below acceptable standards, one solution is to have the team carry extra batteries. Clearly, the more extra batteries they carry, the less likely there is to be a failure in communications due to batteries. Of course, the battery is only one component of the radio system. Others include the antenna, handset, etc. Failure of any one of the essential components causes the system to fail.

This is a relatively simple example of an application of reliability. We will investigate how we can use elementary continuous probability to generate models that can be used to determine the reliability of the police team's equipment.

Component Reliability

We will discuss how to model component reliability. Recall that the reliability function $R(t)$ is defined as $R(t) = P(T > t) = P(\text{component fails after time } t)$. This can also be stated, using T as the component failure time, as

$$R(t) = P(T > t) = 1 - P(T \leq t) = 1 - \int_{-\infty}^{t} f(x)\, dx = 1 - F(t)$$

Thus, if we know the probability density function, $f(t)$ of the time to failure T, we can use probability theory to determine the reliability function $R(t)$. We normally think of these functions as being time dependent; however, this is not always the case. The function might be discrete such as the lifetime of a cell phone since it is dependent on the number of calls made through it (a discrete random variable).

The exponential distribution is a very useful probability distribution in reliability. The exponential's probability density function is

$$f(t) = \begin{cases} \lambda e^{-\lambda t} & t > 0 \\ 0 & \text{otherwise} \end{cases}$$

where the parameter $\lambda > 0$. The mean of the random variable T is $1/\lambda$ and the variance is $1/\lambda^2$. If T denotes the time to failure of a piece of equipment or a system, then $1/\lambda$ is the *mean time to failure*, expressed in units of time. Since $1/\lambda$ is the mean time to failure, λ is the average number of failures per unit time or the failure rate. For example, if a light bulb has time to failure that follows an exponential distribution with a mean time to failure of 50 hours, then its failure rate is 1 light bulb per 50 hours or $1/50$ per hour. So in this case $\lambda = 0.02$ per hour. Note that the mean of the continuous variable T is the mean time to failure of the component and is equal to $1/\lambda$.

Example 6.11. Battery Reliability Problem.
Return to the battery example presented earlier. Let the random variable T be defined as follows: T = time until a randomly selected battery fails. Suppose radio batteries have a time to failure that is exponentially distributed with a mean of 30 hours. We could write $T \sim EXP(\lambda = 1/30)$. Therefore, $\lambda = 1/30$ per hour, so that the PDF $f(t) = (1/30)\, e^{-t/30}$ for $t > 0$, and

$$F(t) = \int_0^t \frac{1}{30}\, e^{-x/30}\, dx = 1 - e^{-t/30}$$

Now we can compute the reliability function for a battery

$$R(t) = 1 - F(t) = 1 - (1 - e^{-t/30}) = e^{-t/30}$$

for $t > 0$.

Recall that in the earlier example, the police must occupy the stake-out for 24 hours. The reliability of the battery for 24 hours is $R(24) = e^{-24/30} = 0.4493$. Thus, the probability that the battery lasts more than 24 hours is 0.4493. This is less than 50%, so we would take a number of extra batteries.

We can easily do these computations in Maple without any new commands.

> $f := x \to \frac{1}{30} \cdot exp\left(-\frac{x}{30}\right)$:

> $F := t \to int(f(x), x = 0..t)$:

> $F(t)$;

$$1 - e^{-t/30}$$

> $R := 1 - F$;

$$R := 1 - F$$

> $R(t)$;

$$e^{-t/30}$$

> $R(24.0)$;

$$0.4493289641$$

Example 6.12. Nickel-Cadmium Battery Proposal.
We have the option to purchase a new nickel cadmium battery for our radios used in the police stakeout operation. Testing has shown that the distribution of the time to failure can be modeled using a parabolic function

$$f(t) = \begin{cases} \dfrac{t}{384} \cdot \left(1 - \dfrac{t}{48}\right) & 0 \le t \le 48 \\ 0 & \text{otherwise} \end{cases}.$$

Let the random variable T be defined as T = (time until a randomly selected battery fails). We can write

$$f(t) = \frac{t}{384} \cdot \left(1 - \frac{t}{48}\right) \quad \text{for } 0 \le t \le 48$$

and

$$F(t) = \int_0^t \frac{x}{384} \cdot \left(1 - \frac{x}{48}\right)\, dx \quad \text{for } 0 \le t \le 48$$

Recall that in the earlier example, the police must operate the stakeout for

24 hours. The reliability of the nickel cadmium battery for 24 hours is

$$R(24) = 1 - F(24) = 1 - \int_0^2 4\frac{x}{384} \cdot \left(1 - \frac{x}{48}\right) dx$$

$$> f := t \to \frac{t}{384} \cdot \left(1 - \frac{t}{48}\right) :$$

$$F := t \to int(f(x), x = 0..t) :$$

$$R := 1 - F :$$

$$> R(24.0);$$

$$0.5000000000$$

The nickel cadmium batteries have a reliability of 0.50 and are an improvement over the original batteries. Thus, we would recommend the new battery.

Models of Large-Scale Systems

In our discussion of reliability in Section 6.4, we discussed series systems, active redundant systems, and standby redundant systems. Unfortunately, things are rarely this simple. The types of systems listed above often appear as subsystems in larger arrangements of components that are called "large-scale systems." Fortunately, if you know how to deal with series systems, active redundant systems, and standby redundant systems, then analyzing system reliabilities for large-scale systems is easy.

Example 6.13. A Large-Scale Network System.
The first and most important step in developing a model to analyze a large-scale system is to draw a picture. Consider the network that appears in Figure 6.14. Subsystem A is a standby redundant system of three components (each with a failure rate of 5 per year) with the decision switch on the left of the figure. Subsystem B1, in the upper right portion of Figure 6.14, is an active redundant system of three components, each with a failure rate of 3 per year, where at least two of the three components must be working for the subsystem to work. Subsystem B2, lower right in the figure, is a two-component parallel system. We define Subsystem B as Subsystems B1 and B2 together in parallel. We assume all components have exponentially distributed times to failure with failure rates as given in Figure 6.14.

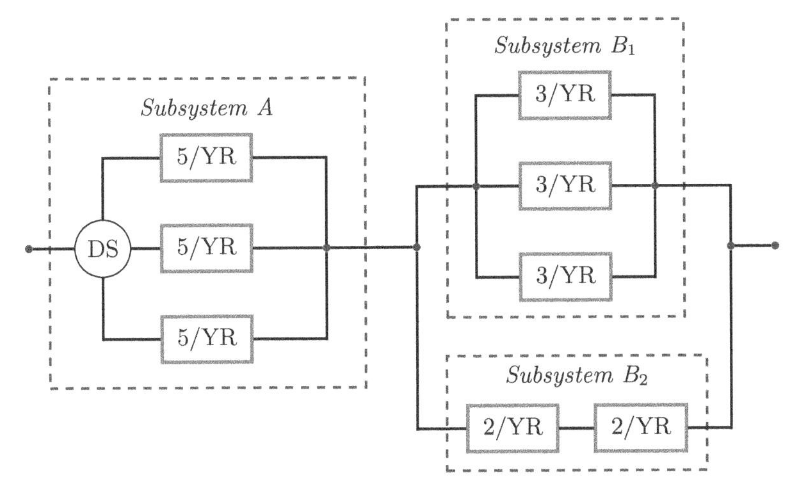

FIGURE 6.14: A Complex Network

We want to know the reliability of the whole system for 6 months. Observe that you already know how to compute the reliabilities for the Subsystems A, B1 and B2. Let's review these computations, and then see how we can use them to simplify our problem.

Subsystem A is a standby redundant system, so we will use the Poisson model (pg. 256). We let $X =$ the number of components which fail in one year. Since 6 months is 0.5 years, we seek $R_A(0.5) = P(x < 3)$, where X follows a Poisson distribution with parameter $\lambda = at = (5)(0.5) = 2.5$. Then,

$$R_A(0.5) = P(X < 3) = P(0 \leq x \leq 2) = 0.5438.$$

Now we consider Subsystem B_1. In our previous sections, we learned how to find individual component reliabilities when the time to failure followed an exponential distribution. The failure rate for Subsystem B_1 is 3 per year, so our individual component reliability is

$$R(0.5) = 1 - F(0.5) = 1 - (1 - e^{-3 \cdot 0.5} = 0.2231.$$

Now, recall that Subsystem B_1 is an active redundant system where two components of the three must work for the subsystem to work. If we let $Y =$ the number of components that function for 6 months, and recognize that Y follows a binomial distribution with $n = 3$ and $P(\text{success}) = 0.2231$, we can quickly compute the reliability of Subsystem B_1 as

$$R_{B_1}(0.5) = P(Y \geq 2) = 1 - P(Y < 2) = 1 - P(Y \leq 1) = 1 - 0.8729 = 0.1271.$$

Finally we can look at Subsystem B_2. Again we use the fact that the failure times follow an exponential distribution. The subsystem consists of two components; obviously, they both need to work for the subsystem to work. The first component's reliability is

$$R(0.5) = 1 - F(0.5) = e^{-2 \cdot 0.5} = 0.3679,$$

and the other component's reliability is

$$R(0.5) = 1 - F(0.5) = e^{-1 \cdot 0.5} = 0.6065,$$

Therefore, the reliability of the subsystem is

$$R_{B_2}(0.5) = 0.3679 \cdot 0.6065 = 0.2231.$$

The network can now be drawn as the reduced system shown in Figure 6.15. From here we determine the reliability of Subsystem B by treating it as a

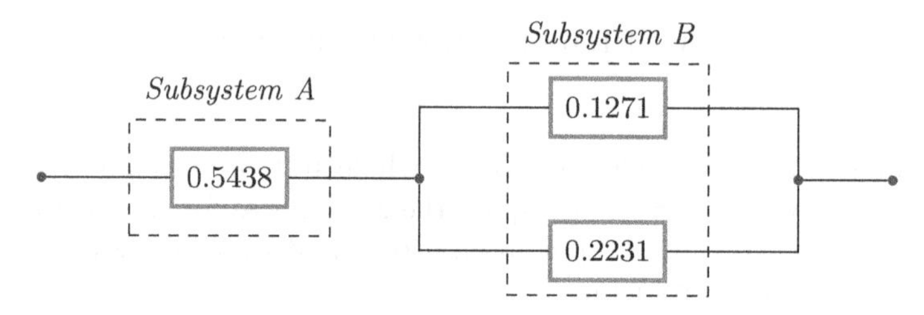

FIGURE 6.15: Simplified Network after Reduction

system of two independent components in parallel where only one component must work. Therefore,

$$R_B(0.5) = 0.1271 + 0.2231 - 0.1271 \cdot 0.2231 = 0.3218.$$

Finally, since Subsystems A and B are in series, we can find the overall system reliability for 6 months by taking the product of the two subsystem reliabilities

$$R_{\text{System}}(0.5) = 0.5438 \cdot 0.3218 = 0.1750.$$

We have used a network reduction approach to determine the reliability for a large-scale system for a given time period. Starting with those subsystems which consist of components independent of other subsystems, we reduced the size of our network by evaluating each subsystem's reliability one at a time. This approach works for any large-scale network consisting of basic subsystems of the type we have studied — series, active redundant, and standby redundant.

We have seen how methods from elementary probability can be used to model reliability problems. The modeling approach presented here is useful in helping to simultaneously improve understanding of both the modeling problems addressed and the mathematics behind these problems.

Exercises

1. A continuous random variable Y representing the time to failure of a 0.50 soda-bottling machine, has a probability density function given by

$$f(y) = \begin{cases} \frac{1}{3}\, e^{-y/3} & y > 0 \\ 0 & \text{otherwise} \end{cases}.$$

   (a) Find the reliability function for Y.

   (b) Find the reliability for 1.2 time periods, $R(1.2)$.

2. The lifetime of a generator engine (measured in hours of operation) is exponentially distributed with a *mean time to failure* (MTTF) of 400 hours. You have received a mission that requires 12 hours of continuous operation after a major storm. Your maintenance book indicates that the generator engine has been operating for 158 hours.

   (a) Find the reliability of your engine for this mission.

   (b) If your vehicle's engine had operated for 250 hours prior to the mission, find the reliability for the mission.

3. You are a project manager for the new system being developed in Huntsville, Alabama. A critical subsystem has two components arranged in a parallel configuration. You have told the contractor that you require this subsystem to be at least 0.995 reliable. One of the subsystems came from an older system and has a known reliability of 0.95. What is the minimum reliability of the other component so that specifications are met?

4. You are in a National Guard Reserve Battalion that has a mission to observe several night operations planned. There is some concern about the reliability of the lighting system for the FEMA Operations Center (TOC). The lights are powered by a 1.5 kW generator that has a MTTF of 7.5 hours.

   (a) Find the reliability of the generator for 10 hours if the generator's reliability is exponential.

   (b) Find the reliability of the power system if two other identical 1.5 kW generators are available. First consider as active redundant, and then as stand-by redundant. Which would improve the reliability the most?

   (c) How many generators would be necessary to insure a 99% reliability?

5. Consider the air traffic control system for a small airport depicted in Figure 6.16 below with the reliability for each component as indicated. Assume all components are independent and the radars are active redundant.

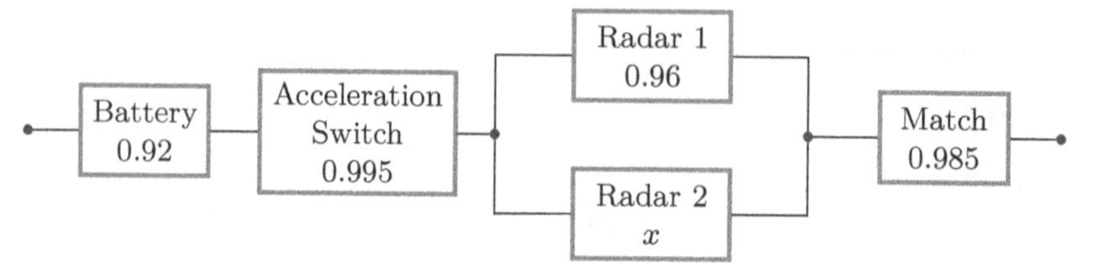

FIGURE 6.16: Air Traffic Control System

(a) Find the system reliability for 6 months when $x = 0.96$.
(b) Find the system reliability for 6 months when $x = 0.939$.

Projects

Project 6.1. Space Program.

The US Space Program is very concerned about the reliability of the space shuttle. "Reliability" of a system is the probability that the system will meet a set of specifications for a given period of time. The failure time of a system is the length of time that the system works according to specification. The failure time distribution for a system is the density function for the failure time t. If we denote the failure time density function by $f(t)$, then the probability that the system will fail before time t_0 is defined by the integral

$$F(t_0) = \int_0^{t_0} f(t)\,dt$$

Then, reliability is defined as $R(t_0) = 1 - F(t_0)$.

Several alternative rocket crafts are being developed for building by Boeing Corporation under a Space X program. Your consultant team has been hired to examine these alternatives and make recommendations to Space X. Your math skills are critical as you need to verify the distribution assumptions as well as find reliabilities to make your recommendations to Space X.

Alternative 1 was built using a Raleigh failure rate. The Raleigh distribution is given by

$$f(x) = \begin{cases} 2\alpha\, x e^{-\alpha x^2} & x > 0 \\ 0 & \text{otherwise} \end{cases}$$

where $\alpha > 0$.

Alternative 2 was built using an exponential failure rate defined by

$$f(x) = \begin{cases} c\, e^{-cx} & x > 0 \\ 0 & \text{otherwise} \end{cases}$$

where $c > 0$.

Alternative 3 was built using a Pareto distribution given by

$$f(x) = \begin{cases} \dfrac{\delta}{x^{\delta+1}} & x > 0 \\ 0 & \text{otherwise} \end{cases}$$

where $\delta > 0$.

Requirements:

1. Validate the assumptions that each distribution is indeed a valid probability distribution; i.e., show that each failure rate follows the requirement of a probability distribution that $\int_{-\infty}^{\infty} f(t)\,dt = 1$. Use improper integrals. Also, show the values for the constants α, c, and δ for which the distributions converge.
2. Given the following estimated constants by Boeing:

   Alternative 1: $\alpha = 0.000016$
   Alternative 2: $c = .0001$
   Alternative 3: $\delta = .0005$

   recommend the alternative that should be selected for the new space craft based upon it lasting 250 hours.

As a minimum, include the following in your report:

- Plots of the failure rates with the estimated constants.
- The expected value (mean time to failure) of each alternative, showing all calculations to support your recommendations.
- Make a short list of possible recommendations for Space X for future operations.

6.7 The Normal Distribution

A continuous random variable X with the probability density function (pdf)

$$f(x;\mu,\sigma) = \frac{1}{\sqrt{2\pi\sigma^2}} \exp\left(-\frac{1}{2}\left(\frac{x-\mu}{\sigma}\right)^2\right)$$

for $-\infty < x < \infty$ is a *normal distribution* with parameters $-\infty < \mu < \infty$ and $\sigma > 0$.

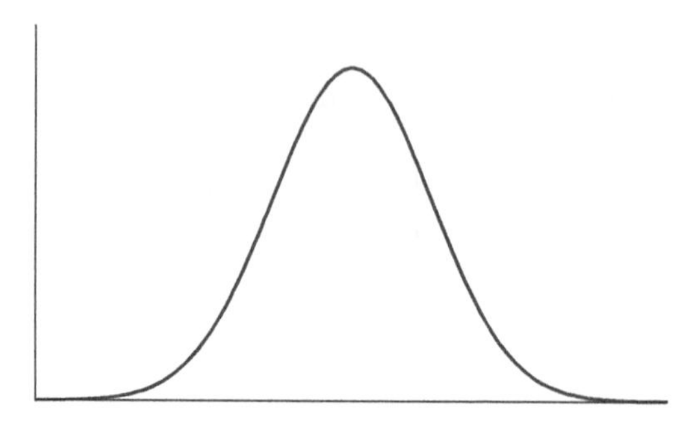

FIGURE 6.17: Bell-Shaped Curve of the Normal Distribution

The plot of the normal distribution is the "bell-shaped curve" in Figure 6.17.

To compute $P(a < x < b)$ when X is a normal random variable, with parameters μ and σ, we must evaluate

$$P(a < x < b) = \int_a^b \frac{1}{\sqrt{2\pi\sigma^2}}\, e^{-\frac{1}{2}\left(\frac{x-\mu}{\sigma}\right)^2} dx$$

There is no elementary algebraic formula for this integral, so the standard normal random variable Z with parameters $\mu = 0$ and $\sigma = 1$ has been numerically evaluated and tabulated for a large number of values. Since most applied problems do not have parameters $\mu = 0$ and $\sigma = 1$, the "standardizing" transformation, often called a *z-transform,* $Z = (x - \mu)/\sigma$ is used. Statistics textbooks illustrate how to convert the random variable X to the standard normal Z. In Maple, we can evaluate the integral numerically to whatever precision is desired.

For an example of the z-transform, consider the following problem. The amount of fluid dispensed into a can of diet soda is approximately a normal random variable with mean $\mu = 11.5$ fluid ounces and standard deviation $\sigma = 0.5$ fluid ounces. We want to determine the probability that between 11 and 12 fluid ounces, $P(11 < x < 12)$, are dispensed. Use the standard transform $Z = (x - 11.5)/(0.5)$. Then $x = 11 \mapsto Z = -1$ and $x = 12 \mapsto Z = 1$, so that

$$P(11 < x < 12) = P(-1 < Z < 1) = \frac{1}{\sqrt{2\pi}} \int_{-1}^{1} e^{-\frac{1}{2}t^2} dt.$$

If we used the normal distribution tables, we can compute this probability to be 0.6826. However, we can use Maple to compute the area between 11 and 12 directly. The *Statistics* package contains the functions we need: X := *RandomVariable*(*Normal*(μ, σ)) creates the normal distribution with mean μ and standard deviation σ naming it X. Then *PDF*(X, t) and

CDF(X, t) calculate the values of the normal's PDF and CDF, respectively, at t.

```
> with(Statistics) :
> X := RandomVariable(Normal(11.5, 0.5)) :
> PDF(X, t);
```

$$0.5641895835\,\sqrt{2}\,e^{-2.000000000(t-11.5)^2}$$

```
> CDF(X, 12) - CDF(X, 11);
```

$$0.6826894920$$

```
> with(plots) :
> Area := plot(PDF(X, t), t = 11..12, filled = true, color = "Maroon") :
  n := plot(PDF(X, t), t = 10..13, thickness = 2) :
> display(Area, n, view = [10..13, 0..0.9], title =
    "Normal(11.5,0.5), Area=0.68269");
```

Normal(11.5,0.5), Area=0.68269

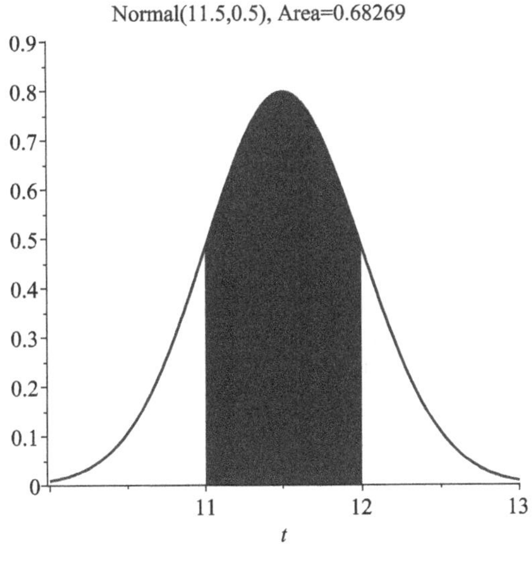

Therefore, 68.27% of the time the cans are filled between 11 and 12 fluid ounces.

We did not need to use a z-transform and the standard normal probability tables since Maple evaluated the integral of the normal function from a to b via the simple statement *CDF*(b) − *CDF*(a).

The Central Limit Theorem

The Central Limit Theorem is one of the most important theorems in probability. One interpretation of the theorem states that if X_1, X_2, ..., X_n are independent random samples from a distribution with a mean μ and a standard deviation σ, and n is sufficiently large (usually $n > 30$) then the distribution of the average or of T (the sum/total) has a normal distribution with parameters

$$\mu_{\bar{X}} = \mu, \quad \sigma^2_{\bar{X}} = \frac{\sigma^2}{n}, \quad \mu_T = n \cdot \mu, \quad \sigma^2_T = n \cdot \sigma$$

For example, when a batch of a certain pharmaceutical is prepared, the amount of natural substance aloe is a random variable with mean value 4.0 grams and standard deviation 1.5 grams. If 50 batches are prepared, what is the probability that the sample average of the aloe is between 3.5 and 3.8 grams?

Since $n = 50$, $(n > 30)$, then the sample average aloe random variable $\bar{X}$ follows a normal distribution with mean $\mu = 4.0$ and standard deviation $\sigma = 1.5/\sqrt{50} = 0.2121$. So, the probability we want, calculated via tables, is

$$P(3.5 < \bar{X} < 3.8) = P\left(\frac{3.5 - 4.0}{0.2121} < Z < \frac{3.8 - 4.0}{0.2121}\right) = 0.1637.$$

Using Maple, we have:

```
> X := RandomVariable(Normal(4.0, 0.2121)) :
> CDF(X, 3.8) − CDF(X, 3.5);
                    0.163678230300331
> Area := plot(PDF(X, t), t = 3.5..3.8, filled = true,
    color = "Maroon") :
  n := plot(PDF(X, t), t = 2.5..5.5, thickness = 2) :
> display(Area, n, view = [3.0..5.5, 0..0.9],
    title = “Normal(4.0,0.212), Area=0.164”);
```

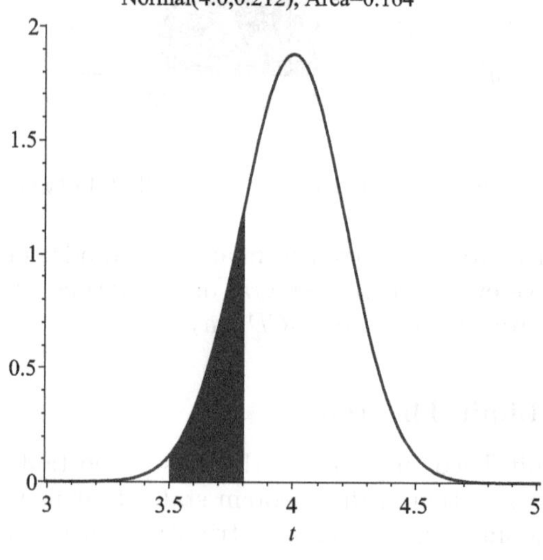

The normal distribution and the central limit (when applicable) are used in many applications of confidence intervals and hypothesis testing.

Exercises

1. Find the following probabilities for the normal distribution $N(\mu, \sigma)$:
   (a) For $X = N(10, 2)$, find $P(X > 6)$
   (b) For $X = N(10, 2)$, find $P(6 < X < 14)$
2. Determine the probability that lies within one standard deviation of the mean, two standard deviations of the mean, and three standard deviations of the mean. Draw a sketch of each region.
   (a) for $N(0, 1)$,
   (b) for $N(\mu, \sigma)$.
3. A tire manufacturer thinks that the amount of wear per normal driving year of the rubber used in their tire follows a normal distribution with mean $\mu = 0.05$ inches and standard deviation $\sigma = 0.05$ inches. If 0.10 inches is considered dangerous, then determine the probability that $P(X > 0.10)$.

6.8 Confidence Intervals and Hypothesis Testing

The basic concepts and properties of *confidence intervals* involve initially understanding and using two assumptions:

1. The population distribution is normal, and
2. the standard deviation σ is known or can be easily estimated.

In its simplest form, we are trying to *estimate* an interval containing μ, thus a *confidence interval* that will likely contain the value of the true parameter of interest. As we are making estimates, μ is not guaranteed to be in the interval. The formula for finding the confidence interval for an unknown population mean from a sample is

$$\left(\bar{X} - Z_{\alpha/2} \cdot \frac{\sigma}{\sqrt{n}}, \bar{X} + Z_{\alpha/2} \cdot \frac{\sigma}{\sqrt{n}}\right)$$

The value of $Z_{\alpha/2}$ is computed from the normality assumption and the level of 'confidence,' $1 - \alpha$, desired.

Let's consider a variation of the diet soda example in the previous section. Suppose the amount of fluid dispensed into a can of diet soda is approximately a normal random variable with unknown mean fluid ounces and a standard deviation of 0.5 fluid ounces. We want to determine a 95% confidence interval for the true mean. A sample of 36 diet cokes was taken and we found a sample mean of $\bar{x} = 11.35$.

Now, we want a 95% confidence interval, so $1 - \alpha = 0.95$; therefore, $\alpha = 0.05$. Since there are two (symmetric) tails, we need each tail to contain only 0.025 probability. We have $\alpha/2 = 0.025$. The value of $Z_{\alpha/2}$ can be found from tables, or we can use Maple.

```
> with(Statistics) :
> Z := RandomVariable(Normal(0, 1)) :
> α := 0.05 :
> fsolve(CDF(Z, t) − CDF(Z, −t) = 1 − α, t);
                              1.959963985
```

So we take $Z_{\alpha/2} = 1.96$. Our confidence interval from the $n = 36$ samples that had sample mean $\bar{x} = 11.35$ and standard deviation $\sigma = 0.5$ is

$$\left(11.35 - 1.96\,\frac{0.5}{\sqrt{36}},\, 11.35 + 1.96\,\frac{0.5}{\sqrt{36}}\right) = (11.187, 11.513)$$

We can write a Maple function to calculate confidence intervals.

```
> ConfidenceInterval := proc(α, μ, σ, n)
    localZ, z, CI;
    use Statistics in
      Z := RandomVariable(Normal(0, 1));
      z := fsolve(CDF(Z, t) − CDF(Z, −t) = 1 − α, t);
      CI := [μ − z · σ/√n, μ + z · σ/√n];
    end use;
  end proc :
> ConfidenceInterval(0.05, 11.35, 0.5, 36);
                     [11.18666967, 11.51333033]
```

The confidence interval as the shaded portion of the normal distribution of the samples is shown in Figure 6.18.

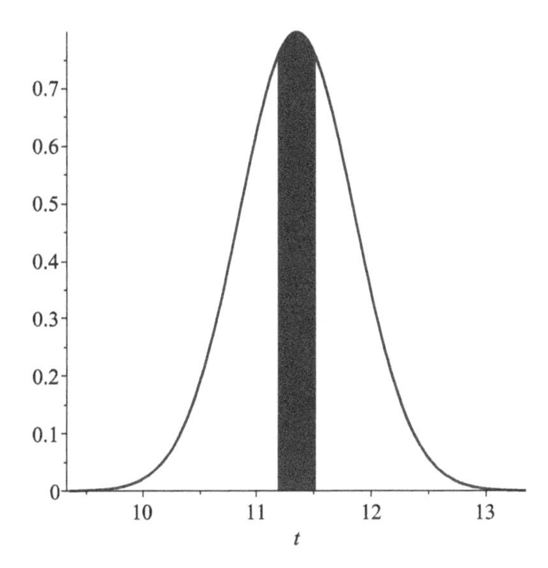

FIGURE 6.18: Confidence Interval for $11.35 \pm 1.96 \cdot 0.5/\sqrt{36}$

Let's interpret the confidence interval. If we took 100 experiments of 36 random samples in each, then calculated the 100 confidence intervals using each sample's mean as $\bar{X} \pm Z_{\alpha/2} \cdot \sigma/\sqrt{n}$, of the 100 confidence intervals, we expect 95 would contain the true population mean, μ. However, we do not know which of the 95 confidence intervals contain the true mean. Thus, to a modeler, each confidence interval built will either contain the true mean or it will not contain the true mean.

Simple Hypothesis Testing

A more powerful technique for inferring information about a parameter is a *hypothesis test.* A statistical hypothesis test is a claim about a single population characteristic or about values of several population characteristics. The *null hypothesis* (which is the claim initially favored or believed to be true) is denoted by H_0. The other hypothesis, the *alternate hypothesis*, is denoted as H_a. We will always keep equality with the null hypothesis. The objective is to decide, based upon sample information, which of the two claims is correct. Typical hypothesis tests can be categorized by three cases and seen in Figures 6.19a–c.

Case 1: H_0: $\mu = \mu_0$ versus H_a: $\mu \neq \mu_0$

Case 2: H_0: $\mu \leq \mu_0$ versus H_a: $\mu > \mu_0$

Case 3: H_0: $\mu \geq \mu_0$ versus H_a: $\mu < \mu_0$

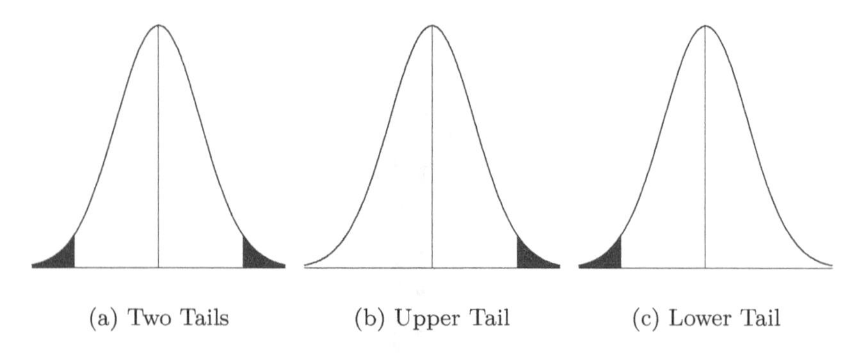

(a) Two Tails (b) Upper Tail (c) Lower Tail

FIGURE 6.19: Hypothesis Testing Regions

There are two types of errors that can be made in hypothesis testing, Type I errors, called α errors, and Type II errors, called β errors. It is important to understand these. Consider the information provided in the table below.

TABLE 6.13: States of Nature

		Reality	
		H_0 is True	H_a is True
Test Conclusion	Accept H_0	$1-\alpha$	β
	Reject H_0	α	$1-\beta$

Several important facts about both α and β follow.

1. $\alpha = P(\text{reject } H_0 \,|\, H_0 \text{ is true}) = P(\text{Type I error})$.
2. $\beta = P(\text{fail to reject } H_0 \,|\, H_0 \text{ is false}) = P(\text{Type II error})$.
3. α is the *level of significance* of the test.
4. $1-\beta$ is the *power* of the test.

Thus, referring to the table we would like α to be small, since it is the probability that we reject H_0 when H_0 is true. We would also want $1-\beta$ to be large since it represents the probability that we reject H_0 when H_0 is false. Part of the modeling process is to determine which of these errors is the most costly, and work to control that error as your primary error of interest.

The following template is provided for hypothesis testing:

STEP 1. Identify the parameter of interest.

STEP 2. Determine the null hypothesis, H_0.

STEP 3. State the alternative hypothesis, H_a.

STEP 4. Give the formula for the test statistic based upon the assumptions that are satisfied.

STEP 5. State the rejection criteria based upon the value of α.

STEP 6. Obtain your sample data and substitute into your test statistic

STEP 7. Determine whether your test statistics lies in the rejection region or the fail-to-reject region.

STEP 8. Make your statistical conclusion. Your choices are to either reject the null hypothesis or fail to reject the null hypothesis. Insure the conclusion is scenario oriented.

Example 6.14. Aviation Transport.
You run a small aviation transport company for a major corporation. You are tired of hearing management complain that your crews rest too much during the day. Aviation rules require a crew to get approximately 9 hours of rest each day. You collect a sample of 36 crew members and determine that their sample average is $\bar{x} = 8.94$ hours with a sample deviation of $s = 0.2$ hours.

The parameter of interest is the true population mean μ. The hypotheses chosen are

$$H_0\colon \mu \geq 9$$
$$H_a\colon \mu < 9$$

The test statistic is $Z = (\bar{x} - \mu)/(s/\sqrt{n})$. This is a "one-tailed test" (Figure 6.19c). We reject H_0, the null hypothesis, if $Z < -1.645$. (In Maple, compute the bound by

```
> Z := RandomVariable(Normal(0, 1)) :
> fsolve(CDF(Z, t) = 0.05, {t});
                    {t = -1.644853627}
```

for $\alpha = 0.05$.)

From our sample of 36 aviators, we find $Z = \dfrac{\bar{x} - \mu}{s/\sqrt{n}} = \dfrac{8.94 - 9}{0.2\sqrt{36}} = -1.8$.

Since $-1.8 < -1.645$. we reject the null hypothesis that aviators rest 9 or more hours per day, and conclude the alternate hypothesis is true, that your aviators rest less than 9 hours per day.

The function *HypothesisTest* in the book's *PSM* package performs these calculations. As usual, load the function via *with*(*PSM*). The syntax is

$$HypothesisTest(\bar{x}, s, n, \mu, \alpha, Case\ \#);$$

where '*Case #*' is 1, 2, or 3 corresponding to the cases in Figure 6.19 and the table above it. To review the Maple code of the *HypothesisTest* function, enter *print*(*HypothesisTest*).

For the aviation pilot example, we have

```
> HypothesisTest(8.94, 0.2, 36, 9.0, 0.05, 3);
  "z = -1.800000000; p = -1.644853627;  ==>  Reject Null Hypothesis"
```

Given our hypothesis test above, the probability of a Type I error, α, is the area under the normal bell-shaped curve centered at μ_0 corresponding to the rejection region shown in Figure 6.20. We chose this value to be 0.05.

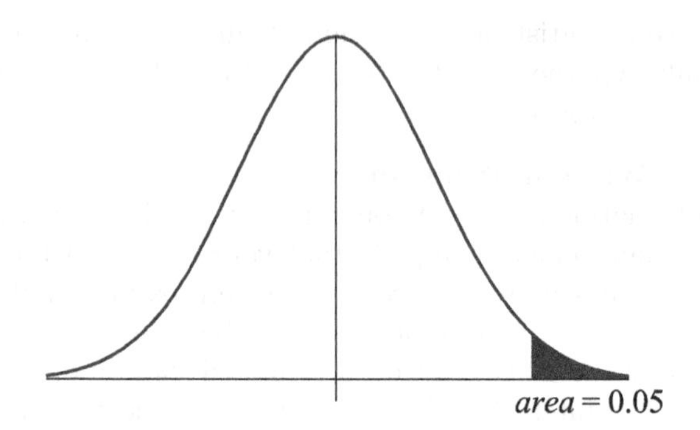

FIGURE 6.20: Type I Error

You must know or suspect another choice of μ to find this problem's Type II error; each new choice of μ will have its own Type II error probability β. Let's continue our example. Assume the new true mean is 8.87 hours of crew rest. Now let's find the probability of a Type II error.

$$\beta = P(\text{Fail to Reject } H_0 \,|\, H_0 \text{ is false})$$

$$\begin{aligned}\beta &= P(\bar{X} > \bar{X}_{cr} \,|\, \mu = 8.87) \\ &= P\left(Z > \frac{\bar{x} - \mu}{s/\sqrt{n}}\right) = P\left(Z > \frac{8.94 - 8.87}{0.2/\sqrt{36}}\right) = P(Z > 2.1) \\ &= p(z < 2.1) = 1 - 0.9821 = 0.01786\end{aligned}$$

This computation shows that over 1.7% of the time, we will fail to reject H_0 when μ really equals 8.87 and not 9.0.

Hypothesis Tests to Compare Sample Means

The following are the hypothesis tests set up for comparing means from two samples.

Tests Comparing Two Sample Means

H_0: $\mu_1 = \mu_2$, we write this as $\mu_1 - \mu_2 = 0$ or $\Delta\mu = 0$

H_a: Any of the $\Delta\mu \neq 0$, $\Delta\mu < 0$, $\Delta\mu > 0$, as required.

Test Statistic: $Z = \dfrac{\bar{x}_1 - \bar{x}_2}{\sqrt{\frac{s_1^2}{n_1} + \frac{s_2^2}{n_2}}}$

Decision: Reject H_0 if and only if $\begin{cases} Z \leq -z_{\alpha/2} \text{ or } Z \geq z_{\alpha/2} & \text{for } \Delta\mu \neq 0 \\ Z \leq -z_\alpha & \text{for } \Delta\mu < 0 \\ z \geq z_\alpha & \text{for } \Delta\mu > 0 \end{cases}$

Tests with Two Population Proportions (Large Sample)

H_0: $p_1 = p_2$ write as $\Delta p = 0$

H_a: Any of the $\Delta p \neq 0$, $\Delta p < 0$, $\Delta p > 0$, as required.

Test Statistic: $z = \dfrac{p_1 - p_2}{\sqrt{\hat{p}\hat{q}\left(\frac{1}{n_1} + \frac{1}{n_2}\right)}}$

Decision: Reject H_0 if and only if $\begin{cases} z \leq -z_{\alpha/2} \text{ or } z \geq z_{\alpha/2} & \text{for } \Delta p \neq 0 \\ z \leq -z_\alpha & \text{for } \Delta p < 0 \\ z \geq z_\alpha & \text{for } \Delta p > 0 \end{cases}$

Return to our example concerning homicides in Hampton and Newport News, VA.

Previously we computed the descriptive statistics and produced displays. The means and boxplots clearly showed that the number of homicides in Newport News was greater than in Hampton. But how much greater? Is it *significant*?

Let's test the hypothesis:

$H_0 : \mu_{\text{Newport News}} = 2 \cdot \mu_{\text{Hampton}}$. Write this as $\Delta\mu = \mu_1 - 2 \cdot \mu_2 = 0$

H_a : We decide that $\Delta\mu > 0$ may be best.

Test Statistic: $Z = \dfrac{\bar{x}_1 - \bar{x}_2}{\sqrt{\frac{s_1^2}{n_1} + \frac{s_2^2}{n_2}}}$

Decision: Reject H_0 if and only if $z \geq z_\alpha$ as we chose $\Delta\mu > 0$

Using Maple makes the computations quite easy; we suggest you always do this. We have previously defined the data frame *HomicidesVA* and loaded the *Statistics* package. Continue by first checking a hypothesis test that the means are equal:

```
> TwoSamplePairedTTest(HomicidesVA["Hampton"],
    HomicidesVA["NewportNews"], 0);
    hypothesis = false,
    confidenceinterval = −12.3818247639853.. − 8.57650856934805,
    distribution = StudentT(47), pvalue = 1.05555468447138 · 10^−14,
    statistic = −11.0799539724117
```

(To see a nicely formatted display, use the option *summarize = embed.*)
We reject the null hypothesis (P-value $= 1.05555468447138 \cdot 10^{-14}$).

Next we test our hypothesis that $\mu_1 - 2 \cdot \mu_2 = 0$. First double the homicides in Hampton (putting that data in *DoubleHampton*) and then run the paired t-test.

```
> DoubleHampton := map(x → 2 · x, HomicidesVA["Hampton"]) :
```

```
> TwoSamplePairedTTest(DoubleHampton,
    HomicidesVA["NewportNews"], 0);
      hypothesis = true,
      confidenceinterval = −2.00748752951105..4.00748752951105,
      distribution = StudentT(47), pvalue = 0.506824908493432,
      statistic = 0.668910675422648
```

Here, we fail to reject the null hypothesis that the number of homicides in Newport News is twice that number of homicides in Hampton.

Exercises

Discuss how to set up each of Exercises 1 to 5 as a hypothesis test.

1. Does drinking coffee increase the risk of getting cancer?
2. Does taking aspirin every day reduce the chance of a heart attack?
3. Which of two gauges is more accurate?
4. Why is a person "innocent until proven guilty"?
5. Is the drinking water safe to drink?
6. Set up a fake trial for a suspected felon. Build a matrix for their innocence or guilt with an appropriate null hypothesis. Which error, Type I or Type II, is the worst error?

Projects

Project 6.1. Numerous complaints have been made that a certain hot coffee machine is not dispensing enough hot coffee into the cup. The vendor claims that on average the machine dispenses at least 8 ounces of coffee per cup. You take a random sample of 36 hot drinks and calculate the mean to be 7.65 ounces with a standard deviation of 1.05 ounces.

(a) Find a 95% confidence interval for the true mean. Next, use Maple to take

30 random samples of size 36 from a normal distribution whose mean is 8 ounces and whose standard deviation is $1.05/\sqrt{30}$. Plot these intervals and determine how many would contain the true mean, $\mu = 8$ ounces.

(b) Set up and conduct a hypothesis test to determine if the vendor's claim is correct. Use an $\alpha = 0.05$ level of significance. Determine the Type II error if the true mean were 7.65 ounces.

Project 6.2. Given data for X_1 and X_2 in Table 6.14, determine the means are the same or not the same at an $\alpha = 0.05$.

TABLE 6.14: X_1 and X_2 Data

X_1	7.6	7.6	7.4	5.7	8.3	6.6	5.6
X_2	8.1	6.6	10.7	9.4	7.8	9.0	8.5

References and Further Reading

[Devore2011] Jay L. Devore, *Probability and Statistics for Engineering and the Sciences*, Cengage Learning, 2011.

[GFH2014] Frank Giordano, William P. Fox, and Steven Horton, *A First Course in Mathematical Modeling*, 5th ed., Nelson Education, 2014.

[Fox2011] William P. Fox, *Mathematical Modeling with Maple*, Nelson Education, 2011.

[FoxHorton1998] William P. Fox and Steven B. Horton, "Military Reliability Modeling" in *Military Mathematical Modeling* M^3, ed. David C. Arney, Vol. 1, USMA, 1998.
https://www.usma.edu/math/Military%20Math%20Modeling/PS1.pdf

[Johnson2012] I. Johnson, *I'll Give You a Definite Maybe, An Introductory Handbook on Probability, Statistics, and Excel* (2000).
https://malvma.viu.ca/~johnstoi/maybe/maybe4.htm

[Walpole2017] Ronald E. Walpole, Raymond H. Myers, Sharon L. Myers, and Keying Ye, *Probability and Statistics for Engineers and Scientists*, 9th ed., Prentice Hall, 2017.

7

Problem Solving with Simulation

Objectives

(1) Understand the concept of Monte Carlo simulation.

(2) Understand random numbers and how to obtain the type random number required for the simulation based upon the distribution needed.

(3) Build simple simulations.

(4) Interpret results of simulations models.

A barber shop has two customers arrive every 30 minutes. The service rate of the barber is three customers every 60 minutes. This implies the time between arrivals is 15 minutes and the mean service time is one customer every 20 minutes. How many customers will be in the queue and what is their average waiting time?

How best to build a model? We could use queueing theory or we could build a simulation model of the system. We'll examine this situation more closely later in the chapter.

7.1 Introduction

Consider an engineering company that conducts vehicle emissions inspections for a specific state within the United States. We have data on when vehicles arrive, when vehicles depart, service times for inspectors under various conditions, number of inspection stations, and penalties levied for failure to meet state inspection standards in terms of waiting time for customers. The company wants to know how it can improve its inspection process throughout the state in order to maximize its profit as well as minimizing the penalties it receives. This type of analysis for a complex system has many variables; we can use a computer simulation to model this operation.

A modeler may encounter situations where the construction of an analytic model is infeasible due to the complexity of the situation. In instances where the behavior cannot be modeled analytically, or data collected directly, the modeler might simulate the behavior indirectly, and then test various alternatives to estimate how each affects the behavior. Data can then be collected to determine which alternative is best. Monte Carlo simulation is a common simulation method that a modeler can use, usually with the aid of a computer. The proliferation of computers in today's society, both in the academic and business worlds, makes Monte Carlo simulation very attractive. It is imperative that we understand, at a minimum, how to use and interpret Monte Carlo simulations as a modeling tool.

There are many forms of simulation ranging from building scale models such as those used by scientists or designers in experimentation to various types of computer simulations. Monte Carlo simulation is a preferred type which is implemented using random numbers. There are many serious mathematical concerns associated with the construction and interpretation of Monte Carlo simulations. We will focus on reinforcing the techniques of simulations with random variates.

A principal advantage of Monte Carlo simulation is the ease with which it can be used to approximate the behavior of very complex systems. Often, simplifying assumptions must be made in order to reduce a complex system into a manageable model. Within the environment forced upon the system, the modeler attempts to represent the real system as closely as possible. This system is probably a stochastic system; however, simulation can allow either a deterministic or stochastic approach; see Figure 7.1. A deterministic model is one that should always provide the same answer based upon the same inputs. Simulating a deterministic event using Monte Carlo simulation, the modeler has introduced randomness into the process and thus answers will not always be the same.

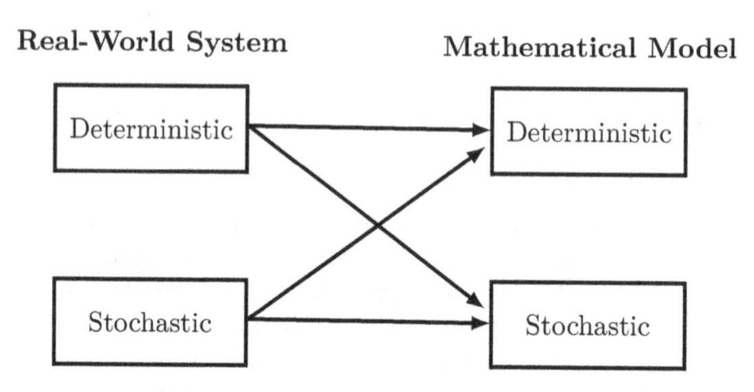

FIGURE 7.1: System versus Model Relationship

Many undergraduate mathematical science, engineering, and operations research programs require or offer a course involving simulation. Typically, such a course will use a high-level language such as C++, Java, Fortran,

SLAM, Prolog, Stella, SIMAN, Python, or GPSS as the software tool to teach simulation. Here we will use Maple to simulate simple modeling scenarios.

Our emphasis is two-fold. First, we want to think in terms of an algorithm, not a specific language. Second, we need to understand that MORE is better in Monte Carlo simulations. The "MORE is better" rule is based on the "Law of Large Numbers" where probabilities are assigned to events in accordance with their limiting relative frequencies.

7.2 Monte Carlo Simulation

A Monte Carlo simulation model is one that uses random numbers to simulate behavior of a situation. Using a known probability distribution (such as uniform, exponential, or normal) or an empirical probability distribution, a modeler assigns a behavior to a specific range of random numbers. The behavior returned from the random number generated is then used in analyzing the problem. For example, if a modeler is simulating the tossing of a fair coin using a uniform random number generator that gives numbers in the range $0 \leq x < 1$, he or she may assign all numbers less than 0.5 to be a "head," while numbers from 0.5 to 1 are "tails." We display this important structure in Figure 7.2.

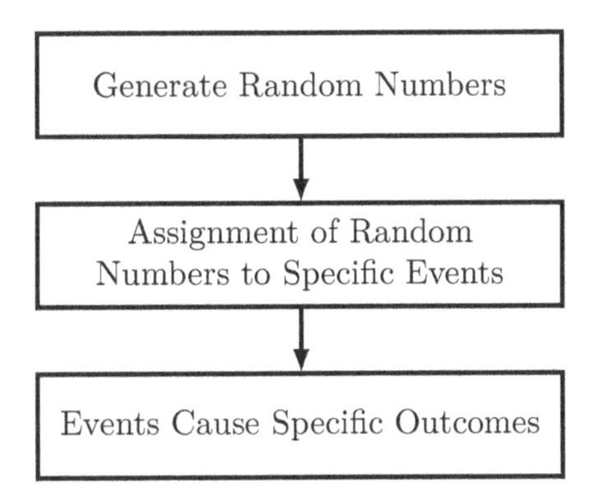

FIGURE 7.2: Relationship between Random Numbers and Outcomes

A Monte Carlo simulation can be used to model either deterministic or stochastic behavior. We can use a Monte Carlo simulation to determine the area under a curve (a deterministic problem) or the probability of winning in Craps (a stochastic problem). In this chapter, we will investigate both: a deterministic problem and a stochastic problem. We discuss how to create algorithms to solve each. We begin with deterministic simulation modeling.

7.2.1 Random Number Generators in Maple

Using random numbers is of paramount importance in performing Monte Carlo simulations, therefore it is imperative that a good random number generator is used. In particular, a modeler must have a method of generating uniform random numbers, $U(0,1)$, that is, numbers that are uniformly distributed between 0 and 1. All other distributions, both known and empirical, can be derived from functions of $U(0,1)$ distributions. At the graduate level, a good deal of class time is spent on the theory behind good and bad random number generators and the tests that can be made on them. There, more is learned about what makes up a "true" random number generator and what does not work well as a random number generator. At the undergraduate level, this is not as necessary if students have access to either random numbers or a good algorithm for generating pseudo-random numbers.

Additionally, most computer languages now use good pseudo-random number generators (although this has not always been the case — the old RANDU[1] generator distributed by IBM was statistically unsound). These generators, called *linear congruence generators* (LCGs), use a recursive sequence $X_i = (aX_{i-1} + c) \mod m$ where a, c, and m determine the statistical quality of the generator. RANDU was an LCG with a particularly bad choice of parameters. Since we will not discuss testing random number generators, we'll "trust" the generators provided by our software packages — modern packages have good generators. Serious study of simulation must, of course, include a study of random number generators since a bad generator will provide output leading a modeler to make poor conclusions.

Maple's *RandomTools* package is devoted to generating random numbers. We will present a few of Maple's basic random number generating commands. The command *rand*() returns a 12-digit non-negative integer, the command *rand*(r) returns a function that will return random integers in the range $0..r$. An excerpt from the *rand* Help Screen follows. (To generate the same values shown in the examples below, start with a new Maple document or execute *restart*.)

[1] See https://csgillespie.wordpress.com/2016/02/16/randu-the-case-of-the-bad-rng/.

rand
random number generator

Calling Sequence
rand(r)

Parameters
r – (optional) numeric range or positive integer

Description

- With no arguments, the call *rand*() returns a random integer sampled uniformly from the range 0 to $10^{12} - 1$.
- If the parameter r is an integer range $a..b$, the call *rand*($a..b$) returns a procedure which, when called, generates random integers in the range $a..b$.
- If r is a numeric range $a..b$ and one of a or b is not an integer, the call *rand*($a..b$) returns a procedure which, when called, generates random floating-point numbers in the range $a..b$.
- If r is a single integer, the call *rand*(r) is the abbreviated form of *rand*($0..r - 1$).
- More than one random number generator may be used at the same time, because *rand*($a..b$) returns a Maple procedure. However, since all random number generators use the same underlying random number sequence, calls to one random number generator will affect the random numbers returned from another.

Examples

```
> rand();
                                395718860534
> rand();
                                193139816415
> roll := rand(1..6) :
> roll();
                                6
> roll();
                                2
```

To generate random real numbers between 0 and 1 use

```
> u := rand(0..1.0) :     # Note the decimal value 1.0
```

```
> u();
                    0.1799302829
```

To obtain random numbers from a particular distribution, we *Sample* the distribution with its associated parameters. *Sample* is from the *Statistics* package. Let's see some examples from a normal(0, 1), a normal(250, 25), an exponential(1/3), and a Poisson(5). Remember to load the *Statistics* package by either using the *Load Package* submenu of the *Tools* menu or the command *with(Statistics)*.

```
> with(Statistics):
> Sample(RandomVariable(Normal(0,1)),1);
                    [-1.07242412799827]
> Sample(RandomVariable(Normal(0,1)),10);
    [ -0.280384023505023, 0.206812415848015, -0.127897916648825,
      -0.389353878496741, 0.871443820742342, 0.724801267308865,
      0.473396924944576, -1.60187232785855, 0.423580977654844,
      -1.95368179160194 ]
> Sample(RandomVariable(Normal(250,25)),4);
  [249.106053654013, 249.837415913434, 248.570424944169, 245.384997634377]
> Sample(RandomVariable(Exponential(1/3)),4);
    [0.0210788932929293, 0.630695315322572, 1.00674288782150,
  0.811960456779717]
> Sample(RandomVariable(Poisson(5)),10);
                    [7., 7., 6., 2., 4., 11., 2., 2., 8., 2.]
> Sample(RandomVariable(Uniform(0,1.0)),4);
    [0.125896635514188, 0.210209077339744, 0.0512164294362061,
  0.0364412455772802]
```

We can use *Uniform*$[0, 1]$ random numbers to generate many distributions that we might need in modeling.

Uniform$[a, b]$

1. Generate a random uniform number U from $[0, 1]$.
2. Return $X = a + (b - a) \cdot U$.

***Exponential* with mean** μ

1. Generate a random uniform number U from $[0, 1]$.
2. Return $X = -\mu \ln(U)$.

Normal$[0, 1]$

1. Generate two random uniform numbers U_1 and U_2 from $[0, 1]$.

2. Set $V_1 = 2U_1 - 1$ and $V_2 = 2U_2 - 1$.
3. Let $W = V_1^2 + V_2^2$.
4. If $W > 1$, go back to Step 1. Otherwise, let $Y = \sqrt{-2\ln(W)/W}$.
5. Put $X_1 = V_1 \cdot Y$ and $X_2 = V_2 \cdot Y$.
6. Return X_1 and X_2. They are both from *Normal*(0, 1).

In a recent article from Maplesoft help for MapleTA, a web-based system for creating and grading tests and assignments, we find the following useful information reproduced below.[2] To use any of the functions in Table 7.2, first load the necessary package via *with*(*MapleTA*[*Builtin*]).

In addition to Maple T.A.'s built-in functions, you can access Maple commands. Maple provides many randomization commands. The following table lists selected randomization commands in Maple. For more information on Maple functions that can be used to generate random numbers, refer to your Maple documentation.

TABLE 7.1: Maple Randomization Commands

Maple Function	Use
rand	Generate a random 12-digit non-negative integer, or a random integer within a specified range (inclusive).
randpoly	Random polynomial generator.
LinearAlgebra[RandomMatrix]	Construct a random Matrix.
LinearAlgebra[RandomVector]	Construct a random Vector.
RandomTools[Generate]	Generate a particular random object. (For example, return a random complex number with real and imaginary parts of a specified flavor, or choose one of the entries in a collection with equal probability.)

Note: When using Maple's random number generator rand, you must include *randomize*() : as the first part of the call. This sets the initial state of the random number generator using a number based on the system clock instead of the default seed in Maple. Without *randomize*, each call to rand will produce the same sequence of values using the default randomization seed.

---

[2] From https://www.maplesoft.com/support/help/MapleTA2017/MapleTAInstructor/ch06s17.aspx.

Generating Random Numbers in Questions

The algorithm syntax in the system allows you to generate random numbers in a variety of ways, depending on your requirements and preferences. Table 7.2 summarizes functions related to random number generation.

TABLE 7.2: Random Number Generators

Function	Range of Operation	Example	Description
rint(n)	$0, \ldots, n-1$	**rint(3)** $= 0$, 1, or 2	Returns a random integer between 0 and $n-1$ (inclusive). Generates n variations.
rint(m,n)	$m, \ldots, n-1$	**rint(3,6)** $= 3$, 4, or 5	Returns a random integer between m and $n-1$ (inclusive). Generates $n-m$ variations.
rint(m,n,k)	m, $m+k$, $m+2k$, $\ldots$, $m+q\cdot k$ (where q is the largest integer such that $m+q\cdot k \leq n-k$)	**rint(3,12,3)** $= 3$, 6, or 9	Returns a random integer between m and $n-1$ (inclusive) in steps of k. Generates approximately $(n-m)/k$ variations.
range(n)	$1, \ldots, n$	**range(3)** $= 1$, 2, or 3	When n is a positive integer, returns a random integer between 1 and n (inclusive). Generates n variations.
range(m,n)	m, $m+1$, $\ldots$, $m+q$ (where q is the floor of $(n-m)/k$)	**range(3,6)** $= 3$, 4, 5, or 6	Returns a random integer between m and n (inclusive). Generates $(n-m)+1$ variations.
range(m,n,k)	m, $m+k$, $m+2k$, $\ldots$, $m+q\cdot k$ (where q is the largest integer such that $m+q\cdot k \leq n-k$)	**range(3,12,3)** $= 3$, 6, 9, or 12	Returns a random number between m and n (inclusive), in steps of k. Generates approximately $(n-m)/k+1$ variations.
rand(m,n)	$m, \ldots, n$	**rand(0.5,9.5)** = all real numbers between 0.5 and 9.5 (inclusive)	Returns a random real number between m and n (inclusive).
rand(m,n,k)	$m, \ldots, n$ expressed to k significant figures	**rand(0.5,9.5,3)** = all real numbers between 0.5 and 9.5 (inclusive), expressed to 3 significant figures	Returns a random real number between m and n (inclusive), *expressed to k significant figures.*

Exercises

Generate 20 random numbers for each distribution given below.

1. *Uniform*$(0, 1)$
2. *Uniform*$(-10, 10)$
3. *Exponential*$(\mu = 0.5)$
4. *Normal*$(0, 1)$
5. *Normal*$(5, 0.5)$
6. Maple's Note above reminds us to use *randomize*() before starting a simulation.

   (a) Execute *restart: rand*(); four times.

   (b) Execute *restart: randomize*() : *rand*(); four times.

   (c) Execute

   restart : *N* :=< *your phone number*>: *randomize*(*N*) : *rand*();

   four times.

   Describe what happened.

7.3 Probability and Monte Carlo Simulation Using Deterministic Behavior

Suppose we need to find the area under the curve, $f(x) = e^{x^2} \cdot \cos(x) \cdot \sqrt{x}$, between 0 and 1.4. This function has no elementary antiderivative, so we cannot use calculus techniques. Let's see how we can use Monte Carlo simulation to evaluate the integral.

One of the keys to good Monte Carlo simulations is understanding the basic axiom of probability: *Probability* is a long-term average. For example, if the probability of an event occurring is $1/5$, then the meaning is "that in the long term, the chance of the event happening is $1/5 = 0.20$," not that it will occur exactly 1 out of every 5 trials; i.e., for a large number of trials, say 1,000, we expect the event will happen approximately, not exactly, $1/5 \cdot 1000 = 200$ times. The event happening 207 times would not surprise us.

Deterministic Simulation Examples

Let's consider deterministic examples of computing the area under a nonnegative curve and the volume under a nonnegative surface.

We will present algorithms for these models as well as produce output from a Monte Carlo simulation to analyze. These algorithms are important to the understanding of simulation as a mathematical modeling tool. The generic framework for the algorithm includes Inputs, Outputs, and the steps required to achieve the desired output. Carefully look for these aspects as we work through the examples.

The algorithm for using Monte Carlo simulation for determining the area under a generic nonnegative curve $y = f(x)$ for $a \leq x \leq b$ is:

Algorithm 1: Monte Carlo Simulation Approximation of Area[3]

INPUTS:

$f(x)$	the nonnegative function
N	total number of random points to generate
$[a, b]$	interval for x
$[0, M]$	interval for y where $M > \max f(x)$ for $x \in [a, b]$

OUTPUTS:

A	the approximate area under the curve $f(x)$ for $x \in [a, b]$

STEPS:

Step 1. Set the counter S to 0

Step 2. For i from 1 to N do Steps 3 - 5

Step 3. Calculate a random point (x_i, y_i) in the rectangle $[a, b] \times [0, M]$

Step 4. Calculate $f(x_i)$

Step 5. If $y_i \leq f(x_i)$, then $S = S + 1$, otherwise do not increment S
Increment i; if $i \leq N$, then go to Step 3. otherwise go to Step 6.

Step 6. Estimate the area A by $A = M \cdot (b - a) \cdot \dfrac{S}{N}$.
Output A and Stop.

We'll begin with an easy function $f(x) = x^3$ over the interval $[0, 2]$. We can easily integrate this function and find the exact answer

$$\int_0^2 x^3 \, dx = 4.$$

[3]Modified from Giordano et al [GFH2014].

Example 7.1. A Simple Area Estimation.
Compute the area under the nonnegative curve $y = x^3$ for $0 \leq x \leq 2$. See Figure 7.3.

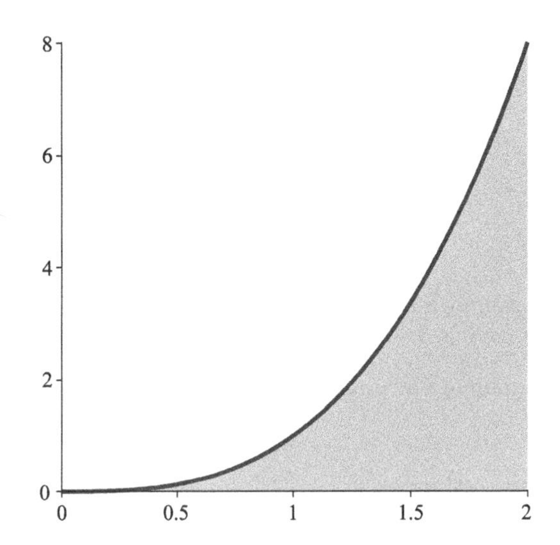

FIGURE 7.3: Graph of $y = x^3$ for $0 \leq x \leq 2$

Now we are ready to *approximate* the area by using Monte Carlo simulation. Remember, the simulation only approximates the solution. Increase the number of trials to get closer to the value. The results are in Table 7.3. Randomness is introduced into the procedure with the Monte Carlo simulation area algorithm. In our Maple program, we provide graphical output as well so that the algorithm may be seen as a process. (Our program *MonteCarloArea* is available in the book's *PSM* Maple package.) In the graphical output, each generated coordinate (x_i, y_i) is a point in the rectangle. Points are randomly generated in our intervals $[a, b]$ for x and $[0, M]$ for y. The curve for the function $f(x)$ is overlaid with the points. The output also gives the approximate area.

First, let's view the graphical outputs from run with $N = 100$ and $N = 5000$ in Figures 7.4a and 7.4b.

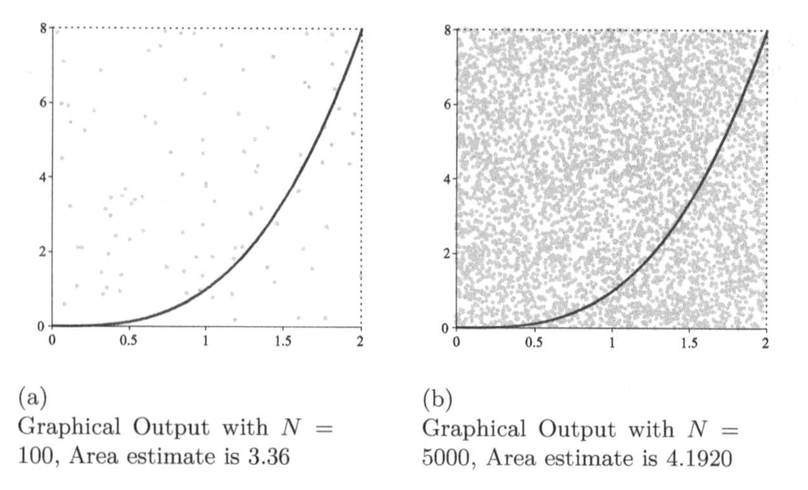

(a)
Graphical Output with $N =$ 100, Area estimate is 3.36

(b)
Graphical Output with $N =$ 5000, Area estimate is 4.1920

FIGURE 7.4: Graphical Output of Monte Carlo Simulation Area Estimates

Now the table of estimates.

TABLE 7.3: Summary of Output for the Area under x^3 from 0 to 2

Number of Trials	Approximate Area	Percent Error
100	3.36	16.00%
500	3.872	3.20%
1,000	4.32	8.00%
5,000	4.1056	2.64%
10,000	4.136	3.40%

We need to stress that in modeling deterministic behavior with stochastic features, we have introduced randomness into the problem — randomness that is not in the original problem. Although more simulation runs are better, it is not always true that as we increase the number of trials, N approaches ∞, that the solution becomes closer to reality. It is generally true that computing more runs gives better results than a small number of runs (16% was the worst error by almost an order of magnitude; it occurred at $N = 100$). In general, more trials are better.

Example 7.2. Estimating the Area under a Non-Integrable Function. Compute the area under the nonnegative curve $f(x) = e^{x^2} \cdot \cos(x) \cdot \sqrt{x}$ for $0 \leq x \leq 1.4$. See Figure 7.5.

This function has no elementary antiderivative, so we can only approximate the value of the area.

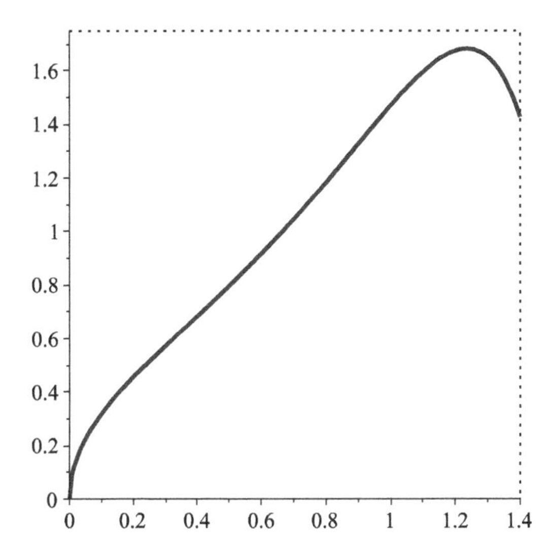

FIGURE 7.5: Graph of $f(x) = e^{x^2} \cdot \cos(x) \cdot \sqrt{x}$ for $0 \le x \le 1.4$

The graphical output using Algorithm 1 with $N = 2000$ appears in Figure 7.6.

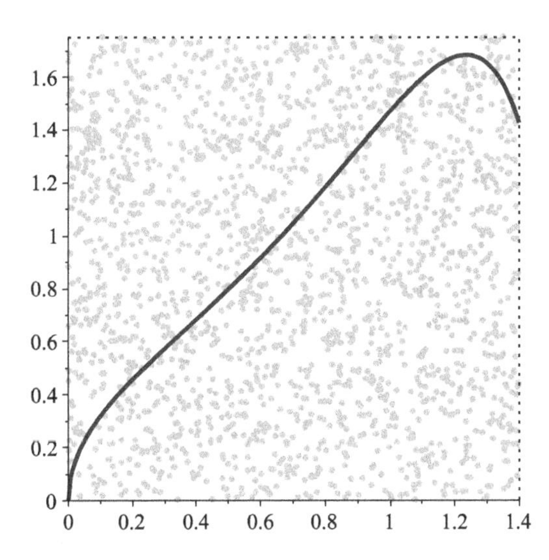

FIGURE 7.6: Graphical Output of Simulation for $N = 2000$, Area ≈ 1.42345

Since this integral has no closed form, we use Maple's integrator to find

```
> int(e^(x^2) · cos(x) · √x, x = 0..1.4);
                                    1.449827600
```

The percent error for this run with $N = 2000$ is only 1.82%.

Keep in mind that every run of the simulation will produce a different answer since the points are randomly generated.

The Monte Carlo simulation technique for finding areas can be easily extended to multiple dimensions. We'll modify Algorithm 1 to find the volume under a surface, like the one shown in Figure 7.7, in the first octant.

Algorithm 2: Monte Carlo Simulation Approximation of Volume

INPUTS:

$f(x, y)$	the nonnegative function
N	total number of random points to generate
$[a, b]$	interval for x
$[c, d]$	interval for y
$[0, M]$	interval for z where $M > \max f(x, y)$ for $(x, y) \in [a, b] \times [c, d]$

OUTPUTS:

V	the approximate volume under the surface $f(x, y)$ over the rectangle $[a, b] \times [c, d]$

STEPS:

Step 1. Set the counter S to 0

Step 2. For i from 1 to N do Steps 3 - 5

  Step 3. Calculate a random point (x_i, y_i, z_i) in the cube $[a, b] \times [c, d] \times [0, M]$

  Step 4. Calculate $f(x_i, y_i)$

  Step 5. If $z_i \leq f(x_i, y_i)$, then $S = S + 1$, otherwise do not increment S
  Increment i; if $i \leq N$, then go to Step 3. otherwise go to Step 6.

Step 6. Estimate the volume V by $V = M \cdot (b - a) \cdot (d - c) \cdot \dfrac{S}{N}$.
Output V and Stop.

Example 7.3. Approximation of a Volume in the First Octant.
Compute the volume in the first octant given by $x^2 + y^2 + z^2 \leq 1$.

The function defining our volume is $f(x, y) = \sqrt{1 - x^2 - y^2}$. The interval $[0, 1]$ works for all three variables x, y, and z.

As the volume is 1/8 of a sphere of radius 1, the exact volume is $\pi/6$.

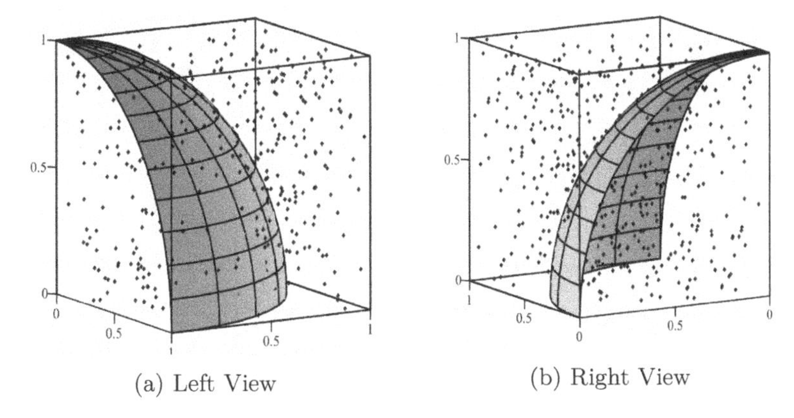

(a) Left View (b) Right View

FIGURE 7.7: Sphere in the First Octant

Table 7.4 provides numerical output from our simulation, and Figure 7.8 provides visual output.[4] (Remember, each run of the algorithm produces different answers because of the randomness inserted using Monte Carlo simulation.) We take $\pi/6$ to four decimals as 0.5236 cubic units for computing percent error.

TABLE 7.4: Output for the Volume of a Unit Sphere in the First Octant

Number of Trials	Approximate Area	Percent Error
100	0.47	10.24%
200	0.595	13.64%
300	0.5030	3.93%
500	0.514	1.833%
1,000	0.518	1.069%
2,000	0.512	2.210%
5,000	0.518	1.069%
10,000	0.5234	0.13368%
20,000	0.5242	0.11459%

We see from the table that generally, though not uniformly, the percent errors become smaller as the number of points N is increased.

[4] Our program *MonteCarloVolume* is available in the book's *PSM* Maple package.

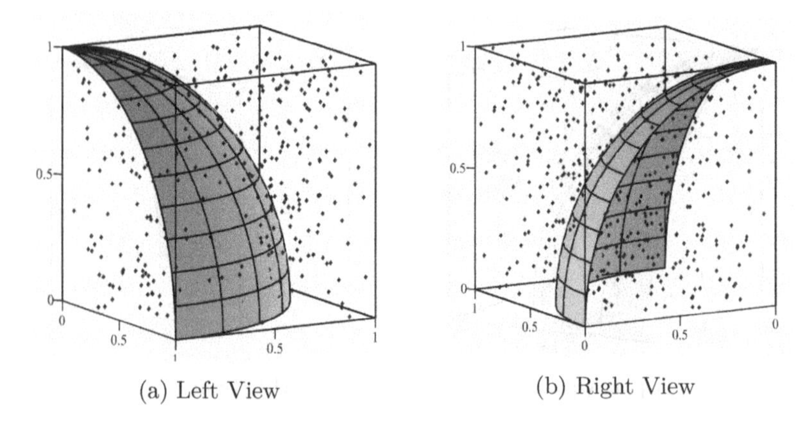

(a) Left View (b) Right View

FIGURE 7.8: Graphical Display of the Algorithm, $N = 500$, $V = 0.518$

Exercises

Use Monte Carlo simulation in each problem below.

1. Approximate the area under the curve $f(x) = 1 + \sin(x)$ over the interval $-\frac{\pi}{2} \leq x \leq \frac{\pi}{2}$.

2. Approximate the area under the curve $f(x) = \sqrt{x}$ over the interval $\frac{1}{2} \leq x \leq \frac{3}{2}$.

3. Approximate the area under the curve $f(x) = \sqrt{1 - x^2}$ over the interval $0 \leq x \leq \frac{1}{2}$.

4. How could you modify Exercise 3 to obtain an approximation to $\pi/2$?

5. Estimate the volume under the surface $f(x, y) = x^2 + y^2$ in the first octant?

7.4 Probability and Monte Carlo Simulation Using Probabilistic Behavior

Unlike deterministic scenarios, probabilistic events inherently incorporate randomness. Let's consider probabilistic examples: computing the probability of getting a "head" or "tail" when flipping a fair coin, or computing the probability of rolling a number from 1 to 6 when rolling a fair die.

Example 7.4. Flipping a Fair Coin.
The algorithm for simulating flipping a coin is relatively straightforward.

Algorithm: Simulating Flipping a Fair Coin

INPUTS:
N total number of random flips to generate

OUTPUTS:
A estimated probabilities for heads and tails

STEPS:

Step 1. Set the counters H and T to 0

Step 2. For i from 1 to N do Steps 3-4

Step 3. Generate a random number x from $Uniform(0, 1)$

Step 4. If $0 \leq x < 0.5$, then $H = H + 1$, otherwise $T = T + 1$
Increment i; if $i \leq N$, then go to Step 3. otherwise go to Step 5.

Step 5. Estimate the probabilities: $P(\text{heads}) \approx H/N$ and $P(\text{tails}) \approx T/N$.
Output the probabilities and Stop.

Implement the coin flipping algorithm in Maple.

```
> Flip := proc (n :: posint)
    local r, H, T, i, x;
    r := rand(0..1.0);
    H, T := 0, 0;
    for i to n do
      x := r();
      if x < 0.5 then H := H + 1 else T := T + 1 end if;
      end do;
    return([evalf(H/n, 4), evalf(T/n, 4)]);
    end proc :
> for N in [2, 10, 1000, 2000, 5000, 10000, 20000] do
    N, Flip(N);
    end do;
                    2          [1., 0.]
                   10    [0.6000, 0.4000]
                 1000    [0.5090, 0.4910]
                 2000    [0.4930, 0.5070]
                 5000    [0.4996, 0.5004]
                10000    [0.5024, 0.4976]
                20000    [0.4957, 0.5043]
```

Example 7.5. Rolling a Fair Die.
Rolling a fair die adds the additional process of multiple assignments (6 for a six-sided die). We can modify the coin flipping algorithm accordingly quite easily. The probability will be (the number of occurrences of each number) / (the total number of trials).

Algorithm: Simulating Rolling a Fair Die

INPUTS:

N total number of random rolls to generate

OUTPUTS:

Die list of estimated probabilities for each possibility

STEPS:

Step 1. Set the counters in *Die* to 0

Step 2. For i from 1 to N do Steps 3-4

Step 3. Generate a random number j from $Integers(1, 6)$

Step 4. Increment $Die_j = Die_j + 1$;

Step 5. Estimate the probabilities: $P(\text{die} = j) \approx Die_j/N$.
Output the probabilities and Stop.

Implement the die tossing algorithm in Maple.

```
> RollDie := proc (n :: posint)
    local roll, Die, i, j;
    roll := rand(1..6);
    Die := [0$6];
    for i to n do
      j := roll();
      Die[j] := Die[j] + 1;
      end do;
    return(evalf(Die/n, 4));
    end proc :
> for N in [10, 100, 1000, 10000] do
    N, Flip(N);
    end do;
              10      [0.5000, 0.2000, 0.1000, 0., 0.1000, 0.1000]
             100      [0.2200, 0.1300, 0.1500, 0.1500, 0.2200, 0.1300]
            1000      [0.1530, 0.1510, 0.1820, 0.1870, 0.1410, 0.1860]
           10000      [0.1671, 0.1662, 0.1760, 0.1646, 0.1604, 0.1657]
```

(*RollDie* is in the *PSM* package.)

The expected probability is 1/6 or 0.1667. Note that as the number of trials increased, the closer our probabilities were to the expected long-run values. We offer the following concluding remark: When you have to run simulations, run them for very large numbers of trials.

Exercises

1. Generate random flips of a fair coin for $n = 20$, 200, 2000, and 20,000.
2. Generate random rolls of a fair die for $n = 20$, 200, 2000, and 20,000.
3. Modify the *Flip* program for a coin that is not fair supposing that $P(\text{heads}) = 9/16$ and $P(\text{tails}) = 7/16$. Compare with Exercise 1.
4. Modify the *RollDie* program to use a 20-sided D&D die. Then generate random rolls of the D&D die for $n = 20$, 200, 2000, and 20,000. What is the expected probability of any particular number appearing?
5. Modify the *RollDie* program to use a fair die having k sides. Test your program with $k = 6$ and $k = 20$, and compare to the previous exercises' results.

7.5 Case Studies: Applied Simulation Models

In this section, we present algorithms and the use of Maple for the following simulations:

- Simulate an aircraft missile attack.
- Given an empirical demand history, simulate the amount of gas a series of gas stations will need.
- Simulate a barber shop.

Case Study 1. A Missile Attack[5]

An analyst develops plans for a missile strike using F-15 aircraft. The F-15s must fly through air defense sites that hold a maximum of 8 missiles. It is vital to ensure success of this attack early in the battle. Each aircraft has a probability of 0.5 of destroying the target, assuming it can get through the air defense systems and then acquire the target. The probability that a single F-15 will acquire a target is approximately 0.9. The protecting air-defense equipment has a probability of stopping the F-15 from either arriving at the target or acquiring the target of 0.4. How many F-15s are needed to have a successful mission assuming we need a 99% success rate?

Begin with an algorithm for simulating the F-15 attack.

[5]Adapted from Meerschaert [MM2007].

Algorithm: Simulating the F-15 Attack

INPUTS:

- N number of F-15s
- M number of air-defense missiles fired
- P probability one F-15 can destroy the target
- Q probability the air-defense can disable an F-15

OUTPUTS:

- S probability of mission success

STEPS:

Step 1. Set S to 0

Step 2. For i from 1 to M do Steps 3-5

Step 3. Calculate $p_i = 1 - (1 - P)^{N-1}$

Step 4. Generate a random b_i from the *Binomial*$(M, i \cdot Q)$ distribution

Step 5. Compute $S = S + p_i \cdot b_i$

Step 6. Output the probability $S = P(\text{mission success})$ and Stop.

We run the simulation letting the number of F-15s vary from 1 to 10, each time calculating the probability of success.

```
> F15Attack := proc(n, m, P, Q)
    local s, B, i, p, b;
    uses Statistics;
    s := 0;
    p := P;
    B := RandomVariable(Binomial(m, Q));
    for i from 0 to m do
      p := 1 - (1 - p)^(n-i);
      b := ProbabilityFunction(B, i);
      s := s + p * b
      end do;
    return(s);
    end proc :
```

```
> for i to 10 do
    i, F15Attack(i, 8, 0.45, 0.40);
  end do;
                         1   0.00755827200
                         2   0.07419703680
                         3    0.3043347039
                         4    0.5924805227
                         5    0.8254837008
                         6    0.9497277101
                         7    0.9912246093
                         8    0.9992039991
                         9    0.9999226475
                        10    0.9999574561
```

We see that the number of F-15s equaling 7 gives us $P(\text{success}) = 0.99122$. Actually, any number of F-15s greater than 7 works to provide a result with the probability of success we desire. We would think the 10 F-15s yielding a $P(\text{success}) = 0.999957$ would virtually guarantee success. Any more would be overkill.

Let's move to a probabilistic inventory problem.

Case Study 2. Gasoline Inventory Simulation[6]

BACKGROUND: You are a consultant to an owner of a chain of gasoline stations along a freeway. The owner wants to maximize their profits and meet consumer demand for gasoline. You decide to look at the following problem.

PROBLEM: Minimize the average daily cost of delivering and storing sufficient gasoline at each station to meet consumer demand.

ASSUMPTIONS: For an initial model, you consider that in the short run the average daily cost is a function of demand rate, storage costs, and delivery costs. You also assume that you need a model for the demand rate. You decide that historical date will assist you.

[6]Adapted from Giordano et al [GFH2014].

TABLE 7.5: Historical Gasoline Demand Data

Demand (gallons)	Number of Occurrences (days)
1000 – 1099	10
1100 – 1199	20
1200 – 1299	50
1300 – 1399	120
1400 – 1499	200
1500 – 1599	270
1600 – 1699	180
1700 – 1799	80
1800 – 1899	40
1900 – 1999	30
Total number of days = 1000	

MODEL FORMULATION: We convert the number of days into probabilities by dividing by the total, and we use the mid-point of the interval of demand as a simplification.

TABLE 7.6: Historical Data as Probabilities

Demand (gallons)	Probability
1000	0.010
1150	0.020
1250	0.050
1350	0.120
1450	0.200
1550	0.270
1650	0.180
1750	0.080
1850	0.040
2000	0.030

Since cumulative probabilities will be more useful, we convert to a CDF, a cumulative distribution function.

TABLE 7.7: Historical Data CDF

Demand (gallons)	Probability
1000	0.000
1050	0.010
1150	0.030
1250	0.080
1350	0.2000
1450	0.400
1550	0.670
1650	0.850
1750	0.930
1850	0.970
2000	1.000

We will use the following algorithm for simulating inventories.

Algorithm: Simulating Gasoline Inventories

INPUTS:

- q quantity delivered (gal)
- d cost per delivery ($)
- T time between deliveries (days)
- s cost of storage ($/gal)
- N number of days in simulation

OUTPUTS:

- C average daily cost ($)

STEPS:

Step 1. Set inventory i to 0 and cost C to 0

Step 2. Begin the next cycle with a delivery (update inventory and cost)

$$i = i + q$$
$$C = C + d$$

Step 3. For j from 1 to T do Steps 4 - 6

Step 4. Generate a demand q_j.
(Use cubic splines to fit a smooth function to the CDF data.)

Step 5. Update inventory $i = i - q_j$

Step 6. Update the cost, $C = C + s \cdot i$ if the Inventory is positive, otherwise set $i = 0$ and go to Step 7.

Step 7. Return to Step 2 until the simulation is complete.

Step 8. Output the average daily cost $\overline{C} = C/N$ and Stop.

The program *GasInventory* below is an implementation of the algorithm in Maple. (*GasInventory* is in the *PSM* package.)

```
> GasInventory := proc(QuantDelivered, DeliveryCost, T, StorageCost, N,
      CDFdata, Galdata)
    local CurrentInventory, Cost, DemandCDF, MinInventory, k, days, x,
      DailyDemand;
    CurrentInventory := 0;
    Cost := 0;
    DemandCDF := unapply(CurveFitting[Spline](CDFdata, Galdata, x), x);
    MinInventory := QuantDelivered;
    for k to ceil(N/T) do
      CurrentInventory := CurrentInventory + QuantDelivered;
      Cost := Cost + DeliveryCost;
      for days to T while days + (k − 1) · T ≤ N do
        x := rand()/10.0^12;
        DailyDemand := evalf(DemandCDF(x));
        CurrentInventory := max(CurrentInventory − DailyDemand, 0);
        MinInventory := min(MinInventory, CurrentInventory);
        Cost := Cost + CurrentInventory · StorageCost;
        end do;
      end do;
    return(fnormal~([CurrentInventory, MinInventory, Cost/N]));
    end proc :
```

Run the simulation three times. For each run, use 20 days and the historical data above where 11,000 gallons are delivered every 7 days at a cost of $500 per delivery and storage cost of $0.50 per gallon.

```
> cdfdata := [0.0, 0.01, 0.03, 0.08, 0.20, 0.40, 0.67, 0.85, 0.93, 0.97, 1.00];
  galdata := [1000, 1050, 1150, 1250, 1350, 1450, 1550, 1650, 1750, 1850, 2000]
> [CurrentInventory, MinInventory, AvgDailyCost];
  Run1 := GasInventory(11000, 500, 7, 0.05, 20, cdfdata, galdata);
  Run2 := GasInventory(11000, 500, 7, 0.05, 20, cdfdata, galdata);
  Run3 := GasInventory(11000, 500, 7, 0.05, 20, cdfdata, galdata);

            [CurrentInventory, MinInventory, AvgDailyCost]
            Run1 := [2862.451664, 922.7349162, 363.0393258]
            Run2 := [1924.447540, 416.2347618, 346.1311072]
            Run3 := [3274.492461, 1067.367251, 377.5502983]
```

The current inventory and minimum inventory values appear to differ quite a bit, but the average daily cost, not so much. The best result was in run 2 where average cost was $346.13. We would run a large number of simulations to gain insight into the dynamics of the problem.

Let's drop the gallons delivered to 10,000.

```
> GasInventory(10000, 500, 7, 0.05, 20, cdfdata, galdata);
                    [1851.769458, 0., 283.8868194]
```

The average daily cost is now only $283.89, but there were lost sales not accounted for. One way to enhance the model is to add a *shortage cost* whenever the inventory is zero, preventing any sales.

Case Study 3. Barbershop Customer Service Problem

A barber shop has two customers arriving every 30 minutes. The service rate of the barber is three customers every 60 minutes. This implies the time between arrivals is 15 minutes and the mean service time is 1 customer every 20 minutes. How many customers will be in the queue and what is their average waiting time?

A queue is a waiting line for service. Examples would be people in line to purchase a movie ticket, the line at a drive-through window to order fast food, or the lines waiting for a cashier at a grocery store. There are two important entities in a queue: customers and servers. There are several important parameters that describe a queue:

a. The number of servers available.

b. Customer average arrival rate: average number of customers arriving to be serviced in a time unit.

c. Server rate: average number of customers processed in a time unit.

d. The number of queues.

In many simple queuing simulations, as well as in theoretical approaches, we assume that arrivals and service times are exponentially distributed with a mean arrival rate of λ_1 and a mean service time of λ_2. We can apply the following theorem.

Theorem. Average Queue Length and Waiting Time.
If the arrival rate is exponential and the service rate is given by any distribution, then the expected number of customers waiting in the queue L_q, and the expected waiting time W_q are given by

$$L_q = \frac{\lambda^2\sigma^2 + \rho^2}{2(1-\rho)}, \quad \text{and} \quad W_q = \frac{L_q}{\lambda}$$

where λ is the mean number of arrival per time period, μ is the mean number of customers serviced per time unit, $\rho = \lambda/\mu$, and σ is the standard deviation of the service time.

BARBER SHOP SIMULATION: We provide an algorithm for simulating the arrival and servicing of customers at the barber shop.

Algorithm: Simulating Barbershop Customers

INPUTS:

T	total time period to simulate
λ	scale parameter for exponentially distributed arrivals
μ	scale parameter for exponentially distributed service time

OUTPUTS:

Report	A dataframe with timings for a randomly generated set of customers

STEPS:

Step 1. Set all times to zero

Step 2. For *Customers* from 1 to n while time is $\leq T$ do Steps 4–6

  Step 3. Generate a random arrival time for the next customer with the *Exponential*$(1/\lambda)$ distribution

  Step 4. If the server is busy add the customer to the queue

  Step 5. Serve the first in the queue: generate random service time with the *Exponential*$(1/\mu)$ distribution

Step 6. Return to Step 2 until the simulation is complete

Step 7. Output the *Report* and the *Total Service Time*

Step 8. Calculate the mean waiting time, maximum waiting time, and server efficiency, then Stop.

The program *BarberShop* is a Maple implementation of the algorithm. The Maple output will be collected in the dataframe *Report* for each person in the queue. Recall that a dataframe is a convenient structure for combining and operating on disparate types of data.

```
> BarberShop := proc(λ, μ, T)
    local Cols, Rows, ArrivalTime, ServeEndTime, Wait, TotalWait,
      TotalService, MaxWait, Server1, serve, Arrival, Report;
    uses Statistics;
    ArrivalTime, ServeEndTime, Wait, TotalWait, TotalService, MaxWait
      := 0$6;
    Cols := ["Arrival Time", "Waiting Time", "Service End",
      "Server's Status"];
    Rows := NULL;
    while ArrivalTime < T do
      Arrival := Sample(RandomVariable(Exponential(1/λ)), 1)[1];
      ArrivalTime := ArrivalTime + Arrival;
      Wait := piecewise(ServeEndTime < ArrivalTime, 0,
           evalf(ServeEndTime − ArrivalTime));
      Server1 := piecewise(ServeEndTime < ArrivalTime, 'Free',
        'Busy');
      TotalWait := evalf(TotalWait + Wait);
      MaxWait := max(Wait, MaxWait);
      ServeEndTime := max(ArrivalTime, ServeEndTime);
      serve := Sample(RandomVariable(Exponential(1/μ)), 1)[1];
      TotalService := TotalService + serve;
      ServeEndTime := evalf(ServeEndTime + serve);
      Rows := Rows, fnormal~([ArrivalTime, Wait, ServeEndTime,
        Server1], 4);
    end do;
    Report := DataFrame(Matrix([Rows]), 'columns' = Cols);
    return[Report, TotalService];
    end proc :
```

We'll show a few rows and the summary analysis. Maple also reduces the display; that default maximum display size of an array is 10×10. To view the entire dataframe, execute *interface(rtablesize* = 500), just as is done for matrices and vectors, then re-execute *TheSimulation*.

```
> λ, μ, T := 3, 5, 60 :
```

```
> BarberShop(λ, μ, T) :
  TheSimulation := %[1];
  TotalServiceTime := %%[2];
```

TheSimulation :=

	"Arrival Time"	"Waiting Time"	"Service End"	"Server's Status"
1	1.373	0.	1.660	*Free*
2	1.524	0.1359	2.326	*Busy*
3	1.676	0.6504	2.339	*Busy*
4	1.724	0.6147	2.381	*Busy*
5	2.931	0.	4.024	*Free*
6	2.952	1.072	4.190	*Busy*
7	3.316	0.8735	4.308	*Busy*
8	3.555	0.7527	4.360	*Busy*
...	...	...	...	...

TotalServiceTime := 29.7811546564664

```
> MaximumWaitingTime = max(TheSimulation["Waiting Time"]);
```

$$MaximumWaitingTime = 1.168$$

```
> TheoreticalServerEfficiency = evalf(λ/μ, 4),
  Simulated = evalf(TotalServiceTime/T, 4);
```

$$TheoreticalServerEfficiency = 0.6000, Simulated = 0.4964$$

$$> TheoreticalAvgWaitingTime = evalf\left(\frac{\lambda}{\mu \cdot (\mu - \lambda)}, 4\right),$$

$$Simulated = evalf(Statistics[Mean](TheSimulation[\text{"Waiting Time"}]), 4);$$

$$TheoreticalAvgWaitingTime = 0.3000, Simulated = 0.1735$$

We see our simulation comes close to the theoretical values. We could use this simulation to try some changes in server training to see if that improves the output. We could alter the code to try more than 1 server to see if that improves the overall service efficiency.

Exercises

1. Modify the missile attack problem if the probability of S is only 0.95 and the probability of an F-15 being deterred by air defense is 0.3. Determine the number of F-15s needed to complete the mission.

2. What if, in the missile attack problem, the air defense units were modified to carry 10 missiles? What impact does that have on the number of F-15s needed?

3. Perform sensitivity analysis on the gasoline inventory problem by modifying the delivery to 11,450 gallons per week. What impact does this have on the average daily cost?
4. Solve for the theoretical L_q and W_q for the barber shop problem.

Projects

Project 7.1. Tollbooths

Heavily traveled toll roads such as the Garden State Parkway, Interstate 95, and so forth, are multilane divided highways that are interrupted at intervals by toll plazas. Because collecting tolls is usually unpopular, it is desirable to minimize motorist annoyance by limiting the amount of traffic disruption caused by the toll plazas. Commonly, a much larger number of tollbooths are provided than the number of travel lanes entering the toll plaza. On entering the toll plaza, the flow of vehicles fans out to the larger number of tollbooths; when leaving the toll plaza, the flow of vehicles is forced to squeeze down to a number of travel lanes equal to the number of travel lanes before the toll plaza. Consequently, when traffic is heavy, congestion increases when vehicles leave the toll plaza. When traffic is very heavy, congestion also builds at the entry to the toll plaza because of the time required for each vehicle to pay the toll.

Construct a mathematical model to help you determine the **optimal number** of tollbooths to deploy in a barrier-toll plaza. Explicitly, first consider the scenario in which there is exactly one tollbooth per incoming travel lane. Then consider multiple tollbooths per incoming lane. Under what conditions is one tollbooth per lane more or less effective than the current practice? The definition of *optimal* is up to you to determine.

Project 7.2. Major League Baseball

Build a simulation to model a baseball game. Use your two favorite teams or favorite all-star players to simulate play of a regulation game.

Project 7.3. NBA Basketball

Build a simulation to model the NBA basketball playoffs.

Project 7.4. Hospital Facilities

Build a simulation to model surgical and recovery rooms for a medium-size hospital in a small city.

Project 7.5. College Final Examination Schedules

(a) Build an algorithm and simulation to create a final exam schedule for classes that have standard starting times 8:00 am, 9:00 am, 10:00 am, ..., 4:00 pm for Monday-Wednesday-Friday classes.

(b) Modify your schedule simulation to add exams for Tuesday-Thursday classes that have standard starting times 8:00 am, 9:30 am, 11:00 am, ..., 5:00 pm.

Project 7.6. Automobile Emissions
Consider a large engineering company that performs emissions control inspections on automobiles for the state. During the peak period, cars arrive at a single location that has four lanes for inspections following exponential arrivals with a mean of 15 minutes. Service times during the same period are uniform: between 15 and 30 minutes. Build a simulation for the length of the queue. If cars wait more than 1 hour, the company pays a penalty of $200 per car. How much money, if any, does the company pay in penalties? Would more inspection lanes help? What costs associated with the inspection lanes need to be considered?

Project 7.7. Recruiting
Monthly demand for recruits is provided in Table 7.8.

TABLE 7.8: Historical Recruiting Data

Demand	Probability	CDF
300	0.05	0.05
320	0.10	0.15
340	0.20	0.35
360	0.30	0.65
380	0.25	0.90
400	0.10	1.00

Additionally, depending on conditions, the average cost per recruit is between $60 and $80 in integer values. Returns from Higher HQ are between 20% and 30% of costs. There is a fixed cost of $2,000/month for the office, phones, etc. Build a simulation model to determine the average monthly costs. Assume

$$Cost = demand \cdot (cost\ per\ recruit) + (fixed\ cost) - (return\ amount),$$

where *return* is given as a percentage of *cost*.

Project 7.8. Inventory Model
The demand for ammunition palettes for resupply on a periodic basis is provided in Table 7.9 below. Lead time if resupply is required is between 1 and 3 days. Currently, we have 7 palettes in stock and no orders due. Fixed cost for placing an order is $20. The cost for holding the unused stock is $0.02 per palette per day. Each time we cannot satisfy a demand the unit goes elsewhere and assumes a loss of $8 to the company. We operate 24/7. Determine

TABLE 7.9: Ammunition Palettes Resupply Demand

Weekly Demand	Frequency	Probability	CDF
0	15	0.05	0.05
1	30	0.10	0.15
2	60	0.20	0.35
3	120	0.40	0.75
4	45	0.15	0.90
5	30	0.10	1.00

the optimum order quantity (# of palettes) and order point (time between orders) to optimize total cost.

Project 7.9. Simple Queuing Problem
The bank manager is trying to improve customer satisfaction by offering better service. The manager wants the average customer to wait less than 2 minutes and the average length of the queue (line) to be 2 or fewer. From collected data, the manager estimates the bank sees about 150 customers per day. Existing service and arrival times are shown in Table 7.10.

TABLE 7.10: Bank Service and Arrival Times

Service Time	Probability	Time Between Arrivals	Probability
1	0.25	0	0.10
2	0.20	1	0.15
3	0.40	2	0.10
4	0.15	3	0.35
		4	0.25
		5	0.05

Determine if the current servers (cashiers) are satisfying the goals. If not, how much improvement is needed in service to accomplish the stated goals.

Project 7.10. Intelligence Gathering (Information Operations)
Table 7.11 shows the likely times (weeks) between intelligence reports[7] arriving at the Operations Center.

[7] For an example of a technology intelligence report, see a *Microsoft Security Intelligence Report* available at https://www.microsoft.com/en-us/security/operations/security-intelligence-report.

TABLE 7.11: Intelligence Report Arrivals

Time between Reports	Probability
1	0.11
2	0.21
3	0.22
4	0.20
5	0.16
6	0.10

The time (weeks) it takes to process these reports is given in Table 7.12.

TABLE 7.12: Intelligence Report Processing Time

Process Time	Probability
1	0.20
2	0.19
3	0.18
4	0.17
5	0.13
6	0.10
7	0.03

Deploying remote sensors allows the reports to come more often changing the arrival times' probabilities as seen in Table 7.13.

TABLE 7.13: Intelligence Report Arrivals with Remote Sensors

Time between Reports	Probability
1	0.22
2	0.25
3	0.19
4	0.15
5	0.12
6	0.07

Advise the director on the current system. Determine utilization and sensor satisfaction. What is an optimal number of report processors needed to insure reports are processed in a timely manner at a minimal cost?

Further Reading and References

[BenIsrael1999] Adi Ben-Isreal, "Probability and Statistics with Maple," ICTCM 12, San Francisco, CA, 1999.

[GFH2014] Frank Giordano, William P. Fox, and Steven Horton, *A First Course in Mathematical Modeling*, 5th ed., Nelson Education, 2014.

[LK2007] Averill M. Law and W. David Kelton. *Simulation Modeling and Analysis*, 4th ed., McGraw Hill, 2007.

[MM2007] Mark Meerschaert, *Mathematical Modeling*, 3rd ed., Academic Press. 2007.

[WW2004] Wayne Winston, *Operations Research: Applications and Algorithms*, 4th ed., Cengage, 2004.

Index